Role of Cancer Stem Cells in Cancer Biology and Therapy

Editors

Thomas Dittmar

Institute of Immunology
Witten/Herdecke University
Witten
Germany

Kurt S. Zänker

Institute of Immunology
Witten/Herdecke University
Witten
Germany

CRC Press
Taylor & Francis Group
Boca Raton London New York

CRC Press is an imprint of the
Taylor & Francis Group, an **informa** business
A SCIENCE PUBLISHERS BOOK

CRC Press
Taylor & Francis Group
6000 Broken Sound Parkway NW, Suite 300
Boca Raton, FL 33487-2742

First issued in hardback 2019

ISBN-13: 978-1-4665-7735-0 (hbk)

Library of Congress Cataloging-in-Publication Data

Role of cancer stem cells in cancer biology and therapy / editors, Thomas Dittmar, Kurt S. Zänker.
p. ; cm.
Includes bibliographical references and index.
ISBN 978-1-4665-7735-0 (hardcover : alk. paper)
I. Dittmar, Thomas. II. Zänker, Kurt S.
[DNLM: 1. Neoplastic Stem Cells. 2. Antineoplastic Agents--therapeutic use. 3. Molecular Targeted Therapy. 4. Neoplasms--drug therapy. QZ 202]
616.99'406--dc23

2012037572

Visit the Taylor & Francis Web site at
http://www.taylorandfrancis.com

and the CRC Press Web site at
http://www.crcpress.com

Preface

To edit a book on the "Role of Cancer Stem Cells in Cancer Biology and Therapy" is not an easy task, but we believe that, if the work is being done with outstanding scientists, who touched upon so many issues, concepts, ideas, proposals—it becomes not a first-rate problem.

PubMed—NCBI lists in 2012 under the heading of "cancer stem cells" more than 30,000 publications—more than a three-fold increase in a 10-years publication period. There is increasing evidence supporting the concept that cancers are organized in a cellular hierarchy with cancer stem cells (CSCs) at the apex. The failure to cure cancer might be due to a misidentification of the target. Current therapy regimens succeed at eliminating bulky disease, but often miss a tumor reservoir that is the source of disease recurrence and metastasis. The CSC hypothesis postulates that tumor growth is driven by a rare subpopulation of tumor cells that show similar biological properties as stem cells. As a result, tumors contain a cellular compartment that retains key stem cell properties including self-renewal, which drives tumorigenesis and differentiation that contributes to cellular heterogeneity. Advances in stem cell biology and technology have led to the identification of stem cells in normal and malignant tissues. Cancer researchers can sequence the genome of single tumor cells, scan the genome for the activity of "good" and "bad" genes and profile the protein expression. But the main question remains still relevant to patients: Why are organ specific tumors formed?

Again, the CSC hypothesis supports the idea that a small subset of cells drives tumor growth and that curing cancer may require those cells to be eliminated. The evolving concept of CSCs will lead to a paradigm shift in the way that chemo- and radiotherapy efficacy is evaluated and how therapeutics are developed. The shrinking of a tumor is not any more the first therapeutic target, but to assess whether a new therapy kills the right sort of tumor cells. The hunt to find ways to kill these cells or push them into a silent niche is now open.

To be successful, we need all the brains and the intelligence of researchers; we need new algorithms to use the trillions of data generated and already stored in the clouds to understand the complexity of a "brain" within a tumor, e.g., why non-stem cells within the tumor are forced to de-differentiate into stem cells.

We have entered the post-genomic era and the master builders that underlie tumor growth are likely CSCs and regulatory phenomena of such cells with other types of cancer cells and their microenvironment.

It is too soon to know whether this fascinating concept of CSC has profound implications for personalized patient care, but we can be again optimistic after a period of relative stagnation in cancer therapy.

This book might be regarded as a further provocative contribution to discuss the role of CSCs in cancer biology and therapy with leading experts and without any boundaries.

Thomas Dittmar
Kurt S. Zänker
Editors
November 2012
Witten, Germany

Introduction

Within the past two decades knowledge about how cancer originates has changed dramatically. What was once believed as a disease that has its origin in somatic cells, which due to chromosomal aberrations, such as mutations and/or aneuploidy events, had undergone a malignant transformation, is now considered as a so-called stem cell-based disease. In 1997 Dick Bonnet and John Dick first discovered a population of acute myeloma leukemia cells exhibiting a leukemia initiating capacity, which means that this particular cell type possesses the differentiative and proliferative capacities and the potential for self-renewal that has been proposed for a leukemic stem cell (Bonnet and Dick 1997). Since then, tumor-initiating cells or so-called cancer stem cells (CSCs) have been identified in a variety of cancers, including chronic myeloid leukemia (CML) (Jamieson et al. 2004), brain tumors (Singh et al. 2004), malignant melanoma (Schatton et al. 2008), breast cancer (Al-Hajj et al. 2003), colon/colorectal cancer (O'Brien et al. 2007) (Dalerba et al. 2007), and pancreatic tumors (Li et al. 2007a). Moreover, the knowledge that cancer is hierarchically organized like normal tissues, namely comprising of a small set of stem/precursor cells and differentiated cells (in the context of cancer this would be the CSCs and the tumor bulk), and that CSCs share similarities with normal stem cells has further changed our understanding of how to treat cancer. Stem cells can be defined as a population of cells exhibiting specific characteristics, such as the capacity to self-renew, to differentiate, to restore tissue function, to exhibit telomerase activity, to express ATP binding cassette (ABC) multidrug resistance transporters and to possess an enhanced and highly efficient DNA repair machinery (Dittmar and Zänker 2008). Under normal circumstances these stem cell characteristics ensure the status quo of stem cells within organ tissues as well as the homeostasis of the organ itself. A loss of stem cells would ultimately result in the destruction of the organ since it can not be longer regenerate itself. This effect can be seen in patients who have received a bone marrow transplant. If the hematopoietic stem cells fail to engraft within the hosts bone marrow the hematopoietic system can not be restored finally leading to the death of the patient. On the other hand, the capability of hematopoietic stem cells to restore the hematopoietic system after transplantation illustrates the potency of stem cells in tissue regeneration.

However, what is desired for stem cells in a physiological context, namely the capactiy to ensure long-term tissue homeostasis and integrity, is unwanted in the pathophysiological context of cancer. Due to their stem cell characteristics CSCs are not only the seeds of primary tumors, but, as per definition only CSCs possess tumor initiation capacity, also of metastatic lesions (Dittmar et al. 2008, Li et al. 2007b). Very recently the importance between epithelial-mesenchymal transition (EMT) and the outgrowth of metastasis linked to self-renewal was conceptualized. Cellular traits of CSCs and EMT programs may play critical roles both early and late in the metastatic cascade (Scheel and Weinberg 2012). Moreover, due to their stem cell characteristics, such as an enhanced resistance towards chemotherapeutic drugs (Dean et al. 2005, Eyler and Rich 2008, Lou and Dean 2007, Nagler et al. 2011) and radiation (Bao et al. 2006, Eyler and Rich 2008, Rich 2007) it is now recognized that CSCs may be able to survive cancer therapy, and as a consequence, that these cells will cause cancer recurrences (Dittmar et al. 2012, Dittmar et al. 2009, Nagler et al. 2011). It is interesting that the heterogeneous phenotypical and molecular traits of human breast cancers are a function of their CSC content (Pece et al. 2010). In this context, current cancer therapies have been compared to cutting a dandelion (or other weeds) off at ground level (Huff et al. 2006, Jones et al. 2004). Even though it may appear to produce the desired effect, only elimination of the root will actually prevent the weed from regrowing (Huff et al. 2006, Jones et al. 2004). The "dandelion phenomenon" illustrates the necessity of a paradigm shift in future cancer therapies. Not the rapidly proliferating tumor cells should be the primary target of newly designed drugs (in fact, in the past the efficacy of new anti-cancer drugs was measured as how fast they block tumor proliferation concomitant with tumor bulk shrinking—the faster, the better), but rather should be the CSCs. There is a hope to inhibit tumor relapse caused by cancer cells with stem cell properties by combined treatment modalities of hyperthermia and chemo- and/or radiotherapy (Pelicci et al. 2011). As mentioned above, the loss of stem cells would ultimately result in the destruction of the organ since it can not be longer regenerate itself. Thus, elimination of CSCs would be a valid option to ultimately cure cancer (Reya et al. 2001). But what are the current concepts to selectively eradicate CSCs? CSCs share several similarities with normal stem cells since they most likely derive from normal stem cells and/or their progenitor cells (Houghton et al. 2004, Jaiswal et al. 2003, Jamieson et al. 2004). Thus, molecular targets that would be suitable to eliminate CSCs might also eliminate normal stem cells.

Moreover, recent data indicate that CSC biology is much more complex then previously assumed. Wright and colleagues demonstrated that Brca1 breast tumors contain distinct $CD44^{+}/CD24^{-}$ and $CD133^{+}$ cells with CSC characteristics indicating that Brca1 mouse mammary tumors harbor a heterogenous stem cell population (Wright et al. 2008). Phenotypic subpopulation of CSCs have also been identified in metastatic colon cancer (Botchkina et al. 2009), breast cancer cell lines (Charafe-Jauffret et al. 2009), and acute lymphoblastic leukemia (ALL) (Mullighan et al. 2007, Mullighan et al. 2008, Notta et al. 2011, Shackleton et al. 2009). Particularly the data obtained from ALL initiating cells revealed that CSCs consist of a pool of diverse subclones, which derived in an evolutive manner (Mullighan et al. 2008, Notta et al. 2011). Moreover, the finding that the primary pool of ALL initiating cells already harbors CSC subclones that were responsible for relapses (Mullighan et al. 2008) illustrates the complexity of CSC biology as well as underlines the challenge to physicians and scientists to find targets and compounds that will selectively kill all CSC clones. The search has begun and the first CSC eradicating drugs/overall strategies have been developed and are currently tested clinically. Some of these data will be presented in this book and they will provide a hopeful outlook for future anti-cancer strategies.

We are glad that so many internationally recognized experts accepted our invitation to contribute to this exciting book and to share their results with the scientific community working in the field of CSC research. We thank them all for their interest in this important topic and that they found the time to present excellent and comprehensive overviews of addressing the most important and recent findings in their particular field of research.

We further would like to thank Raju Primlani from Science Publishers for his helpful assistance, his excellent collaboration and for giving us the opportunity to realize this challenging project.

We hope that this book may encourage new scientific approaches within the field of CSC biology as well as closer interdisciplinary collaborations on fascinating and important issues on cancer research.

November 2012
Thomas Dittmar
Kurt S. Zänker
Witten, Germany

References

Al-Hajj, M., M.S. Wicha, A. Benito-Hernandez, S.J. Morrison, and M.F. Clarke. 2003. Prospective identification of tumorigenic breast cancer cells. Proc Natl Acad Sci USA 100: 3983–3988.

Bao, S., Q. Wu, R.E. McLendon, Y. Hao, Q. Shi, A.B. Hjelmeland, M.W. Dewhirst, D.D. Bigner, and J.N. Rich 2006. Glioma stem cells promote radioresistance by preferential activation of the DNA damage response. Nature 444: 756–760.

Bonnet, D., and J.E. Dick 1997. Human acute myeloid leukemia is organized as a hierarchy that originates from a primitive hematopoietic cell. Nat Med 3: 730–737.

Botchkina, I.L., R.A. Rowehl, D.E. Rivadeneira, M.S. Karpeh, Jr., H. Crawford, A. Dufour, J. Ju, Y. Wang, Y. Leyfman, and G.I. Botchkina 2009. Phenotypic subpopulations of metastatic colon cancer stem cells: genomic analysis. Cancer Genomics Proteomics 6: 19–29.

Charafe-Jauffret, E., C. Ginestier, F. Iovino, J. Wicinski, N. Cervera, P. Finetti, M.H. Hur, M.E. Diebel, F. Monville, J. Dutcher, M. Brown, P. Viens, L. Xerri, F. Bertucci, G. Stassi, G. Dontu, D. Birnbaum, and M.S. Wicha 2009. Breast cancer cell lines contain functional cancer stem cells with metastatic capacity and a distinct molecular signature. Cancer Res 69: 1302–1313.

Dalerba, P., S.J. Dylla, I.K. Park, R. Liu, X. Wang, R.W. Cho, T. Hoey, A. Gurney, E.H. Huang, D.M. Simeone, A.A. Shelton, G. Parmiani, C. Castelli, and M.F. Clarke 2007. Phenotypic characterization of human colorectal cancer stem cells. Proc Natl Acad Sci USA 104: 10158–10163.

Dean, M., T. Fojo, and S. Bates 2005. Tumour stem cells and drug resistance. Nat Rev Cancer 5: 275–284.

Dittmar, T., C. Heyder, E. Gloria-Maercker, W. Hatzmann, and K.S. Zanker 2008. Adhesion molecules and chemokines: the navigation system for circulating tumor (stem) cells to metastasize in an organ-specific manner. Clin Exp Metastasis 25: 11–32.

Dittmar, T., C. Nagler, B. Niggemann, and K.S. Zänker 2013. The dark side of stem cells: triggering cancer progression by cell fusion. Curr Mol Med accepted.

Dittmar, T., C. Nagler, S. Schwitalla, G. Reith, B. Niggemann, and K.S. Zanker 2009. Recurrence cancer stem cells—made by cell fusion? Med Hypotheses 73: 542–547.

Dittmar, T., and K.S. Zänker Targeting Cancer Stem Cells. pp. 189–197. In: Dittmar, T. and K.S. Zänker (eds.). 2008. Cancer and Stem Cells, Nova Publishers. Hauppauge.

Eyler, C.E., and J.N. Rich 2008. Survival of the fittest: cancer stem cells in therapeutic resistance and angiogenesis. J Clin Oncol 26: 2839–2845.

Houghton, J., C. Stoicov, S. Nomura, A.B. Rogers, J. Carlson, H. Li, X. Cai, J.G. Fox, J.R. Goldenring, and T.C. Wang 2004. Gastric cancer originating from bone marrow-derived cells. Science 306: 1568–1571.

Huff, C.A., W. Matsui, B.D. Smith, and R.J. Jones 2006. The paradox of response and survival in cancer therapeutics. Blood 107: 431–434

Jaiswal, S., D. Traver, T. Miyamoto, K. Akashi, E. Lagasse, and I.L. Weissman 2003. Expression of BCR/ABL and BCL-2 in myeloid progenitors leads to myeloid leukemias. Proc Natl Acad Sci USA 100: 10002–10007.

Jamieson, C.H., L.E. Ailles, S.J. Dylla, M. Muijtjens, C. Jones, J.L. Zehnder, J. Gotlib, K. Li, M.G. Manz, A. Keating, C.L. Sawyers, and I.L. Weissman 2004. Granulocyte-macrophage progenitors as candidate leukemic stem cells in blast-crisis CML. N Engl J Med 351: 657–667.

Jones, R.J., W.H. Matsui, and B.D. Smith 2004. Cancer stem cells: are we missing the target? J Natl Cancer Inst 96: 583–585.

Li, C., D.G. Heidt, P. Dalerba, C.F. Burant, L. Zhang, V. Adsay, M. Wicha, M.F. Clarke, and D.M. Simeone 2007a. Identification of pancreatic cancer stem cells. Cancer Res 67: 1030–1037.

Li, F., B. Tiede, J. Massague, and Y. Kang 2007b. Beyond tumorigenesis: cancer stem cells in metastasis. Cell Res 17: 3–14.

Lou, H., and M. Dean 2007. Targeted therapy for cancer stem cells: the patched pathway and ABC transporters. Oncogene 26: 1357–1360.

Mullighan, C.G., S. Goorha, I. Radtke, C.B. Miller, E. Coustan-Smith, J.D. Dalton, K. Girtman, S. Mathew, J. Ma, S.B. Pounds, X. Su, C.H. Pui, M.V. Relling, W.E. Evans, S.A. Shurtleff, and J.R. Downing 2007. Genome-wide analysis of genetic alterations in acute lymphoblastic leukaemia. Nature 446: 758–764.

Mullighan, C.G., L.A. Phillips, X. Su, J. Ma, C.B. Miller, S.A. Shurtleff, and J.R. Downing 2008. Genomic analysis of the clonal origins of relapsed acute lymphoblastic leukemia. Science 322: 1377–1380.

Nagler, C., K.S. Zanker, and T. Dittmar 2011. Cell Fusion, Drug Resistance and Recurrence CSCs. Adv Exp Med Biol 714: 173–182.

Notta, F., C.G. Mullighan, J.C. Wang, A. Poeppl, S. Doulatov, L.A. Phillips, J. Ma, M.D. Minden, J.R. Downing, and J.E. Dick. 2011. Evolution of human BCR-ABL1 lymphoblastic leukaemia-initiating cells. Nature 469: 362–367.

O'Brien, C.A., A. Pollett, S. Gallinger, and J.E. Dick 2007. A human colon cancer cell capable of initiating tumour growth in immunodeficient mice. Nature 445: 106–110.

Pece, S., D. Tosoni, S. Confalonieri, G. Mazzarol, M. Vecchi, S. Ronzoni, L. Bernard, G. Viale, P.G. Pelicci, and P.P. Di Fiore 2010. Biological and molecular heterogeneity of breast cancers correlates with their cancer stem cell content. Cell 140: 62–73.

Pelicci, P.G., P. Dalton, and R. Orecchia 2011. Heating cancer stem cells to reduce tumor relapse. Breast Cancer Res 13: 305.

Reya, T., S.J. Morrison, M.F. Clarke, and I.L. Weissman 2001. Stem cells, cancer, and cancer stem cells. Nature 414: 105–111.

Rich, J.N. 2007. Cancer stem cells in radiation resistance. Cancer Res 67: 8980–8984.

Schatton, T., G.F. Murphy, N.Y. Frank, K. Yamaura, A.M. Waaga-Gasser, M. Gasser, Q. Zhan, S. Jordan, L.M. Duncan, C. Weishaupt, R.C. Fuhlbrigge, T.S. Kupper, M.H. Sayegh, and M.H. Frank 2008. Identification of cells initiating human melanomas. Nature 451: 345–349.

Scheel, C., and R.A. Weinberg 2012. Cancer stem cells and epithelial-mesenchymal transition: Concepts and molecular links. Semin Cancer Biol 22: 396–403.

Shackleton, M., E. Quintana, E.R. Fearon, and S.J. Morrison 2009. Heterogeneity in cancer: cancer stem cells versus clonal evolution. Cell 138: 822–829.

Singh, S.K., C. Hawkins, I.D. Clarke, J.A. Squire, J. Bayani, T. Hide, R.M. Henkelman, M.D. Cusimano, and P.B. Dirks 2004. Identification of human brain tumour initiating cells. Nature 432: 396–401.

Wright, M.H., A.M. Calcagno, C.D. Salcido, M.D. Carlson, S.V. Ambudkar, and L. Varticovski 2008. Brca1 breast tumors contain distinct CD44+/CD24– and CD133+ cells with cancer stem cell characteristics. Breast Cancer Res 10: R10.

Contents

Chapter 1

Cancer Stem Cells—Fact or Fiction?

Yasunari Kanda

Introduction

Tumors are heterogeneous in their cellular morphology, proliferation rate, differentiation grade, genetic lesions, and therapeutic response (Heppner and Miller 1983). Cellular and molecular mechanisms that cause tumor heterogeneity remain largely unknown.

According to a conventional "clonal evolution model", tumors arise from normal cells that mutate and generate abnormal offsprings which also mutate, forming a mass of genetically varied cancer cells (Fearon and Vogelstein 1990). Most tumor cells have an ability to contribute to tumor maintenance. All cells within a tumor have an equal chance of acquiring genetic mutations necessary for driving tumor growth.

Growing evidence suggests that tumors are organized in a hierarchy of heterogeneous cell populations and are formed and maintained from a small population of stem/stem-like cells, known as "cancer stem cell (CSC) model". CSCs are defined on the basis of characteristics such as high tumorigenicity, self-renewal, and differentiation that contribute to heterogeneity. Only CSCs caused by genetic mutations acquire an ability to contribute to tumor progression.

This review focuses on the current protocols employed to identify and characterize CSCs from established cancer cell lines. Based on these protocols, this review aims to summarize most of the evidences that support the CSC model and its apparent contradictions. These protocols can provide new insights into the targeting of CSCs and their implications for cancer therapy.

Section of cell signaling, Division of Pharmacology, National Institute of Health Sciences, 1-18-1 Kamiyoga, Setagaya, 158-8501 JAPAN.
E-mail: kanda@nihs.go.jp

Historical Background

The original idea of a CSC was proposed many decades ago. In 1937, a single leukemic cell was shown to be able to transmit a systemic disease when transplanted into a mouse (Furth and Kahn 1937). Subsequently, the frequency of tumor-initiating cells in leukemia was assessed by the development of quantitative methods, such as the spleen colony-forming assay (Makino 1956, Bruce and Van Der Gaag 1963). The resultant tumors showed similar morphological heterogeneity to that of the original tumor. Collectively, these pioneering studies formed the basis of the CSC concept.

Because of the development of technologies, such as flow cytometry and xenotransplantation, proof for the applicability of CSC concept was provided in the mid-1990s. Dick and colleagues found that most subtypes of acute myeloid leukemia (AML) could be engrafted in immunodeficient mice; however, the engraftment could be initiated from only $CD34^{+}CD38^{-}$ fractions (Lapidot et al. 1994). Moreover, this xenotransplantation assay revealed the frequency of CSCs, which was found to be in the order of one per million tumor cells (Bonnet and Dick 1997). Therefore, rare populations are believed to have the ability to self-renew and to produce heterogeneous tumors.

Based on a similar approach, many researchers applied the CSC concept to solid tumors. CSCs in human in breast cancer cells were identified using $CD44^{+}CD24^{-/low}$ fractions (Al-Hajj et al. 2003). To date, CSCs have been found in various solid tumors of other organs such as brain, colon, pancreas, prostate, and ovary (Collins et al. 2005, Curley et al. 2009, Dalerba et al. 2007, Ponti et al. 2005, Singh et al. 2004). Because of technical difficulties involved in purification of rare CSCs, the current methods used for enrichment of CSCs using surface markers, side populations, and the ALDEFLUOR assay are described below.

Isolation and Characterization of CSCs

Sphere formation

The sphere culture technique was originally used for isolation of normal neural stem cells. Neural stem cells and their derived progenitor cells can be enriched and expanded *in vitro* by virture of their ability to form floating aggregates, termed "neurospheres" (Reynolds and Weiss 1992). These non-adherent spheres were enriched in stem/progenitor cells and were capable of differentiating into neurons and glia. In these spheres, 4%–20% of cells were stem cells, whereas the other cells represented progenitor cells in different phases of differentiation (Weiss et al. 1996), suggesting that stem cells have been successfully enriched. The markers and receptors that regulate neural stem cell growth have been identified using this cell culture system (Hiramoto et al. 2007, Holmberg et al. 2005, Nagato et al. 2005). These non-adherent culture conditions were also adapted to other normal stem cells. Mammary stem cells grown in suspension form mammospheres, which are equivalent to neurospheres (Dontu et al. 2003).

Subsequently, the sphere culture technique was applied to expand CSCs from a variety of cancer cell lines and clinical samples (Galli et al. 2004, Singh et al. 2003, 2004). For example, it was demonstrated that CD133^{+} cells could grow as spheres, whereas CD133^{-} cells could not. Glioblastoma-derived spheres have been shown to express a number of neural stem cell markers and are highly tumorigenic (Lee et al. 2006).

Similar findings have been obtained from other types of tumors. We applied this approach to the formation of non-adherent spheres using established human cancer cell lines such as MCF-7 (Hirata et al. 2010). As shown in Fig. 1A, cancer cell lines from various tissues have an ability to grow as spheres in stem cell medium supplemented with growth factors such as basic FGF and N_2. Compared with adherent cells, sphere cells exhibited a higher expression of stemness markers (Oct 3/4, Nanog, and Sox2), suggesting that they have self-renewal properties (Fig. 1B). In addition, sphere cells from MCF-7 cells exhibited high tumorigenicity when injected in Balb/c nude mice (Fig. 1C).

Thus, sphere assay may represent a potentially valid and useful technique for enhancement of the frequency of CSCs from cell lines and clinical samples. However, stem cell populations cannot be purified entirely by the sphere technique. When a single cell from MCF-7 spheres was seeded in ultra-low dishes, some cells formed secondary spheres, whereas others did not (data not shown). CSCs from clinical tumors have been shown to be highly variable (Visvader and Lindeman 2008), which may be because the stem cell population was contaminated with more differentiated cells. Taken together, further purification and characterization of CSCs is required.

Surface markers

CSCs have been isolated using cell surface markers, which are similar to their normal counterparts. As shown in Table 1, CSCs have been isolated by various markers from many types of cancers. The choice of markers varies among tissues and species.

CD44^{+}/CD24^{-}

CSCs in human solid tumors were discovered using CD44^{+}CD24$^{-/low}$ in breast cancer cells (Al-Hajj et al. 2003). As few as 200 cells from this

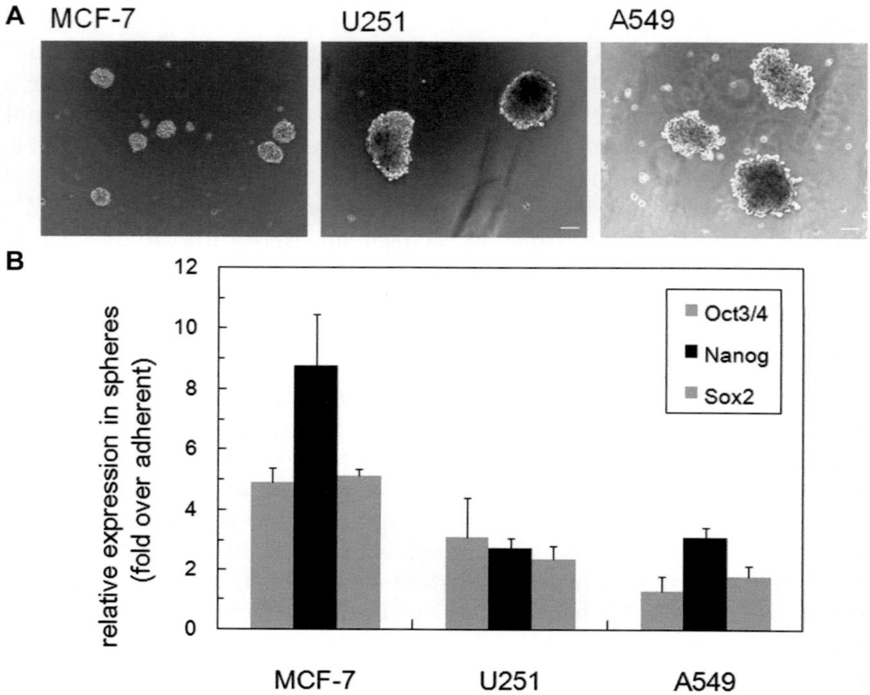

C

in vivo tumor formation

tumor / injections

control	sphere
1 / 10	4 / 10

Fig. 1. Sphere culture in established cancer cell lines. (A) Sphere formation in ultra-low attachment dishes (bar=100 μm). (B) Expression of stemness markers in sphere culture. (C) Tumor incidence indicates the number of tumors/the number of injections.

subpopulation were capable of forming tumors when injected into NOD/SCID mice, whereas many other cells did not form tumors (Al-Hajj et al. 2003). The tumors derived from this subpopulation acquired the phenotypic heterogeneity of the initial tumor, which contained a minority

Table 1. Various surface markers for isolation of CSCs.

Surface marker	Tumor type	References	Year
$CD34^+/CD38^-$	Acute myeloid leukemia	Bonnet and Dick	1997
$CD44^+/CD24^-$	Breast	Al-Hajj et al.	2003
$CD133^+$	Brain	Singh et al.	2003
$CD133^+$	Prostate	Miki et al.	2004
$CD133^+$	Colon	O'Brien et al.	2005
$CD20^+$	Melanoma	Fang et al.	2005
$CD44^+$	Head and neck	Prince et al.	2007
$CD133^+/CXCR4^+$	Pancreas	Hermann et al.	2007
$CD44^+/CD117^+$	Ovary	Zhang et al.	2008
CD90	Liver	Yang et al.	2008
CD105	Kidney	Bussolati et al.	2008
$CD49f^+$	Glioblastoma	Lathia et al.	2010
CD271	Melanoma	Boiko et al.	2010

of $CD44^+CD24^{-/low}$ cells. The $CD44^+CD24^{-/low}$ phenotype has been used to identify and isolate CSCs from breast cancer specimens and cell lines (Fillmore and Kuperwasser 2008, Ponti et al. 2005). In addition to breast cancer, CSCs in ovarian cancer cells have been isolated using $CD44^{+/}CD24^-$ (Zhang et al. 2008).

With regard to breast cancer cell lines, the MCF-7 cell line had little $CD44^+/CD24^{-/low}$ population, with most cells positive for CD44 and CD24 (Fig. 2). In contrast, almost the entire MDA-MB231 cell line showed $CD44^+/$

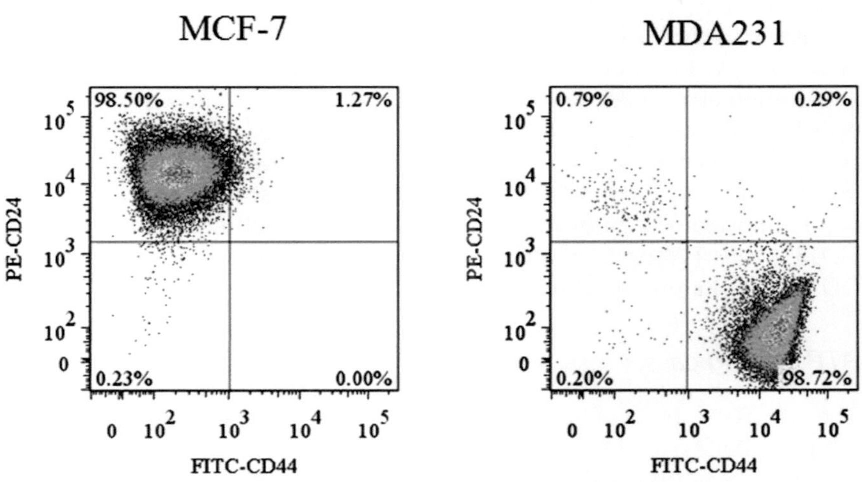

Fig. 2. Expression of CD44 and CD24 in established breast cancer cell lines.

Color image of this figure appears in the color plate section at the end of the book.

$CD24^{-/low}$ population. Although the $CD44^{+}/CD24^{-/low}$ phenotype was found to correlate with the more aggressive basal-like subtype of breast cancer (Honeth et al. 2008), this marker can not be generalized for breast cancers. It is possible that $CD44^{+}/CD24^{-/low}$ cells are still heterogeneous and that only a subgroup within $CD44^{+}CD24^{-/low}$ cells is self-renewing.

$CD34^{+}/CD38^{-}$

As described in section 2, the $CD34^{+}/CD38^{-}$ cell population has been identified as a cell surface marker of leukemic CSCs. This $CD34^{+}/CD38^{-}$ cell population had the capacity to initiate leukemia in NOD-SCID mice when compared with the $CD34^{-}$ and $CD34^{+}/CD38^{+}$ cell populations (Bonnet and Dick 1997). Although the identification of leukemic CSCs marked a major advancement in the field of stem cell research, this subpopulation is still considered heterogeneous (Sarry et al. 2011). A strict definition of leukemic CSCs is necessary to further target these cells.

CD133

CD133, also known as PROML1 or prominin, is a transmembrane glycoprotein originally identified in mouse neuroepithelial stem cells and human hematopoietic stem cells (Miraglia et al. 1997, Weigmann et al. 1997). CD133 was also found in other progenitor cells, such as endothelial cells, lymphangiogenic cells, and myoangiogenic cells (Peichev et al. 2000, Salven et al. 2003, Shmelkov et al. 2005). Although its biological function is still unclear, CD133 has been recognized as a putative CSC marker for cancers of the brain, colon, prostate and pancrea (Hermann et al. 2007, Miki et al. 2007, O'Brien et al. 2007, Ricci-Vitiani et al. 2007, Richardson et al. 2004, Singh et al. 2004). However, several reports have shown that $CD133^{-}$ cells have self-renewal properties. For example, the $CD133^{-}$ population in colonic cancer cells was shown capable of self-renewal and tumorigenicity (Shmelkov et al. 2008). $CD133^{-}$ cells derived from several glioma patients were tumorigenic in nude rats and several of the resulting tumors contained $CD133^{+}$ cells (Wang et al. 2008). Taken together, these data suggest that CD133 is not a universal marker for CSC isolation.

ATP-binding cassette sub-family B member 5 (ABCB5)

The $ABCB5^{+}$ subpopulation of melanoma cells has been shown to exhibit a high capacity for re-establishing malignancy after xenotransplantation into mice (Schatton et al. 2008). In addition, these authors reported that the systemic administration of monoclonal antibodies against ABCB5 induced antibody-dependent cell-mediated cytotoxicity in $ABCB5^{+}$ malignant melanoma-initiating cells and exerted tumor-inhibitory effects in a xenograft model.

CD49f (Integrin α6)

CD49f, also known as Integrin α6, was expressed in the stem cells of several tissues including epidermal, keratinocyte, and mammary stem cells (Fortunel et al. 2003, Jones and Watt 1993, Li et al. 1998). A small subpopulation of mouse mammary stem cells, sorted as $CD45^{-}/Ter119^{-}/CD31^{-}/Sca\text{-}1^{low}/CD24^{med}/CD49f^{high}$, was used to purify a rare subset of adult mouse mammary stem cells that were capable of regenerating an entire mammary gland *in vivo* (Stingl et al. 2006). CSCs in glioblastomas was shown to express high levels of CD49f and targeting CD49f inhibited self-renewal, proliferation, and tumor formation (Lathia et al. 2010). Another study in mice showed that knockdown of CD49f caused mammosphere-derived cells to lose their ability to grow as mammospheres and abrogated their tumorigenicity (Cariati et al. 2008). Taken together, CD49f may be a potential therapeutic target for some cancers.

CD90

CD90, also known as Thy-1, is a glycosylphosphatidylinositol-anchored protein that plays a role in T-cell activation (Haeryfar and Hoskin 2004). CD90 is also expressed in primitive cells such as mesenchymal stem cells (Dennis et al. 2007) and hepatic stem cells (La'zaro et al. 2003). CD90 has shown potential in identifying hepatic CSCs from hepatocellular carcinoma cell lines, tumor specimens, and blood samples from liver cancer patients (Yang et al. 2008).

CD271

CD271, also known as the p75 nerve growth factor receptor, is a marker for neural crest stem cells (Morrison et al. 1999). Using $NOD/SCID/IL2R\gamma c^{null}$ mice, melanoma cells expressing CD271 had a higher tumor-initiation capacity than CD271-negative cells. However, the negative fraction was also able to generate tumors in this mouse model (Boiko 2010). The association with CD271 suggests a possible developmental link between neural crest cells and melanoma.

As described above, surface markers are very useful tools with regards to the enrichment of CSCs from a variety of cancer cell lines and clinical specimens. However, there are currently no universal surface markers for a pure population of CSCs. Since many surface markers are also expressed on normal stem cells, normal stem cells can be a contaminant for the CSC population. More specific markers need to be determined.

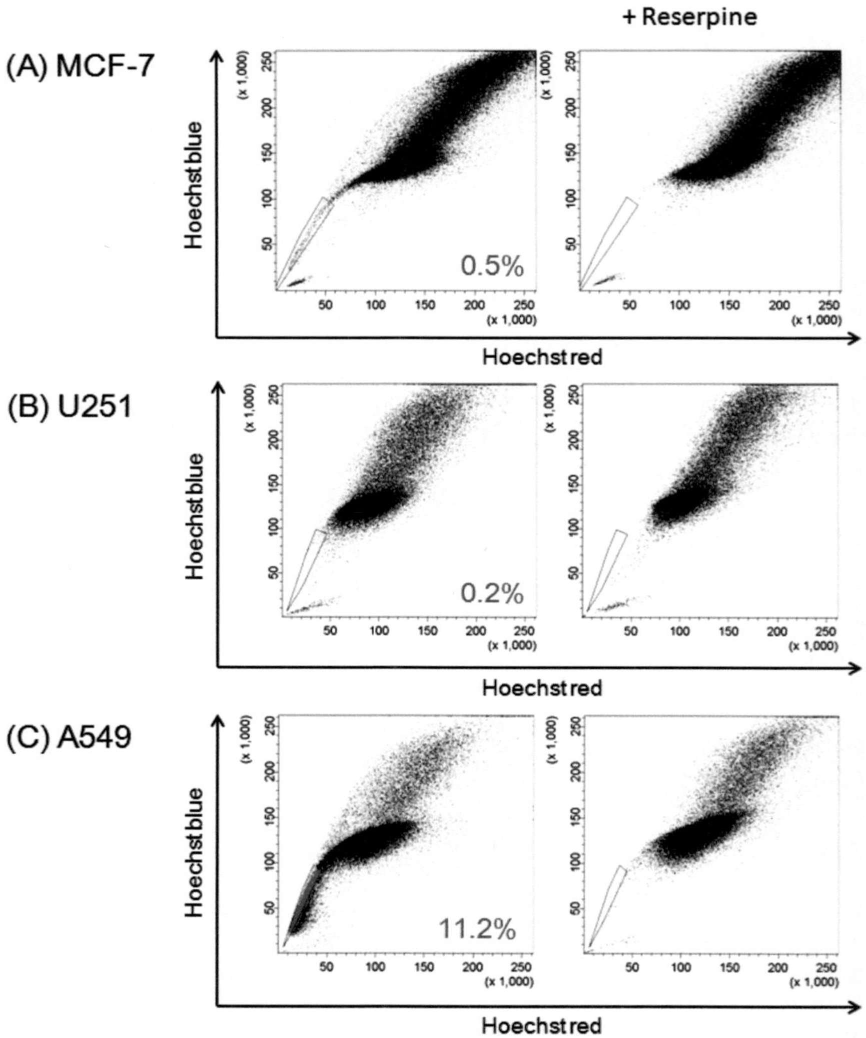

Fig. 3. Side population assay in human cancer cell lines. MCF-7 (A), U251 (B), and A549 (C) cancer cell lines were stained with Hoechst 33342 dye in the absence or presence of reserpine, an inhibitor of the ABC transporter, and were analyzed by flow cytometry.

Color image of this figure appears in the color plate section at the end of the book.

Side population

The use of Hoechst 33342 dye to identify and isolate CSCs as a side population (SP) overcomes the barrier of diverse surface markers and replaces it with more direct functional markers (Hadnagy et al. 2006). SP cells were originally identified in normal murine hematopoietic stem cells, this method being based on the efficient and specific efflux of the fluorescent DNA-binding dye Hoechst 33342 by an ATP-binding cassette (ABC) transporter (Goodell et al. 1996). Kondo et al. were the first to report SP cells in rat C6 glioma cell line (Kondo et al. 2004). Because SP cells have been identified in various cell lines and tumor specimens and exhibit higher tumorigenicity than non-SP cells (Hirschmann-Jax et al. 2004, Kondo et al. 2004, Nakanishi et al. 2010), the SP phenotype is believed to detect CSCs.

Consistent with previous reports, we also detected SP cells in cancer cell lines representing multiple tissues of origin: breast (MCF-7 cells), brain (U251 cells) and lung (A549 cells) (Fig. 3). To determine the SP gate, it is important to use an ABC transporter-blocking agent such as verapamil or reserpine as a control. Reserpine was more effective than verapamil in inhibiting efflux of Hoechst 33342 by SP cells in cancer cells. In contrast, SP cells have not been observed in human MDA-MB-231 breast cancer cells (data not shown). Some reports suggest that SP cells do not distinguish stem cells (Triel et al. 2004). Although SP cells may not be universal, the SP technique could help to identify more specific CSC markers by comparing the expression profiles of SP and non-SP cells.

ALDEFLUOR assay

Similar to the SP assay, ALDEFLUOR assay was developed as a direct functional marker of CSCs. This assay is based on the finding that human hematopoietic stem cells have increased aldehyde dehydrogenase (ALDH) activity, which is responsible for oxidizing intracellular aldehydes and results in the oxidation of retinol to retinoic acid (Hess et al. 2004).

The ALDEFLUOR assay uses an ALDH fluorescent substrate, BODIPY-aminoacetaldehyde (BAAA). BAAA is transported into living cells through passive diffusion and is converted into the reaction product BODIPY-aminoacetate (BAA^-) by intracellular ALDH. BAA^- is retained within cells and becomes brightly fluorescent (Christ et al. 2007). A specific ALDH inhibitor, diethylaminobenzaldehyde (DEAB), is used to determine background fluorescence. Thus, cells with high ALDH activity can be detected by standard flow cytometry.

As shown in Fig. 4A, we found that MCF-7 cancer cells contain approximately 1% of ALDEFLUOR-positive cells. This is consistent with the hypothesis that CSCs are a small subpopulation. CSCs with high ALDH activity generated breast tumors in NOD/SCID mice, with phenotypic

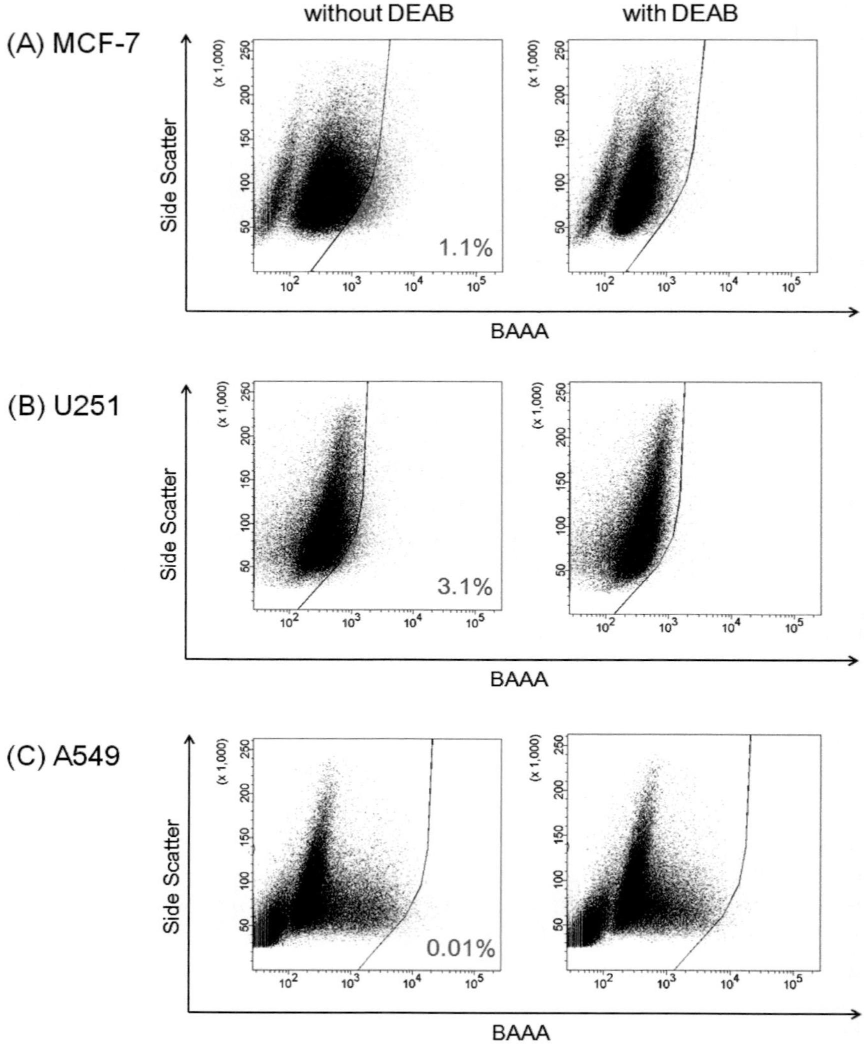

Fig. 4. ALDEFLUOR assay in human breast cancer cell lines. MCF-7 (A), U251 (B), and A549 (C) cancer cell lines were incubated with ALDEFLUOR substrate BAAA alone (left) or in the presence of the ALDH inhibitor DEAB (right), and then analyzed by flow cytometry. DEAB was used to establish the baseline fluorescence of these cells (blue) and to define the ALDEFLUOR-positive region (red).

Color image of this figure appears in the color plate section at the end of the book.

characteristics resembling those of the parental tumor (Ginestier et al. 2007). In addition, ALDH expression is associated with poor prognosis in breast cancer patients (Ginestier et al. 2007, Marcato et al. 2011). This method has been used to isolate CSCs not only from breast cancer, but also from multiple myeloma and leukemia cells (Matsui et al. 2004, Pearce et al. 2005). We also found that U251 glioma cell line contained approximately 1% of ALDEFLUOR-positive cells (Fig. 4B). These data suggest that the ALDEFLUOR assay is a potential protocol for CSC isolation.

There are several limitations of this technique. First, it is not suitable for use in certain tumors. For example, both ALDEFLUOR^{br} and ALDEFLUOR^{low} cells from the H522 lung carcinoma cell line were capable of initiating tumors after transplantation into NOD/SCID mice, and tumors generated from ALDEFLUOR^{low} cells grew faster and were larger than tumors from ALDEFLUOR^{br} cells (Ucar et al. 2009). In addition, we did not detect ALDEFLUOR-positive cells in A549 lung caercinoma cell line (Fig. 4C).

Secondly, the stem cell population identified by the ALDEFLUOR assay is presumably heterogeneous and must be dissected using additional surface markers. In breast cancer cell lines, cell selection using the CD44^{+}/CD24^{-}/ALDH^{+} phenotype increased the tumorigenicity of breast cancer cells in comparison with CD44^{+}/CD24^{-} or ALDH^{+} cells (Ginestier et al. 2007). This suggests that CSCs obtained with a given marker can be further divided into distinct metastatic or non-metastatic subpopulations using additional markers.

Self-renewal Pathways of CSCs

The CSC model provides therapeutic strategies beyond traditional anti-proliferative agents (Zhou et al. 2009). A potential approach to eliminating CSCs is by blocking the essential self-renewal signaling pathway for CSC survival. Since self-renewal is critical for both normal stem cells and CSCs, common self-renewal pathways presumably exist among them. In addition, these self-renewal pathways may be more conserved than surface markers among CSCs. These observations suggest that the search for drugs that target this common mechanism would be a powerful strategy in the field of novel drug discovery.

It has been suggested that specific signaling pathways such as Notch, Wnt/β-catenin, Hedgehog and Bmi-1 play a role in the self-renewal and differentiation of normal stem cells. Alterations in genes that encode signaling molecules belonging to these pathways have been found in human tumor samples (Lobo et al. 2007, Sánchez-García et al. 2007), suggesting that they are likely involved in CSC regulation.

Notch

The Notch signaling pathway plays an important role in the maintenance of a variety of adult stem cells, including breast, neural and intestinal stem cells, by promoting self-renewal (Dontu et al. 2004, Fre et al. 2005, Hitoshi et al. 2002). Components of the Notch pathway reportedly act as oncogenes in a wide range of human tumors including breast cancers and gliomas (Kanamori et al. 2007, Reedijk et al. 2005, Stylianou et al. 2006). In addition, breast CSCs have been shown to exist within breast cell lines and primary samples and to self-renew through the Notch pathway (Harrison et al. 2010, Hirata et al. 2010). Neurospheres derived from human glioblastoma specimens have been shown to grow through the Notch-dependent pathway (Fan et al. 2010). γ-Secretase inhibitors, which inhibit cleavage of activated Notch receptors and thereby prevent Notch signaling, may be a promising approach for clinical trial. MK-0752, γ-Secretase inhibitors, is currently undergoing clinical trials as a target for breast cancer stem cells following chemotherapy (www.ClinicalTrials.gov, number NCT00645333) and recurrent CNS malignancies (Fouladi et al. 2011).

Wnt/β-catenin

The Wnt/β-catenin signaling pathway plays an important role in embryonic development (Clevers 2006). This pathway is considered a master switch that controls proliferation versus differentiation in both stem cell and cancer cell maintenance and growth in intestinal, epidermal, and hematopoietic tissues (Reya and Clevers 2005, Van der Wetering et al. 2002). Wnt pathways are commonly hyperactivated in tumors and are required for sustained tumor growth (Reya and Clevers 2005). Several small molecule inhibitors of Wnt/β-catenin signaling have been developed. ICG-001 selectively antagonizes interactions between β-catenin and the cyclic AMP response element-binding protein (CBP), which is a transcriptional co-activator essential for β-catenin-mediated transcription (Emami et al. 2004). NSC668036 binds to the PDZ domain of the Wnt-pathway signaling molecule Disheveled and mimics the endogenous Wnt inhibitor Dapper1 (Zhang et al. 2006). XAV939 selectively inhibits β-catenin-mediated transcription. This inhibitor stimulates β-catenin degradation by stabilizing axin, which is a member of the destruction complex that induces ubiquitin-mediated degradation of β-catenin (Huang et al. 2009). Among these Wnt inhibitors, an ICG-001 analog (known as PRI-721) is currently being tested in a clinical trial in patients with gastrointestinal cancer.

Hedgehog

The Hedgehog signaling pathway was initially identified in Drosophila as a mediator of segmental patterning during development (Nusslein-Volhard and Wieschaus 1980). This pathway is also essential for maintaining the normal adult stem cell population (Ingham and McMahon 2001). Xu et al.

identified a Hedgehog-dependent subset of brain tumor stem cells (Xu et al. 2008). Inhibition of Hedgehog signaling has been shown to be effective in a pancreatic cancer xenograft model (Jimeno et al. 2009). Moreover, the Hedgehog pathway has also been implicated in maintaining human leukemic CSCs (Dierks et al. 2008, Zhao et al. 2009). Loss of smoothened, which is a Hedgehog pathway component, resulted in depletion of chronic myeloid CSCs. Based on these data, many inhibitors of this pathway are currently under development (Mahindroo et al. 2009). For example, GDC-0449 was originally identified as a smoothened antagonist in a chemical compound screen (Robarge et al. 2009) and has been used in a clinical trial in patients with solid tumor (Von Hoff et al. 2009). BMS-833923 (XL139) was also used in a clinical trial for uncontrolled basal cell nevus syndrome (Siu et al. 2010).

Bmi-1

Bmi-1 is one of the polycomb group of genes, which have been shown to play a role in self-renewal of hematopoietic and neural stem cells (Molofsky et al. 2005, Park et al. 2004). Aberrant expression of Bmi-1 contributes to cancer development, such as that of the breast, head and neck, and myeloid leukemia (Sawa et al. 2005, Silva et al. 2007, Song et al. 2006). Bmi-1 has been reported to express in head and neck CSCs (Prince et al. 2007). Knockdown of Bmi-1 inhibited CSC self-renewal capacity in breast and hematopoietic cancer (Liu et al. 2006, Rizo et al. 2009).

Source of CSCs in Novel Drug Discovery

According to the CSC model, new cancer therapies should focus on targeting and eliminating CSCs, which requires characterization of CSCs from clinical specimens or established cancer cell lines.

Clinical CSC samples are difficult to obtain and expand *in vitro*. Our experience with breast cancer cells obtained from clinical tumors indicates that a common, distinctive feature of breast CSCs is currently not available. Since a large number of cells would be required for high-throughput screening of lead compounds, clinical specimens are not suitable for drug development. In contrast, CSCs in human cancer cell lines could serve as a suitable model for both drug discovery and elucidation of the mechanism of disease. CSCs in human cancer cell lines are easily accessible and provide a simple model for obtaining reproducible results.

We recently reported the validity of CSCs in MCF-7 cells (Hirata et al. 2010). A concentration of nicotine closely related to the blood concentration in cigarette smokers (10nM–10μM) increases the proportion of the $ALDH^+$ population in MCF-7 cells. This population, which forms mammospheres and was characterized with respect to Notch signaling, is related to breast cancer development. Another study showed that the estrogen/FGF/Tbx3 signaling axis regulates CSC numbers both *in vitro* and *in vivo* by using a proportion of the $CD44^+$/$CD24^-$/ESA^+ population in MCF-7 cells (Fillmore et al. 2010). In addition, identification of novel anti-cancer drugs by the use of drug-screening platforms that target CSCs have begun (Gupta et al. 2009, Pollard et al. 2009). Considerable efforts should be made to assess the validity, optimal experimental conditions, and genetic stability of such a screening system (van Staveren et al. 2009).

Controversy Regarding CSCs

As described above, growing evidence has been accumulated to support the CSC model. Nevertheless, there is still controversy regarding CSCs.

It is still unclear how CSCs are generated. Do CSCs really exist within cancer cells? Is there a hierarchy in CSCs for all tumors? It has been speculated that normal stem cells in various tissues are malignantly transformed by multiple steps such as genetic and epigenetic mutations (Visvader and Lindeman 2008). It is not understood whether tumor progression-driving genetic events accumulate only in CSCs. Although most solid tumors show extensive genomic instability (Charames and Bapat 2003), there is no information regarding genomic stability in CSCs. A recent study suggests conversion between CSCs and non-CSCs (Chaffer et al. 2011, Gupta et al. 2011). These authors demonstrated the possibility that the dedifferentiation of transformed malignant cells results in the production of CSCs. This plasticity may account for the current inconsistencies observed in the CSC model.

Another inconsistency widely observed is that CSCs are always rare, this view is based on the original data on AML CSCs (Bonnet and Dick 1997). However, the validity of xeno-transplantation is questionable, as the microenvironment in mice is not suited for tumorigenicity with regard to human cancer cells. Although rare CSCs are considered to cause disease relapse, tumor relapse is difficult to study by xeno-transplantation of mice, whose life-spans is a maximum of two years. Several studies have reported means to overcome the initial problem of xeno-transplantation. In a mouse model of hepatocellular carcinoma, only the $CD133^+CD45^-$ subpopulation was found to generate tumors in nude mice (Rountree et al. 2009). Another study showed the existence of CSCs expressing both myoepithelial and luminal markers ($CD29^{high}CD24^{high}$) in a p53-null mouse mammary tumor model (Zhang et al. 2008). The frequency of CSCs was only 1:300, supporting the notion that CSCs are a minor subpopulation. However, the frequency of CSCs may depend on the experimental conditions of xeno-transplantation

(Quintana et al. 2008). Further investigation is required to establish an ideal animal model for the elucidation of CSC phenotype.

Conclusion

CSCs play a central role in the field of cancer biology, and evidence is accumulating that CSCs are involved in tumorigenesis. CSCs in established cancer cell lines are considered good *in vitro* models. These CSCs can be easily isolated by the protocols described above and are useful in chemical screening. Considering the variation of CSCs in established cancer cell lines, it is essential to optimize the protocol of CSC isolation and characterize CSCs. Moreover, it currently remains unknown how CSCs function in patients. Future studies are required to investigate whether the CSCs identified in cancer cell lines have the same properties as those in clinical specimens.

Although the contribution of CSCs to cancer development remains unclear, it is an attractive notion that conventional chemotherapies eliminate differentiated or differentiating cells forming the bulk of the tumor, although they are not able to eliminate CSCs. Targeting of CSCs will hopefully lead to the development of new drugs that can eliminate cancer cells and prevent recurrence and metastasis.

Acknowledgments

This work was supported by a grant from Health Sciences of the National Institute of Biomedical Innovation (No. 09-02), a Grant-in-Aid for Scientific Research from the Ministry of Education, Culture, Sports, Science, and Technology, Japan (No. 23590322) and a Health and Labour Sciences Research Grant from the Ministry of Health, Labour and Welfare, Japan, and a grant from the Smoking Research Foundation.

References

Al-Hajj, M., M.S. Wicha, A. Benito-Hernandez, S.J. Morrison, and M.F. Clarke. 2003. Prospective identification of tumorigenic breast cancer cells. Proc Natl Acad Sci USA 100: 3983–3988.

Boiko, A.D., O.V. Razorenova, M. van de Rijn, S.M. Swetter, D.L. Johnson, D.P. Ly, P.D. Butler, G.P. Yang, B. Joshua, M.J. Kaplan, M.T. Longaker, and I.L. Weissman. 2010. Human melanoma-initiating cells express neural crest nerve growth factor receptor CD271. Nature 466: 133–137.

Bonnet, D., and J.E. Dick. 1997. Human acute myeloid leukemia is organized as a hierarchy that originates from a primitive hematopoietic cell. Nat Med 3: 730–737.

Bruce, W.R., and H. Van Der Gaag. 1963. A quantitative assay for the number of murine lymphoma cells capable of proliferation *in vivo*. Nature 199: 79–80.

Bussolati, B., S. Bruno, C. Grange, U. Ferrando, and G. Camussi. 2008. Identification of a tumorinitiating stem cell population in human renal carcinomas. FASEB J 22: 3696–3705.

Cariati, M., A. Naderi, J.P. Brown, M.J. Smalley, S.E. Pinder, C. Caldas, and A.D. Purushotham. 2008. α-6 integrin is necessary for the tumourigenicity of a stem cell-like subpopulation within the MCF7 breast cancer cell line. Int J Cancer 122: 298–304.

Chaffer, C.L., I. Brueckmann, C. Scheel, A.J. Kaestli, P.A. Wiggins, L.O. Rodrigues, M. Brooks, F. Reinhardt, Y. Su, K. Polyak, L.M. Arendt, C. Kuperwasser, B. Bierie, and R.A. Weinberg. 2011. Normal and neoplastic nonstem cells can spontaneously convert to a stem-like state. Proc Natl Acad Sci USA 108: 7950–7955.

Charames, G.S., and B. Bapat. 2003. Genomic instability and cancer. Curr Mol Med 3: 589–596.

Christ, O., K. Lucke, S. Imren, K. Leung, M. Hamilton, A. Eaves, C. Smith, and C. Eaves. 2007. Improved purification of hematopoietic stem cells based on their elevated aldehyde dehydrogenase activity, Haematologica 92: 1165–1172.

Clevers, H. 2006. Wnt/β-catenin signaling in development and disease. Cell 127: 469–480.

Collins, A.T., P.A. Berry, C. Hyde, M.J. Stower, and N.J. Maitland. 2005. Prospective identification of tumorigenic prostate cancer stem cells. Cancer Res 65: 10946–10951.

Curley, M.D., V.A. Therrien, C.L. Cummings, P.A. Sergent, C.R. Koulouris, A.M. Friel, D.J. Roberts, M.V. Seiden, D.T. Scadden, B.R. Rueda, and R. Foster. 2009. Cd133 expression defines a tumor initiating cell population in primary human ovarian cancer. Stem Cells 27: 2875–2883.

Dalerba, P., S.J. Dylla, I.K. Park, R. Liu, X. Wang, R.W. Cho, T. Hoey, A. Gurney, E.H. Huang, D.M. Simeone, L. Ricci-Vitiani, D.G. Lombardi, E. Pilozzi, M. Biffoni, M. Todaro, C. Peschle, and R. De Maria. 2007. Identification and expansion of human colon-cancer-initiating cells. Nature 445: 111–115.

Dennis, J.E., K. Esterly, A. Awadallah, C.R. Parrish, G.M. Poynter, and K.L. Goltry. 2007. Clinical-scale expansion of a mixed population of bone Marrow derived stem and progenitor cells for potential use in bone tissue regeneration. Stem Cells 25: 2575–2582.

Dierks, C., R. Beigi, G.R. Guo, K. Zirlik, M.R. Stegert, P. Manley, C. Trussell, A. Schmitt-Graeff, K. Landwerlin, H. Veelken, and M. Warmuth. 2008. Expansion of Bcr-Abl-positive leukemic stem cells is dependent on Hedgehog pathway activation. Cancer Cell 14: 238–249.

Dontu, G., W.M. Abdallah, J.M. Foley, K.W. Jackson, M.F. Clarke, M.J. Kawamura, and M.S. Wicha. 2003. *In vitro* propagation and transcriptional profiling of human mammary stem/progenitor cells. Genes Dev 17: 1253–1270.

Dontu, G., K.W. Jackson, E. McNicholas, M.J. Kawamura, W.M. Abdallah, and M.S. Wicha. 2004. Role of Notch signaling in cell-fate determination of human mammary stem/progenitor cells. Breast Cancer Res 6: R605–615.

Emami, K.H., C. Nguyen, H. Ma, D.H. Kim, K.W. Jeong, M. Eguchi, R.T. Moon, J.L. Teo, H.Y. Kim, S.H. Moon, J.R. Ha, and M. Kahn. 2004. A small molecule inhibitor of β-catenin/CREB-binding protein transcription. Proc Natl Aca Sci USA 101: 12682–12687.

Fan X., L. Khaki, T.S. Zhu, M.E. Soules, C.E. Talsma, N. Gul, C. Koh, J. Zhang, Y.M. Li, J. Maciaczyk, G. Nikkhah, F. Dimeco, S. Piccirillo, A.L. Vescovi, and C.G. Eberhart. 2010. Notch pathway blockade depletes CD133-positive glioblastoma cells and inhibits growth of tumor neurospheres and xenografts. Stem Cells 28: 5–16.

Fang, D., T.K. Nguyen, K. Leishear, R. Finko, A.N. Kulp, S. Hotz, P.A. Van Belle, X. Xu, D.E. Elder, and M. Herlyn. 2005. A tumorigenic subpopulation with stem cell properties in melanomas. Cancer Res 65: 9328–9337.

Fearon, E.R., and B. Vogelstein. 1990. A genetic model for colorectal tumorigenesis. Cell 61: 759–767.

Fillmore, C.M., P.B. Gupta, J.A. Rudnick, S. Caballero, P.J. Keller, E.S. Lander, and C. Kuperwasser. 2010. Estrogen expands breast cancer stem-like cells through paracrine FGF/Tbx3 signaling. Proc Natl Acad Sci USA 107: 21737–21742.

Fillmore, C.M., and C. Kuperwasser. 2008. Human breast cancer cell lines contain stem-like cells that self-renew, give rise to phenotypically diverse progeny and survive chemotherapy. Breast Cancer Res 10: R25.

Fortunel, N.O., H.H. Otu, H.H. Ng, J. Chen, X. Mu, T. Chevassut, X. Li, M. Joseph, C. Bailey, J.A. Hatzfeld, A. Hatzfeld, F. Usta, V.B. Vega, P.M. Long, T.A. Libermann, and B. Lim. 2003. Comment on " 'Stemness': transcriptional profiling of embryonic and adult stem cells" and "a stem cell molecular signature". Science 302: 393.

Fre, S., M. Huyghe, P. Mourikis, S. Robine, D. Louvard, and S. Artavanis-Tsakonas. 2005. Notch signals control the fate of immature progenitor cells in the intestine. Nature 435: 964–968.

Fouladi, M., C.F. Stewart, J. Olson, L.M. Wagner, A. Onar-Thomas, M. Kocak, R.J. Packer, S. Goldman, S. Gururangan, A. Gajjar, T. Demuth, L.E. Kun, J.M. Boyett, and R.J. Gilbertson. 2011. Phase I trial of MK-0752 in children with refractory CNS malignancies: a pediatric brain tumor consortium study. J Clin Oncol 29: 3529–3534.

Furth, J., and M. Kahn. 1937. The transmission of leukemia of mice with a single cell. Am J Cancer 31: 276–282.

Galli, R., E. Binda, U. Orfanelli, B. Cipelletti, A. Gritti, S. De Vitis, R. Fiocco, C. Foroni, F. Dimeco, and A. Vescovi. 2004. Isolation and characterization of tumorigenic, stem-like neural precursors from human glioblastoma. Cancer Res 64: 7011–7021.

Ginestier, C., M.H. Hur, E. Charafe-Jauffret, F. Monville, J. Dutcher, M. Brown, J. Jacquemier, P. Viens, C.G. Kleer, S. Liu, A. Schott, D. Hayes, D. Birnbaum, M.S. Wicha, and G. Dontu. 2007. ALDH1 is a marker of normal and malignant human mammary stem cells and a predictor of poor clinical outcome. Cell Stem Cell 1: 555–567.

Goodell, M.A., K. Brose, G. Paradis, A.S. Conner, and R.C. Mulligan. 1996. Isolation and functional properties of murine hematopoietic stem cells that are replicating *in vivo*. J Exp Med 183: 1797–1806.

Gupta, P.B., T.T. Onder, G. Jiang, K. Tao, C. Kuperwasser, R.A. Weinberg, and E.S. Lander. 2009. Identification of selective inhibitors of cancer stem cells by high-throughput screening. Cell 138: 645–659.

Gupta, P.B., C.M. Fillmore, G. Jiang, S.D. Shapira, K. Tao, C. Kuperwasser, and E.S. Lander. 2011. Stochastic state transitions give rise to phenotypic equilibrium in populations of cancer cells. Cell 146: 633–644.

Hadnagy, A., L. Gaboury, R. Beaulieu, and D. Balicki. 2006. SP analysis may be used to identify cancer stem cell populations. Exp Cell Res 312: 3701–3710.

Haeryfar, S.M., and D.W. Hoskin. 2004. Thy-1: more than a mouse pan-T cell marker. J Immunol 173: 3581–3588.

Harrison, H., G. Farnie, S.J. Howell, R.E. Rock, S. Stylianou, K.R. Brennan, N.J. Bundred, and R.B. Clarke. 2010. Regulation of breast cancer stem cell activity by signaling through the Notch4 receptor. Cancer Res 70: 709–718.

Heppner, G.H., and B.E. Miller. 1983. Tumor heterogeneity: biological implications and therapeutic consequences, Cancer Metastasis Rev 2: 5–23.

Hermann, P.C., S.L. Huber, T. Herrler, A. Aicher, J.W. Ellwart, M. Guba, C.J. Bruns, and C. Heeschen. 2007. Distinct populations of cancer stem cells determine tumor growth and metastatic activity in human pancreatic cancer. Cell Stem Cell 1: 313–323.

Hess, D.A., T.E. Meyerrose, L. Wirthlin, T.P. Craft, P.E. Herrbrich, M.H. Creer, and J.A. Nolta. 2004. Functional characterization of highly purified human hematopoietic repopulating cells isolated according to aldehyde dehydrogenase activity. Blood 104: 1648–1655.

Hiramoto, T., Y. Kanda, Y. Satoh, K. Takishima, and Y. Watanabe. 2007. Dopamine D2 receptor stimulation promotes the proliferation of neural progenitor cells in adult mouse hippocampus. Neuroreport 18: 659–664.

Hirata N., Y. Sekino, and Y. Kanda. 2010. Nicotine increases cancer stem cell population in MCF-7 cells, Biochem Biophys Res Commun 403: 138–143.

Hirschmann-Jax, C., A.E. Foster, G.G. Wulf, J.G. Nuchtern, T.W. Jax, U. Gobel, M.A. Goodell, and M.K. Brenner. 2004. A distinct "side population" of cells with high drug efflux capacity in human tumor cells. Proc Natl Acad Sci USA 101: 14228–14233.

Hitoshi, S., T. Alexson, V. Tropepe, D. Donoviel, A.J. Elia, J.S. Nye, R.A. Conlon, T.W. Mak, A. Bernstein, and D. van der Kooy. 2002. Notch pathway molecules are essential for the maintenance, but not the generation, of mammalian neural stem cells. Genes Dev 16: 846–858.

Holmberg, J., A. Armulik, K.A. Senti, K. Edoff, K. Spalding, S. Momma, R. Cassidy, J.G. Flanagan and J. Frisén. 2005. Ephrin-A2 reverse signaling negatively regulates neural progenitor proliferation and neurogenesis. Genes Dev 19: 462–471.

Honeth, G., P.O. Bendahl, M. Ringnér, L.H. Saal, S.K. Gruvberger-Saal, K. Lövgren, D. Grabau, M. Fernö, A. Borg and C. Hegardt. 2008. The CD44+/CD24– phenotype is enriched in basal-like breast tumors. Breast Cancer Res 10: R53.

Huang, S.M., Y.M. Mishina, S. Liu, A. Cheung, F. Stegmeier, G.A. Michaud, O. Charlat, E. Wiellette, Y. Zhang, S. Wiessner, M. Hild, X. Shi, C.J. Wilson, C. Mickanin, V. Myer, A. Fazal, R. Tomlinson, F. Serluca, W. Shao, H. Cheng, M. Shultz, C. Rau, M. Schirle, J. Schlegl, S. Ghidelli, S. Fawell, C. Lu, D. Curtis, M.W. Kirschner, C. Lengauer, P.M. Finan, J.A. Tallarico, T. Bouwmeester, J.A. Porter, A. Bauer, and F. Cong. 2009. Tankyrase inhibition stabilizes axin and antagonizes Wnt signalling. Nature 461: 614–620.

Ingham, P.W., and A.P. McMahon. 2001. Hedgehog signaling in animal development: paradigms and principles. Genes Dev 15: 3059–3087.

Jimeno, A., G. Feldmann, A. Suárez-Gauthier, Z. Rasheed, A. Solomon, G.M. Zou, B. Rubio-Viqueira, E. García-García, F. López-Ríos, W. Matsui, A. Maitra, and M. Hidalgo. 2009. A direct pancreatic cancer xenograft model as a platform for cancer stem cell therapeutic development. Mol Cancer Ther 8: 310–314.

Jones, P.H., and F.M. Watt. 1993. Separation of human epidermal stem cells from transit amplifying cells on the basis of differences in integrin function and expression. Cell 73: 713–724.

Kanamori, M., T. Kawaguchi, J.M. Nigro, B.G. Feuerstein, M.S. Berger, L. Miele, and R.O. Pieper. 2007. Contribution of Notch signaling activation to human glioblastoma multiforme. J Neurosurg 106: 417–427.

Kondo, T., T. Setoguchi, and T. Taga. 2004. Persistence of a small subpopulation of cancer stem-like cells in the C6 glioma cell line. Proc Natl Acad Sci USA 101: 781–786.

Lapidot, T., C. Sirard, J. Vormoor, B. Murdoch, T. Hoang, J. Caceres-Cortes, M. Minden, B. Paterson, M.A. Caligiuri, and J.E. Dick. 1994. A cell initiating human acute myeloid leukaemia after transplantation into SCID mice. Nature 367: 645–648.

Lathia, J.D., J. Gallagher, J.M. Heddleston, J. Wang, C.E. Eyler, J. Macswords, Q. Wu, A. Vasanji, R.E. McLendon, A.B. Hjelmeland, and J.N. Rich. 2010. Integrin α6 Regulates Glioblastoma Stem Cells. Cell Stem Cell 6: 421–432.

La'zaro, C.A., E.J. Croager, C. Mitchell, J.S. Campbell, C. Yu, J. Foraker, J.A. Rhim, G.C. Yeoh, and N. Fausto. 2003. Establishment, characterization, and long-term maintenance of cultures of human fetal hepatocytes. Hepatology 38: 1095–1106.

Lee, J., S. Kotliarova, Y. Kotliarov, A. Li, Q. Su, N.M. Donin, S. Pastorino, B.W. Purow, N. Christopher, W. Zhang, J.K. Park, and H.A. Fine. 2006. Tumor stem cells derived from glioblastomas cultured in bFGF and EGF more closely mirror the phenotype and genotype of primary tumors than do serum-cultured cell lines. Cancer Cell 9: 391–403.

Liu, S., G. Dontu, I.D. Mantle, S. Patel, N.S. Ahn, K.W. Jackson, P. Suri, and M.S. Wicha. 2006. Hedgehog signaling and Bmi-1 regulate self-renewal of normal and malignant human mammary stem cells. Cancer Res 66: 6063–6071.

Lobo, N.A., Y. Shimono, D. Qian, and M.F. Clarke. 2007. The biology of cancer stem cells. Annu Rev Cell Dev Biol 23: 675–699.

Mahindroo, N., C. Punchihewa, and N. Fujii. 2009. Hedgehog-Gli signaling pathway inhibitors as anticancer agents. J Med Chem 52: 3829–3845.

Makino, S. 1956. Further evidence favoring the concept of the stem cell in ascites tumors of rats. Ann. NY Acad Sci 63: 818–830.

Marcato, P., C.A. Dean, D. Pan, R. Araslanova, M. Gillis, M. Joshi, L. Helyer, L. Pan, A. Leidal, S. Gujar, C.A. Giacomantonio, and P.W. Lee. 2011. Aldehyde dehydrogenase activity of breast cancer stem cells is primarily due to isoform ALDH1A3 and its expression is predictive of metastasis. Stem Cells 29: 32–45.

Matsui, W., C.A. Huff, Q. Wang, M.T. Malehorn, J. Barber, Y. Tanhehco, B.D. Smith, C.I. Civin, and R.J. Jones. 2004. Characterization of clonogenic multiple myeloma cells. Blood 103: 2332–2336.

Miki, J., B. Furusato, H. Li, Y. Gu, H. Takahashi, S. Egawa, I.A. Sesterhenn, D.G. McLeod, S. Srivastava, and J.S. Rhim. 2007. Identification of putative stem cell markers, CD133 and CXCR4, in hTERT-immortalized primary nonmalignant and malignant tumor-derived human prostate epithelial cell lines and in prostate cancer specimens. Cancer Res 67: 3153–3161.

Miraglia, S., W. Godfrey, A.H. Yin, K. Atkins, R. Warnke, J.T. Holden, R.A. Bray, E.K. Waller, and D.W. Buck. 1997. A novel five-transmembrane hematopoietic stem cell antigen: isolation, characterization, and molecular cloning. Blood 90: 5013–5021.

Molofsky, A.V., S. He, M. Bydon, S.J. Morrison, and R. Pardal. 2005. Bmi-1 promotes neural stem cell self-renewal and neural development but not mouse growth and survival by repressing the p16Ink4a and p19Arf senescence pathways. Genes Dev 19: 1432–1437.

Morrison, S.J., P.M. White, C. Zock, and D.J. Anderson. 1999. Prospective identification, isolation by flow cytometry, and *in vivo* self-renewal of multipotent mammalian neural crest stem cells. Cell 96: 737–749.

Nagato, M., T. Heike, T. Kato, Y. Yamanaka, M. Yoshimoto, T. Shimazaki, H. Okano, and T. Nakahata. 2005. Prospective characterization of neural stem cells by flow cytometry analysis using a combination of surface markers. J Neurosci Res 80: 456–466.

Nakanishi, T., S. Chumsri, N. Khakpour, A.H. Brodie, B. Leyland-Jones, A.W. Hamburger, D.D. Ross, and A.M. Burger. 2010. Side-population cells in luminal-type breast cancer have tumour-initiating cell properties, and are regulated by HER2 expression and signalling. Br J Cancer 102: 815–826.

Nusslein-Volhard, C., and E. Wieschaus. 1980. Mutations affecting segment number and polarity in Drosophila. Nature 287: 795–801.

O'Brien, C.A., A. Pollett, S. Gallinger, and J.E. Dick. 2007. A human colon cancer cell capable of initiating tumour growth in immunodeficient mice. Nature 445: 106–110.

Park, I.K., S.J. Morrison, and M.F. Clarke. 2004. Bmi1, stem cells, and senescence regulation. J Clin Invest 113: 175–179.

Pearce, D.J., D. Taussig, C. Simpson, K. Allen, A.Z. Rohatiner, T.A. Lister, and D. Bonnet. 2005. Characterization of cells with a high aldehyde dehydrogenase activity from cord blood and acute myeloid leukemia samples. Stem Cells 23: 752–760.

Peichev, M., A.J. Naiyer, D. Pereira, Z. Zhu, W.J. Lane, M. Williams, M.C. Oz, D.J. Hicklin, L. Witte, M.A. Moore, and S. Rafii. 2000. Expression of VEGFR-2 and AC133 by circulating human CD34$^+$ cells identifies a population of functional endothelial precursors. Blood 95: 952–958.

Pollard, S.M., K. Yoshikawa, I.D. Clarke, D. Danovi, S. Stricker, R. Russell, J. Bayani, R. Head, M. Lee, M. Bernstein, J.A. Squire, A. Smith, and P. Dirks. 2009. Glioma stem cell lines expanded in adherent culture have tumor-specific phenotypes and are suitable for chemical and genetic screens. Cell Stem Cell 4: 568–580.

Ponti, D., A. Costa, N. Zaffaroni, G. Pratesi, G. Petrangolini, D. Coradini, S. Pilotti, M.A. Pierotti, and M.G. Daidone. 2005. Isolation and *in vitro* propagation of tumorigenic breast cancer cells with stem/progenitor cell properties. Cancer Res 65: 5506–5511.

Prince, M.E., R. Sivanandan, A. Kaczorowski, G.T. Wolf, M. Kaplan, P. Dalerba, I.L. Weissman, M.F. Clarke, and L.E. Ailles. 2007. Identification of a subpopulation of cells with cancer stem cell properties in head and neck squamous cell carcinoma. Proc Natl Acad Sci USA 104: 973–978.

Quintana, E., M. Shackleton, M.S. Sabel, D.R. Fullen, T.M. Johnson, and S.J. Morrison. 2008. Efficient tumor formation by single human melanoma cells. Nature 456: 593–598.

Reedijk, M., S. Odorcic, L. Chang, H. Zhang, N. Miller, D.R. McCready, G. Lockwood, and S.E. Egan. 2005. High-level coexpression of JAG1 and NOTCH1 is observed in human breast cancer and is associated with poor overall survival. Cancer Res 65: 8530–8537.

Reya, T., and H. Clevers. 2005. Wnt signalling in stem cells and cancer. Nature 434: 843–850.

Reynolds, B.A., and S. Weiss. 1992. Generation of neurons and astrocytes from isolated cells of the adult mammalian central nervous system. Science 255: 1707–1710.

Ricci-Vitiani, L., D.G. Lombardi, E. Pilozzi, M. Biffoni, M. Todaro, C. Peschle, and R. De Maria. 2007. Identification and expansion of human colon-cancer-initiating cells. Nature 445: 111–115.

Richardson, G.D., C.N. Robson, S.H. Lang, D.E. Neal, N.J. Maitland, and A.T. Collins. 2004. CD133, a novel marker for human prostatic epithelial stem cells. J Cell Sci 117: 3539–3545.

Rizo, A., S. Olthof, L. Han, E. Vellenga, G. de Haan, and J.J. Schuringa. 2009. Repression of BMI1 in normal and leukemic human $CD34^+$ cells impairs selfrenewal and induces apoptosis. Blood 114: 1498–1505.

Robarge, K.D., S.A. Brunton, G.M. Castanedo, Y. Cui, M.S. Dina, R. Goldsmith, S.E. Gould, O. Guichert, J.L. Gunzner, J. Halladay, W. Jia, C. Khojasteh, M.F. Koehler, K. Kotkow, H. La, R.L. Lalonde, K. Lau, L. Lee, D. Marshall, J.C. Marsters Jr, L.J. Murray, C. Qian, L.L. Rubin, L. Salphati, M.S. Stanley, J.H. Stibbard, D.P. Sutherlin, S. Ubhayaker, S. Wang, S. Wong, and M. Xie. 2009. GDC-0449-a potent inhibitor of the Hedgehog pathway. Bioorg Med Chem Lett 19: 5576–5581.

Rountree, C.B., W. Ding, L. He, and B. Stiles. 2009. Expansion of CD133-expressing liver cancer stem cells in liver-specific phosphatase and tensin homolog deleted on chromosome 10-deleted mice. Stem Cells 27: 290–299.

Salven, P., S. Mustjoki, R. Alitalo, K. Alitalo, and S. Rafii. 2003. VEGFR-3 and CD133 identify a population of $CD34^+$ lymphatic/vascular endothelial precursor cells. Blood 101: 168–172.

Sánchez-García, I., C. Vicente-Dueñas, and C. Cobaleda. 2007. The theoretical basis of cancer-stem-cell based therapeutics of cancer: can it be put into practice? Bioessays 29: 1269–1280.

Sarry, J.E., K. Murphy, R. Perry, P.V. Sanchez, A. Secreto, C. Keefer, C.R. Swider, A.C. Strzelecki, C. Cavelier, C. Récher, V. Mansat-De Mas, E. Delabesse, G. Danet-Desnoyers, and M. Carroll. 2011. Human acute myelogenous leukemia stem cells are rare and heterogeneous when assayed in NOD/SCID/IL2Rγc-deficient mice. J Clin Invest 121: 384–395.

Sawa, M., K. Yamamoto, T. Yokozawa, H. Kiyoi, A. Hishida, T. Kajiguchi, M. Seto, A. Kohno, K. Kitamura, Y. Itoh, N. Asou, N. Hamajima, N. Emi, and T. Naoe. 2005. BMI-1 is highly expressed in M0-subtype acute myeloid leukemia. Int J Hematol 82: 42–47.

Schatton, T., G.F. Murphy, N.Y. Frank, K. Yamaura, A.M. Waaga-Gasser, M. Gasser, Q. Zhan, S. Jordan, L.M. Duncan, C. Weishaupt, R.C. Fuhlbrigge, T.S. Kupper, M.H. Sayegh, and M.H. Frank. 2008. Identification of cells initiating human melanomas. Nature 451: 345–349.

Shmelkov, S.V., S. Meeus, N. Moussazadeh, P. Kermani, W.K. Rashbaum, S.Y. Rabbany, M.A. Hanson, W.J. Lane, R. St Clair, K.A. Walsh, S. Dias, J.T. Jacobson, B.L. Hempstead, J.M. Edelberg, and S. Rafii. 2005. Cytokine preconditioning promotes codifferentiation of human fetal liver $CD133^+$ stem cells into angiomyogenic tissue. Circulation 111: 1175–1183.

Shmelkov, S.V., J.M. Butler, A.T. Hooper , A. Hormigo, J. Kushner, T. Milde., R. St Clair, M. Baljevic, I. White, D.K. Jin, A. Chadburn, A.J. Murphy, D.M. Valenzuela, N.W. Gale, G. Thurston, G.D. Yancopoulos, M. D'Angelica, N. Kemeny, D. Lyden, and S. Rafii. 2008. CD133 expression is not restricted to stem cells, and both $CD133^+$ and $CD133^-$ metastatic colon cancer cells initiate tumors. J Clin Invest 118: 2111–2120.

Silva, J., V. García, J.M. García, C. Peña, G. Domínguez, R. Díaz, Y. Lorenzo, A. Hurtado, A. Sánchez, and F. Bonilla. 2007. Circulating Bmi-1 mRNA as a possible prognostic factor for advanced breast cancer patients. Breast Cancer Res 9: R55.

Singh, S.K., I.D. Clarke, M. Terasaki, V.E. Bonn, C. Hawkins, J. Squire, and P.B. Dirks. 2003. Identification of a cancer stem cell in human brain tumors. Cancer Res 63: 5821–5828.

Singh, S.K., C. Hawkins, I.D. Clarke, J.A. Squire, J. Bayani, T. Hide, R.M. Henkelman, M.D. Cusimano, and P.B. Dirks. 2004. Identification of human brain tumour initiating cells. Nature 432: 396–401.

Siu, L.L., K. Papadopoulos, S.R. Alberts, R. Kirchoff-Ross, B. Vakkalagadda, L. Lang, C.M. Ahlers, K.L. Bennett, and J.M. Van Tornout. 2010. A first-in-human, phase I study of an oral hedgehog (HH) pathway antagonist, BMS-833923 (XL139), in subjects with advanced or metastatic solid tumors. J Clin Oncol (ASCO) abstract no 2501.

Song, L.B., M.S. Zeng, W.T. Liao, L. Zhang, H.Y. Mo, W.L. Liu, J.Y. Shao, Q.L. Wu, M.Z. Li, Y.F. Xia, L.W. Fu, W.L. Huang, G.P. Dimri, V. Band, and Y.X. Zeng. 2006. Bmi-1 is a novel molecular marker of nasopharyngeal carcinoma progression and immortalizes primary human nasopharyngeal epithelial cells. Cancer Res 66: 6225–6632.

Stingl, J., P. Eirew, I. Ricketson, M. Shackleton, F. Vaillant, D. Choi, H.I. Li, and C.J. Eaves. 2006. Purification and unique properties of mammary epithelial stem cells. Nature 439: 993–997.

Stylianou, S., R.B. Clarke, and K. Brennan. 2006. Aberrant activation of notch signalling in human breast cancer. Cancer Res 66: 1517–1525.

Triel, C., M.E. Vestergaard, L. Bolund, T.G. Jensen, and U.B. Jensen. 2004. Side population cells in human and mouse epidermis lack stem cell characteristics. Exp Cell Res 295: 79–90.

Ucar, D., C.R. Cogle, J.R. Zucali, B. Ostmark, E.W. Scott, R. Zori, B.A. Gray, and J.S. Moreb. 2009. Aldehyde dehydrogenase activity as a functional marker for lung cancer. Chem Biol Interact 178: 48–55.

van de Wetering, M., E. Sancho, C. Verweij, W. de Lau, I. Oving, A. Hurlstone, K. van der Horn, E. Batlle, D. Coudreuse, A.P. Haramis, M. Tjon-Pon-Fong, P. Moerer, M. van den Born, G. Soete, S. Pals, M. Eilers, R. Medema, and H. Clevers. 2002. The beta-catenin/TCF-4 complex imposes a crypt progenitor phenotype on colorectal cancer cells. Cell 111: 241–250.

van Staveren, W.C., D.Y. Solís, A. Hébrant, V. Detours, J.E. Dumont, and C. Maenhaut. 2009. Human cancer cell lines: Experimental models for cancer cells *in situ*? For cancer stem cells? Biochim Biophys Acta 1795: 92–103.

Visvader, J.E., and G.J. Lindeman. 2008. Cancer stem cells in solid tumours: accumulating evidence and unresolved questions. Nat Rev Cancer 8: 755–768.

Von Hoff, D.D., P.M. LoRusso, C.M. Rudin, J.C. Reddy, R.L. Yauch, R. Tibes, G.J. Weiss, M.J. Borad, C.L. Hann, J.R. Brahmer, H.M. Mackey, B.L. Lum, W.C. Darbonne, J.C. Marsters Jr, F.J. de Sauvage, and J.A. Low. 2009. Inhibition of the Hedgehog pathway in advanced basal-cell carcinoma. N Engl J Med 361: 1164–1172.

Wang, J., P.Ø. Sakariassen, O. Tsinkalovsky, H. Immervoll, S.O. Boe, A. Svendsen, L. Prestegarden, G. Rosland, F. Thorsen, L. Stuhr, A. Molven, R. Bjerkvig, and P.Ø. Enger. 2008. CD133 negative glioma cells form tumors in nude rats and give rise to CD133 positive cells. Int J Cancer 122: 761–768.

Weigmann, A., D. Corbeil, A. Hellwig, and W.B. Huttner. 1997. Prominin, a novel microvilli-specific polytopic membrane protein of the apical surface of epithelial cells, is targeted to plasmalemmal protrusions of non-epithelial cells. Proc Natl Acad Sci USA 94: 12425–12430.

Weiss, S., C. Dunne, J. Hewson, C. Wohl, M. Wheatley, A.C. Peterson, and B.A. Reynolds. 1996. Multipotent CNS stem cells are present in the adult mammalian spinal cord and ventricular neuroaxis. J Neurosci 16: 7599–7609.

Xu, Q., X. Yuan, G. Liu, K.L. Black, and J.S. Yu. 2008. Hedgehog signaling regulates brain tumor-initiating cell proliferation and portends shorter survival for patients with PTEN-coexpressing glioblastomas. Stem Cells 26: 3018–26.

Yang, Z.F., D.W. Ho, M.N. Ng, C.K. Lau, W.C. Yu, P. Ngai, P.W. Chu, C.T. Lam, R.T. Poon, S.T. Fan. 2008. Significance of $CD90^+$ cancer stem cells in human liver cancer. Cancer Cell 13: 153–166.

Zhang, L., X. Gao, J. Wen, Y. Ning, and Y.G. Chen. 2006. Dapper 1 antagonizes Wnt signaling by promoting dishevelled degradation. J Biol Chem 281: 8607–8612.

Zhang, M., F. Behbod, R.L. Atkinson, M.D. Landis, F. Kittrell, D. Edwards, D. Medina, A. Tsimelzon, S. Hilsenbeck, J.E. Green, A.M. Michalowska, and J.M. Rosen. 2008. Identification of tumor-initiating cells in a p53-null mouse model of breast cancer. Cancer Res 68: 4674–4682.

Zhang, S., C. Balch, M.W. Chan, H.C. Lai, D. Matei, J.M. Schilder, P.S. Yan, T.H. Huang, and K.P. Nephew. 2008. Identification and characterization of ovarian cancer-initiating cells from primary human tumors. Cancer Res 68: 4311–4320.

Zhao, C., A. Chen, C.H. Jamieson, M. Fereshteh, A. Abrahamsson, J. Blum, H.Y. Kwon, J. Kim, J.P. Chute, D. Rizzieri, M. Munchhof, T. Vanarsdale, P.A. Beachy, and T. Reya. 2009. Hedgehog signalling is essential for maintenance of cancer stem cells in myeloid leukaemia. Nature 458: 776–779.

Zhou, B.B., H. Zhang, M. Damelin, K.G. Geles, J.C. Grindley, and P.B. Dirks. 2009. Tumour-initiating cells: challenges and opportunities for anticancer drug discovery. Nat Rev Drug Discov 8: 806–823.

Chapter 2

Cancer Stem Cells of Sarcoma

Tomohiro Fujiwara,[1,a,2,4] Akira Kawai,[2] Akihiko Yoshida,[3] Toshifumi Ozaki[4] and Takahiro Ochiya[1,b,*]

Introduction

Sarcomas are malignant mesenchymal tumors. Mesenchymal tissue is defined as a complex of nonepithelial structures of the body exclusive of reproductive tissue, glia, and hematopoietic and lymphoid tissue. Embryonically, these nonepithelial structures are derived from the mesoderm and, to a lesser degree, from the neuroectoderm and are grouped into fibrous tissue, adipose tissue, skeletal muscle, blood and lymph vessels, and peripheral nervous tissue (Aogi et al. 2000). The word "sarcoma" is derived from the Greek word *sarkoma* meaning "fleshy outgrowth"; sarcomas can present as either a bone sarcoma or a soft tissue sarcoma (Misra et al. 2009).

Bone Sarcoma

Bone tissue is composed of many kinds of cells, such as osteoblasts, osteocytes, chondrocytes, and periosteal cells, and, in bone marrow, hematopoietic cells, adipocytes, neurocytes, fibroblasts, and vascular

[1]Division of Molecular and Cellular Medicine, National Cancer Center Research Institute, Tokyo, Japan.
[a]E-mail: tofujiwa@ncc.go.jp; tomomedvn@gmail.com
[b]E-mail: tochiya@ncc.go.jp
[2]Department of Musculoskeletal Tumor Surgery, National Cancer Center Hospital, Tokyo, Japan.
[3]Department of Pathology and Clinical Laboratories, National Cancer Center Hospital, Tokyo, Japan.
[4]Department of Orthopedic Surgery, Okayama University Graduate School of Medicine, Dentistry, and Pharmaceutical Sciences, Okayama, Japan.
*Corresponding author

smooth myocytes. Since the origin of sarcoma has not been identified, bone sarcoma classification is based on morphological findings, such as cell type, architecture, and matrix production. The morphological features of benign and malignant as well as non-neoplastic conditions may overlap. The classification of the World Health Organization (WHO) is presented in Table 1 (Fletcher et al. 2002).

Table 1. WHO classification of bone sarcoma.
(adapted from Fletcher, C.D.M., K.K. Unni., and F. Mertens. 2002.
Pathology and genetics of tumors of soft tissue and bone. p. 226. IARC Press. Lyon)

CARTILAGE TUMORS	EWING SARCOMA/PRIMITIVE
Osteochondroma	NEUROECTODERMAL TUMOR
Chondroma	Ewing sarcoma
Enchondroma	HEMATOPOIETIC TUMORS
Periosteal chondroma	Plasma cell myeloma
Multiple chondromatosis	Malignant lymphoma, NOS
Chondroblastoma	GIANT CELL TUMOR
Chondromyxoid fibroma	Giant cell tumor
Chondrosarcoma	Malignancy in giant cell tumor
Central, primary, and secondary	NOTOCHORDAL TUMORS
Peripheral	Chordoma
Dedifferentiated	VASCULAR TUMORS
Mesenchymal	Hemangioma
Clear cell	Angiosarcoma
OSTEOGENIC TUMORS	SMOOTH MUSCLE TUMORS
Osteoid osteoma	Leiomyoma
Osteoblastoma	Leiomyosarcoma
Osteosarcoma	LIPOGENIC TUMORS
Conventional	Lipoma

Table 1. contd....

Table 1. contd....

OSTEOGENIC TUMORS	LIPOGENIC TUMORS
chondroblastic	Liposarcoma
fibroblastic	NEURAL TUMORS
osteoblastic	Neurilemmoma
Telangiectatic	MISCELLANEOUS TUMORS
Small cell	Adamantinoma
Low grade central	Metastatic malignancy
Secondary	MISCELLANEOUS LESIONS
Parosteal	Aneurysmal bone cyst
Periosteal	Simple cyst
High grade surface	Fibrous dysplasia
FIBROGENIC TUMORS	Osteofibrous dysplasia
Desmoplastic fibroma	Langerhans cell histiocytosis
Fibrosarcoma	Erdheim-Chester disease
FIBROHISTIOCYTIC TUMORS	Chest wall hamartoma
Benign fibrous histiocytoma	JOINT LESIONS
Malignant fibrous histiocytoma	Synovial chondromatosis

Excluding myeloma and lymphoma, malignant primary bone sarcomas constitute 0.2% of all malignancies in adults and approximately 5% of childhood malignancies, for which data were obtained in one large series (Surveillance, Epidemiology, and End-Results (SEER) study). In North America and Europe, the incidence rate for bone sarcomas in males is approximately 0.8 new cases per 100,000 population a year (Fletcher et al. 2002). Cancer registry data with histological stratification indicate that osteosarcoma is the most common primary malignant tumor of bone, accounting for approximately 35% of cases, followed by chondrosarcoma for 25%, Ewing sarcoma for 16%, and chordoma for 8% (Table 3) (Dorfman and Czerniak 1995).

Soft tissue sarcoma

Soft tissue is defined as the supportive tissue of various organs and the nonepithelial, extraskeletal structures exclusive of lymphohematopoietic tissues. It includes fibrous connective tissue, adipose tissue, skeletal muscle, blood/lymph vessels, and the peripheral nervous system. Embryologically, most of it is derived from the mesoderm, with a neuroectodermal contribution in the case of peripheral nerves. Soft tissue sarcomas are malignant mesenchymal neoplasms that share a common embryological and connective tissue origin. Since their origin has not been clarified, the classification system commonly used in soft tissue sarcoma is also based on histopathology. The generally accepted basis for soft tissue tumor classification is also the WHO system (Table 2) (Fletcher et al. 2002).

According to the SEER study, which included 26,758 cases for 1978–2001, leiomyosarcoma was the most common sarcoma, accounting for 23% of cases. Other major histological types included malignant fibrous histiocytoma (MFH; 17%), liposarcoma (11%), dermatofibrosarcoma (10%), and rhabdomyosarcoma (4%). Together, these 6 histological types accounted for 67% of all cases (Table 3) (Toro et al. 2006). In this report, MFH was the second most common soft tissue sarcoma in their series. However, it is accepted that MHF does not show true histiocytic differentiation and its morphological pattern is shared by a variety of poorly differentiated malignancies. As a result, the diagnostic term MFH is now reserved for pleomorphic sarcomas without defined differentiation. Therefore, the decline in MFH incidence rates identified since 1990 is probably due to changes in diagnostic criteria that parallel changes in the understanding of MFH.

Problems with the current treatments

According to the histological type, treatment options for most sarcoma patients include surgical resection followed by limb or trunk reconstruction, pre-operative (neoadjuvant) and/or post-operative (adjuvant) chemotherapy, and radiotherapy. Although surgical resection is the mainstay of treatment for musculoskeletal sarcomas, chemotherapy also has a proven role in the primary therapy of certain types of bone sarcomas and a potential role for some patients with soft tissue sarcomas (Wesolowski and Budd 2010). In osteosarcoma cases, for example, recruitment of chemotherapy in the 1970s drastically improved the prognosis of patients (Ferrari et al. 2009). More recent randomized trials have shown that treatment of osteosarcoma patients with modern multiagent chemotherapy regimens, which include doxorubicin (DOX), cisplatin (CDDP), methotrexate (MTX), and ifosfamide (IFO), results in a 5-year survival rate of approximately 60–80%. Furthermore, response to neoadjuvant (preoperative) treatment has

Table 2. WHO classification of soft tissue sarcoma.
(adapted from Fletcher, C.D.M., K.K. Unni., and F. Mertens. 2002. Pathology and genetics of tumors of soft tissue and bone. p. 10-11. IARC Press. Lyon)

ADIPOCYTIC TUMORS	SO-CALLED FIBROHISTIOCYTIC TUMORS	CHONDRO-OSSEOUS TUMORS
Benign	Benign	Soft tissue chondroma
Lipoma	Giant cell tumor of tendon sheath	Mesenchymal chondrosarcoma
Lipomatosis	Diffuse-type giant cell tumor	Extraskeletal osteosarcoma
Lipomatosis of nerve	Deep benign fibrous histiocytoma	
Lipoblastoma / Lipoblastomatosis	Intermediate (rarely metastasizing)	TUMORS OF UNCERTAIN DIFFERENTIATION
Angiolipoma	Plexiform fibrohistiocytic tumor	Benign
Myolipoma	Giant cell tumor of soft tissues	Intramuscular myxoma
Chondroid lipoma	Malignant	(incl. cellular variant)
Extrarenal angiomyolipoma	Pleomorphic 'MFH' / Undifferentiated	Juxta-articular myxoma
Extra-adrenal myelolipoma	pleomorphic sarcoma	Deep ('aggressive') angiomyxoma
Spindle cell	Giant cell 'MFH' / Undifferentiated	Pleomorphic hyalinizing
Pleomorphic lipoma	pleomorphic sarcoma	angiectatic tumor
Hibernoma	with giant cells	Ectopic hamartomatous thymoma
Intermediate (locally aggressive)	Inflammatory 'MFH' / Undifferentiated	Intermediate (rarely metastasizing)

Table 2. contd....

Table 2. contd....

ADIPOCYTIC TUMORS	SO-CALLED FIBROHISTIOCYTIC TUMORS	TUMORS OF UNCERTAIN DIFFERENTIATION
Atypical lipomatous tumor/	pleomorphic sarcoma with	Angiomatoid fibrous histiocytoma
Well differentiated liposarcoma	prominent inflammation	Ossifying fibromyxoid tumor
Malignant		(incl. atypical / malignant)
Dedifferentiated liposarcoma	SMOOTH MUSCLE TUMORS	Mixed tumor/
Myxoid liposarcoma	Angioleiomyoma	Myoepithelioma/
Round cell liposarcoma	Deep leiomyoma	Parachordoma
Pleomorphic liposarcoma	Genital leiomyoma	Malignant
Mixed-type liposarcoma	Leiomyosarcoma (excluding skin)	Synovial sarcoma
Liposarcoma, not otherwise specified		Epithelioid sarcoma
	PERICYTIC (PERIVASCULAR) TUMORS	Alveolar soft part sarcoma
FIBROBLASTIC / MYOFIBROBLASTIC	Glomus tumor (and variants)	Clear cell sarcoma of soft tissue
TUMORS	malignant glomus tumor	Extraskeletal myxoid chondrosarcoma
Benign	Myopericytoma	("chordoid" type)
Nodular fasciitis		PNET / Extraskeletal Ewing tumor
Proliferative fasciitis	SKELETAL MUSCLE TUMORS	pPNET

Proliferative myositis	Benign	extraskeletal Ewing tumor
Myositis ossificans	Rhabdomyoma	Desmoplastic small round cell tumor
fibro-osseous pseudotumor of digits	adult type	Extra-renal rhabdoid tumor
Iischemic fasciitis	fetal type	Malignant mesenchymoma
Elastofibroma	genital type	Neoplasms with perivascular epithelioid
Fibrous hamartoma of infancy	Malignant	cell differentiation (PEComa)
Myofibroma / Myofibromatosis	Embryonal rhabdomyosarcoma	clear cell myomelanocytic tumor
Fibromatosis colli	(incl. spindle cell,	Intimal sarcoma
Juvenile hyaline fibromatosis	botryoid, anaplastic)	
Inclusion body fibromatosis	Alveolar rhabdomyosarcoma	
Fibroma of tendon sheath	(incl. solid, anaplastic)	
Desmoplastic fibroblastoma	Pleomorphic rhabdomyosarcoma	
Mammary-type myofibroblastoma		
Calcifying aponeurotic fibroma	VASCULAR TUMORS	
Angiomyofibroblastoma	Benign	
Cellular angiofibroma	Hemangiomas of	
Nuchal-type fibroma	subcut/deep soft tissue	

Table 2. contd....

Table 2. contd....

FIBROBLASTIC/MYOFIBROBLASTIC TUMORS	VASCULAR TUMORS
Gardner fibroma	capillary
Calcifying fibrous tumor	cavernous
Giant cell angiofibroma	arteriovenous
Intermediate (locally aggressive)	venous
Superficial fibromatoses (palmar / plantar)	intramuscular
Desmoid-type fibromatosis	synovial
Lipofibromatosis	Epithelioid hemangioma
Intermediate (rarely metastasizing)	Angiomatosis
Solitary fibrous tumor	Lymphangioma
and hemangiopericytoma	Intermediate (locally aggressive)
(incl. lipomatous hemangiopericytoma)	Kaposiform hemangioendothelioma
Inflammatory myofibroblastic tumor	Intermediate (rarely metastasizing)
Low grade myofibroblastic sarcoma	Retiform hemangioendothelioma
Myxoinflammatory	Papillary intralymphatic angioendothelioma
fibroblastic sarcoma	

Infantile fibrosarcoma	Composite hemangioendothelioma
Malignant	Kaposi sarcoma
Adult fibrosarcoma	Malignant
Myxofibrosarcoma	Epithelioid hemangioendothelioma
Low grade fibromyxoid sarcoma	Angiosarcoma of soft tissue
hyalinizing spindle cell tumor	
Sclerosing epithelioid fibrosarcoma	

Table 3. Incidence of bone and soft tissue sarcomas.

(A) Relative Frequencies of Bone Sarcomas by Histologic Type: SEER Data 1973–1987

	Cases	**Percent**
Osteosarcoma	922	35.1
Chondrosarcoma	677	25.8
Ewing sarcoma	420	16.0
Chordoma	221	8.4
Malignant fibrous histiocytoma (MFH)	149	5.7
Angiosarcoma	36	1.4
Unspecified	32	1.2
Other	170	6.4
Total	*2627*	*100.0*

(adapted from Dorfman, H.D. and B. Czerniak. 1995. Bone cancers. Cancer. 75: 203–210)

(B) Soft tissue sarcomas* diagnosed during 1978–2001 in the 9 SEER program areas by histologic type

	Cases	**Percent**
Leiomyosarcoma	6393	23.9
Malignant fibrous histiocytoma (MFH)	4577	17.1
Liposarcoma	3086	11.5
Dermatofibrosarcoma	2810	10.5
Rhabdomyosarcoma	1218	4.6
Angiosarcoma	1092	4.1
Nerve sheath tumor/MPNST	1061	4.0
Fibrosarcoma	964	3.6
Sarcoma, NOS	3424	12.8
Other specified soft tissue sarcomas	2133	8.0
Synovial sarcoma	615	2.3
Total	*26758*	*100.0*

MPNST, malignant peripheral nerve sheath tumor, NOS, not otherwise specified.

*Excluding sarcomas of bones and joints and Kaposi sarcoma.

(adapted from Toro, J.R. and L.B. Travis, H.J. Wu, K. Zhu, C.D. Fletcher, and S.S. Devesa. 2006. Incidence patterns of soft tissue sarcomas, regardless of primary site, in the surveillance, epidemiology and end results program, 1978–2001: An analysis of 26,758 cases. Int J Cancer 15; 119: 2922–2930)

become the most important predictor of outcome, as the median survival of osteosarcoma patients who have greater than 90% necrosis in the resected specimen following neoadjuvant chemotherapy is about 90% at 5 years.

However, one of the current problems is that the prognosis of osteosarcoma patients began to plateau about 20 years ago and many patients develop resistance in the standard therapies and tumor recurrence. The prognosis is much worse for patients with metastases (Longhi et al.

2006, Ta et al. 2009), most of which are lung metastases. Targeting molecules associated with sarcomagenesis, "targeted therapy," has been an exciting development in sarcoma treatment in the past 10 years. However, such therapies are currently limited for many kinds of sarcoma. Furthermore, there are no fewer cases involving metastases long after initial treatments (Halldorsson et al. 2009). Many problems remain to be solved before the prognosis of sarcoma patients improves.

Considering the characteristics and heterogeneity of sarcoma described below, it is possible that a subset of sarcoma cells might resist various stresses producing recurrences or metastases; this is known as the "cancer stem cell hypothesis." Therefore, there is a great need to identify, characterize, and target sarcoma cancer stem cells for the improvement of sarcoma treatments.

Origin of Sarcoma

Sarcomas constitute a large and heterogeneous group of neoplasms in terms of pathophysiology and molecular oncology. Considering the heterogeneity of sarcomas, a possible speculation is that pluripotent cells, such as tissue stem cells, could be involved in sarcomagenesis (Fig. 1A).

Tissue stem cells

Mesenchymal stem cell (MSC). Speculation had been that the pluripotent cells exist in bone morrow. Friedenstein et al. were the first to demonstrate this idea in an experimental approach (Friedenstein et al. 1976). Currently, MSCs from not only bone marrow stroma but also alternative sources, such as adipose tissue, have provided exciting prospects for cell-based tissue engineering and regeneration (Tuan et al. 2003, Banas et al. 2007). However, in most studies, it remains to be determined whether true stem cells are present or whether the population is instead a diverse mixture of lineage-specific progenitors (Toguchida 2009). Inconsistency in the published reports of the growth characteristics and differentiation potential of MSCs underscores the need for a functional definition of these cells. At present, one of the criteria that the International Society for Cellular Therapy (ISCT) proposed to define an MSC population is that the cells must be "greater than or equal to 95% positive for CD73 (ecto-50-nucleotidase), CD90 (Thy-1), and CD105 (endoglin), and no more than 2% of the cells may express CD34 (hematopoietic progenitor and endothelial cell marker), CD45 (pan-leukocyte marker), CD11b or CD14 (monocyte and macrophage markers), CD19 or CD79α (B cell markers), and HLA-DR (marker of stimulated MSCs)" (Kuhn and Tuan 2010). Other proposed MSC markers include CD44, CD49a, STRO-1, CD200, CD271, and CD146 (Kuhn and Tuan 2010).

It is assumed that MSC or MSC-like cells are localized not only in bone marrow but also in other tissues, such as adipose tissue, for example, and can be isolated (Fibbe 2002).

Neural crest stem cell (NCSC). Other multipotent stem cells that differentiate into bone and cartilage include neural crest stem cells. Originally, in the vertebrate embryo, neuroectodermal neural crest cells (NCC) have remarkably broad potencies, giving rise, after a migratory phase, to neurons and glial cells in the peripheral nervous system and to skin melanocytes, all being designated here as "neural" derivatives (Le Douarin et al. 2008). NC-derived cells also include non-neural, "mesenchymal" cell types, such as chondrocytes and bone cells, myofibroblasts and adipocytes, which largely contribute to the head structures in amniotes. A multipotent progenitor cells isolated from neural crest have the capacity to self-renew and to generate neurons, glia, and smooth muscle and have therefore been termed neural crest stem cells (NCSCs) (Morrison et al. 1999, Nagoshi et al. 2008). NCSCs are highly migratory and invasive and, during embryogenesis, travel to numerous tissues throughout the body (von Levetzow et al. 2011). As they are also identified in bone marrow (Nagoshi et al. 2009), the relationship between MSC and NCSC has been a matter of interest.

Clinical implications

There are some clinical features supporting the hypothesis that sarcomas originate from these tissue stem cells (Toguchida 2009).

Sarcomas containing different mesenchymal components

Three types of sarcomas, described below, represent a mixture of distinct histological subtypes in one sarcoma tissue.

Malignant mesenchymoma. Malignant mesenchymomas are rare soft tissue tumors that contain two or more distinct histological subtypes of sarcoma within the same tumor. They are generally considered high-grade neoplasms and are associated with a poor prognosis (Brady et al. 1996).

Malignant Triton tumor. Malignant triton tumors (Fig. 1Ba) are malignant peripheral nerve sheath tumors with rhabdomyosarcomatous differentiation (James et al. 2008). This entity was originally described by Masson and Martin in 1938, who suggested that the neural elements induced skeletal muscle differentiation, as was believed to occur in skeletal muscle of the Triton salamander.

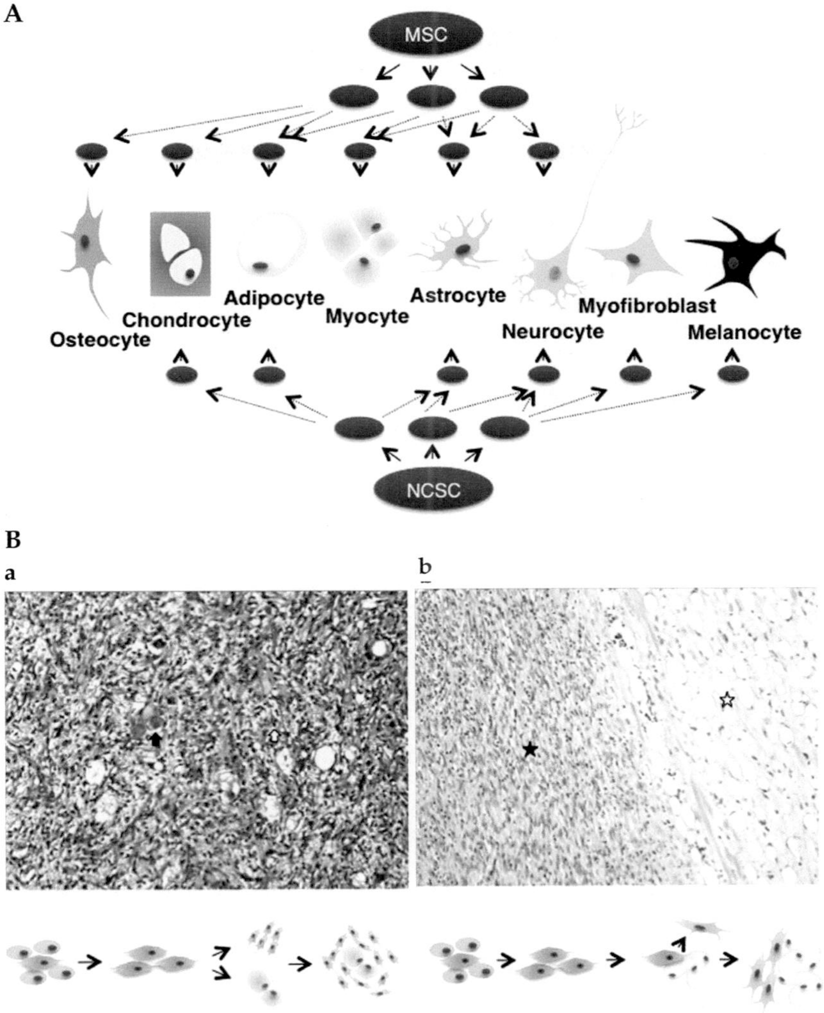

Fig. 1. Considerations for the origin of sarcoma. (A) Multi-differentiating potential of the mesenchymal stem cell (MSC) and neural crest stem cell (NCSC) (adapted from Toguchida J. Sarcoma-initiating cells and tissue stem cells. pp. 65–71. *In:* H. Esumi, N. Takakura, K. Miyazono, M. Mori [eds.] 2009. Experimental Medicine. 27. Yodosha, Tokyo, Japan). (B) Schemes of sarcomagenesis suggesting that tissue stem cells might be the origin of sarcomas. a. Malignant Triton tumor. Microscopically, the tumor cells exhibited pleomorphic (Schwannian) spindle tumor cells (white arrow) with focal rhabdomyosarcomatous differentiation (black arrow). b. Dedifferentiated liposarcoma. An abrupt transition from conventional well-differentiated liposarcoma (☆) to dedifferentiated liposarcoma composed of pleomorphic spindle cells (★).

Color image of this figure appears in the color plate section at the end of the book.

Ectomesenchymoma. Ectomesenchymoma is a rare malignant neoplasm usually consisting of rhabdomyosarcoma (RMS) with a neural component (Kawamoto et al. 1987).

These tumors are speculated to be composed of a subpopulation of cells that have two or more specific lineages of differentiation.

Dedifferentiated sarcoma

Dedifferentiation is the progression of cells toward a less differentiated state in which the original line of differentiation is no longer evident (Katoh et al. 2004). The concept of dedifferentiation was first proposed by Dahlin and Beabout in 1971, when they described dedifferentiated chondrosarcoma as a distinct clinicopathological entity characterized by a low-grade chondrosarcoma juxtaposed to a histologically different high-grade sarcoma (Dahlin and Beabout 1971).

Dedifferentiated liposarcoma. Dedifferentiated liposarcoma (DDLS) (Fig. 1Bb) is a term that was first introduced by Evans in 1979 to describe liposarcomas containing a mixture of atypical lipomatous tumor (ALT)/ well-differentiated liposarcoma (WDLS) and a high-grade nonlipogenic sarcomatous component, usually with an abrupt transition between the two components (Cha 2011). Dedifferentiated areas exhibit a wide morphological spectrum. Most cases have areas of high-grade poorly differentiated sarcoma resembling pleomorphic malignant fibrous histiocytoma, fibrosarcoma, malignant hemangiopericytoma, or high-grade myxofibrosarcoma.

Dedifferentiated chondrosarcoma. Dedifferentiated chondrosarcoma (DCS) is a high-grade, aggressive anaplastic sarcoma that progresses from a previous low-grade chondrosarcoma. DCSs have been described as "osteosarcomatous" or "fibrosarcomatous" transformations of low-grade chondrogenic neoplasms (Wick et al. 1987).

These sarcomas are speculated to include a subpopulation of the pluripotent cells with acquired novel genetic alterations that transform into completely different lineages.

Experimental implications

Some experimental results also indicate that tissue stem cells are the origins of sarcomas. If the hypothesis is right, cell lines originating from sarcoma tissues might have a potential of multi-directional differentiation (Toguchida 2009).

Multi-differentiation potential of osteosarcoma. Osteosarcoma is defined as a malignant tumor composed of neoplastic mesenchymal cells synthesizing osteoid or immature bone. On the other hand, the histological findings can be extremely variable. Osteosarcoma can be subdivided into several histological subtypes; osteoblastic, chondroblastic, and fibroblastic osteosarcoma. Aoyama et al. demonstrated that the cell line established from chondroblastic osteosarcoma expressed not only osteoblastic but also chondroblastic genetic markers and represented both histological types under a differentiation environment *in vitro* and *in vivo* (Aoyama et al. 2004). Furthermore, this cell line could differentiate into adipose, neural, muscular, and vascular lineages, suggesting ability similar to that of MSCs.

Multi-differentiation potential of synovial sarcoma. Synovial sarcoma (SS) is a mesenchymal spindle cell tumor that displays variable epithelial differentiation, including glandular formation, and has a specific chromosomal translocation t(X; 18) (p11; q11). The name "synovial" comes from the morphological similarity with joint synovium; however, it does not arise from or differentiate toward synovium, which lacks epithelial differentiation and has different histochemistry. No origin from or continuity with pre-existing epithelium has been identified. Nagayama et al. examined the genome-wide gene expression profiles of 13 SS cases and 34 other spindle cell sarcoma cases by cDNA microarray consisting of 23,040 genes (Nagayama et al. 2002). A hierarchical clustering analysis grouped SS and malignant peripheral nerve sheath tumor (MPNST) into the same category, and these two types of tumors shared expression patterns of numerous genes relating to neural differentiation. Several genes were up-regulated in almost all SS cases, and the presumed functions of known genes among them were related to migration or differentiation of neural crest cells, suggesting the possibility of the neuroectodermal origin of SS. On the other hand, Naka et al. demonstrated that SS cells, on SS18-SSX silencing with siRNAs, exhibited morphological transition from spherical growth in suspension to adherent growth in the monolayer, additional expression of later mesenchymal and hematopoietic lineage genes, and broader differentiation potentials into osteocytes, chondrocytes, adipocytes, and macrophages in appropriate differentiation environments. These data suggest that SS is a stem cell malignancy (Naka et al. 2010).

Sarcomagenesis of MSC

Several publications have questioned the ability of MSCs to undergo malignant transformation (Shima et al. 2007, Mohseny et al. 2009). Shima et al. reported the spontaneous transformation of bone marrow-derived human MSC (hMSC), isolated and expanded independently in two laboratories. They tried to immortalize hMSC by inactivating the *p16*^{INK4A} gene using the *Bmi1* gene, established immortalized human MSC (ihMSC), which retained the potential for the multi-directional differentiation of the original cells, and tested the feasibility of using ihMSC as presarcomatous cells. The transformation of ihMSC by the *H-ras* gene showed the phenotype of fully transformed cells and retained the adipogenic and chondrogenic, but not osteogenic, potential (Shima et al. 2007).

Cancer Stem Cells of Sarcoma

The cancer stem cell hypothesis proposes that, within a heterogeneous tumor, there is a small subpopulation of cells called "cancer stem cells (CSCs)" that are responsible for forming the bulk of the tumor (Clarke et al. 2006, Visvader and Lindeman 2008). These cells are considered to be similar to stem cells and may arise from the transformation of stem cells or the dedifferentiation of non-stem cells (Visvader 2011). The common consensus is that they are capable of both self-renewal and differentiation into all of the cells within a tumor (Clarke et al. 2006).

The first evidence of the existence of CSCs came from studies of hematological malignancies (Clevers 2011). Initial attempts to characterize CSCs were accomplished using cell surface molecules in acute myeloid leukemia. Several groups demonstrated that CSCs capable of initiating leukemia were found in the $CD34^+CD38^-$ fraction (Lapidot et al. 1994, Warner et al. 2004). Recently, CSCs have been isolated from several human solid tumors that have markers for putative normal stem cells, including breast cancer ($CD44^+CD24^-ESA^+$) (Al-Hajj et al. 2003), brain cancer ($CD133^+$) (Singh et al. 2004), prostate cancer ($CD44^+/\alpha 2\beta 1^{high}/CD133^+$) (Collins et al. 2005), hepatocellular carcinoma ($CD133^+$) (Yin et al. 2007), pancreatic cancer ($CD44^+CD24^-ESA^+$, $CD133^+CXCR4^+$) (Hermann et al. 2007, Li et al. 2007) and colon cancer ($CD133^+$) (Ricci-Vitiani et al. 2007).

Gibbs et al. were the first to demonstrate CSC in sarcomas in 2005. They demonstrated that spheres from osteosarcoma cell line possessed the CSC phenotype as described below. To date, many reports have been published as to CSCs or TICs in both bone and soft tissue sarcomas using various methods as follows:

1) Sphere formation
2) Side population (SP)
3) Cell surface markers
4) Self-renewal marker genes

Regardless of these methods, the common phenotype has been strong tumorigenicity *in vivo*. This might be regarded as a common consensus of CSCs. The implications of sarcoma CSCs have been reported most frequently in osteosarcoma, followed by Ewing sarcoma. These reports are summarized in Tables 4 and 5.

Bone sarcoma

Osteosarcoma-Sarcosphere, SP, ALDH, CD133, CD117, Stro-1, Oct4, and Sox2. Osteosarcoma is a primary mesenchymal tumor that is characterized histologically by the production of osteoid by malignant cells. It is the most common primary malignancy of bone, with approximately 900 new cases reported in the United States annually (Geller 2010). It represents less than 1% of cancers reported within the United States, with a peak incidence of 4.4 cases per million annually in the adolescent and young adult population. The most common primary sites are the distal femur, proximal tibia, and proximal humerus, with approximately 50% of cases originating in the vicinity of the knee.

The WHO classification recognizes additional histological variants in addition to the conventional osteosarcoma (osteoblastic, chondroblastic, and fibroblastic types): telangiectatic osteosarcoma, small cell osteosarcoma, low-grade central osteosarcoma, secondary osteosarcoma, parosteal osteosarcoma, and periosteal osteosarcoma according to the dominant histological feature (Fletcher et al. 2002).

The standard treatment of patients with conventional osteosarcoma consists of neoadjuvant chemotherapy, resection, and adjuvant chemotherapy (Marina et al. 2004). In the past, treatment of the primary tumor was amputation, whereas a high percentage of patients are currently being treated by limb salvage surgery (Bacci et al. 2000, Bielack et al. 1999, Weis 1999, Lindner et al. 1999). With combined treatment (neoadjuvant chemotherapy, surgery, and adjuvant chemotherapy), the 5-year survival for patients with no metastatic disease at diagnosis has been 60% to 80% (Provisor et al. 1997, Bacci et al. 2000, Rytting et al. 2000, Meyers et al. 2008). However, for patients who present with metastatic disease, outcomes are far worse at <30% survival (Ferguson and Goorin 2001). Pulmonary metastasis is the predominant site of osteosarcoma recurrence and the most common cause of death. The survival rate has not improved for 20 years despite multiple clinical trials with increased intensity; therefore, new therapeutic targets and approaches must be sought to suppress pulmonary metastasis of osteosarcoma for better prognosis.

Table 4. Multiple phenotypes of sarcoma CSCs in biopsies and cell lines of human/mouse bone sarcoma according to the CSC markers.

Sarcoma type	CSC marker	Cell line (No.)	Frequency, %[a]	*In vitro*				*In vivo*		Year	Refs (et al.)
				Asymmetric division	Stem cell marker	Chemo-resistance	Invasion	Tumorigenicity	Metastasis		
Osteosarcoma	Sarcosphere	Biopsy (5)	0.1–1	✓	✓					2005	Gibbs
		Human (3)	0.025–0.25		✓					2009	Wang, Fujii
		Mouse (3)	0.1–1.5	✓	✓	✓				2008	Wilson
	SP	Biopsy (5)	NA					✓		2007	Wu
		Human (2)	0.17–0.31		✓	✓	✓	✓		2008, 2009	Tsuchida, Murase
	CD133	Biopsy (2, 18)	5–7.8	✓	✓			✓		2009, 2011	Veselska, Tirino
		Human (3)	0.92–10.86		✓	✓				2008, 2009	Tirino, Kelly
	CD117/ Stro-1	Biopsy (1)	NA					✓			Adhikari
		Human (2)	1					✓		2010	
		Mouse (3)	1–3	✓	✓	✓		✓	✓		
	ALDH	Human (4)	0.59–45.07	✓	✓	✓		✓		2010	Wang, Honoki
	Oct3/4	Biopsy (1)	NA	✓	✓			✓	✓	2009	Levings
	Sox2	Biopsy (18)	>50 (IHC)		✓					2011	Basu-Roy
		Human (7)	NA		✓						
		Mouse (4)	45–74.5[b]	✓	✓		✓	✓			
Ewing sarcoma	Sarcosphere	Human (1)	0.78		✓	✓				2009	Fujii
	CD133	Biopsy (4, 48)	0–99[c]		✓	✓		✓		2009, 2010	Suva, Jiang
		Human (1)									
	ALDH	Human (5)	2	✓		✓		✓		2010	Awad

Chondrosarcoma	Sarcosphere	Biopsy (4)	0.01–0.1							2005	Gibbs
	SP	Biopsy (4)	NA							2007	Wu
	CD133	Biopsy (6)	0.21–3.5					✓		2011	Tirino
Chordoma	SP	Biopsy (1)	NA							2007	Wu
	CD133	Biopsy (1)	0.8							2011	Tirino

SP, side population. NA, not available. [a]analyzed by flow cytometry. [b]Sca-1^{high}Sox2^{high} population. [c]analyzed by immunohistochemistry.

Table 5. Multiple phenotypes of sarcoma CSCs in biopsies and cell lines of human/mouse soft tissue sarcoma according to the CSC markers.

Sarcoma type	CSC marker	Cells (No.)	Frequency, %[a]	*In vitro*				*In vivo*		Year	Refs (et al.)
				Asymmetric division	Stem cell marker	Chemo-resistance	Invasion	Tumorigenicity	Metastasis		
Leiomyo-sarcoma	SP	Biopsy (1)	3							2007	Wu
	CD133	Biopsy (1)	0.9							2011	Tirino
MFH	SP	Biopsy (4) Human (1)	NA 5.28	✓	✓			✓		2011 2009	Tirino Murase
Liposarcoma	SP	Biopsy (2)	NA							2007	Wu
	CD133	Human (3)	0.31–5.2							2011	Tirino
	ALDH/ CD133	Human (1)	NA	✓				✓		2011	Stratford
Rhabdomyo-sarcoma	SP	Human (1)	1.5–2			✓	✓	✓		2008	Tsuchida
	CD133	Human (3)	20	✓		✓		✓		2011	Walter
	FGFR-4	Human (3)	1.6–2.6	✓	✓			✓		2009	Hirotsu
Synovial sarcoma	SP	Biopsy (3)	NA					✓		2007	Wu
	CD133	Biopsy (8) Human (2)	0.84–7.23 1.6–20.5[b] 0.1–1.7							2010, 2011	Jefferson, Tirino

Fibrosarcoma	Sarcosphere	Human (1)	0.25	✓	✓					2009	Fujii
	CD133	Biopsy (4) Human (1)	0.42–1.23		✓			✓		2011	Tirino
	ALDH	Human (1)	9							2010	Honoki

SP, side population. NA, not available. [a]analyzed by flow cytometry. [b]analyzed by immunohistochemistry.

Sarcosphere. Reynolds and Weiss first cultured cells that exhibit stem cell properties, such as free-floating spheres, called neurospheres, from the adult brain (Reynolds and Weiss 1992). They dissected striatal tissue to single cells and plated them in nonadherent conditions in serum-free medium in the presence of an epidermal growth factor (EGF). Spheres from sarcoma cell lines with the similar procedures, called "sarcospheres," have been reported to possess CSC phenotypes (Gibbs et al. 2005, Wilson et al. 2008).

Gibbs et al. reported that all of the several cell lines established from patient biopsies and the MG63 cell line formed spheres at a frequency of 0.1–1% under an anchorage-independent environment with a serum-free N2 medium with growth factors, human EGF, and human FGF. The marker genes of pluripotent embryonic stem cells, Oct3/4 and Nanog, were greater in sarcospheres than in adherent cells. These spheres could form secondary spheres after dissociated single cells, suggesting that they contained a small subpopulation that had self-renewing ability.

Similarities between human and canine osteosarcoma cell lines were identified in view of the CSC phenotype of the sarcosphere (Wilson et al. 2008). Sarcosphere formation was identified in all of 3 canine cell lines, D-17, UWOS-1, and UWOS-2, and 1 human cell line, MG63, expressing Nanog, Oct4, and STAT3 at a frequency of 0.1–1.5%. Their ability to reproduce consistently after multiple passages was also confirmed. Wang et al. also identified sarcosphere formation from human cell lines OS99-1, MG63, HuO9, and SaOS2 and confirmed the expression of Oct3/4 A, Oct3/4 B, and Nanog by RT-PCR, immunohistochemistry, and flow cytometry (Wang et al. 2009).

Fujii et al. investigated drug resistance between sarcosphere and adherent cells (Fujii et al. 2009). MG63 spheres showed strong resistance to DOX and CDDP and increased expression of DNA repair enzyme genes MLH1 and MSH2 compared to adherent cells, indicating that a DNA repair inhibitor had the potential to enhance the efficacy of chemotherapeutics.

Side population (SP). In the analysis of hematopoietic stem cells, a subpopulation that effluxes the DNA-binding dye Hoechst 33342 out of the cell membrane through an ATP-binding cassette (ABC) transporter was recognized as a stem cell population (Goodell et al. 1996, Zhou et al. 2001, Ibrahim et al. 2007). This cell population expressing the ABC transporter is defined as side population (SP) cells, which are distinguished from cells of the other population (main population; MP) (Murase et al. 2009). SP cells showed CSC phenotypes characterized by asymmetric cell division, drug resistance, and tumorigenicity (Huber et al. 2005, Chiba et al. 2008). However, the following problems have been reported regarding SP cells as a CSC fraction (Wu and Alman 2008). First, cells resistant to Hoechst 33342 dye do not necessarily show tumorigenicity and metastatic ability, as CSCs do (Patrawala et al. 2005). Second, cytometry gating strategies used to isolate SP cells lack the consistency of gating strategies used in marker staining (Chiba et al. 2008). These problems might lead to cross-contamination of the SP and the non-SP cells, resulting in controversial data.

Wu et al. demonstrated that SP cells were identified in 5 surgically excised osteosarcoma samples; the frequencies were not proven (Wu et al. 2007). The SP cells from one sample were analyzed for tumorigenicity in NOD/SCID mice, showing higher tumorigenicity than MP cells. Murase et al. screened the frequency of SP cells in 7 osteosarcoma cell lines (OS2000, KIKU, NY, HuO9, HOS, U2OS, and SaOS2). As a result, only NY included SP cells, at a small percentage (0.38%; Murase et al. 2009). No further study was done to confirm the phenotype of SP cells of NY. Tsuchida et al. have shown that CDDP treatment could transform a non-tumorigenic osteosarcoma SP fraction to a highly tumorigenic phenotype, which is described later in this chapter.

The enrichment of the CSC fraction as a result of chemotherapeutics have described. Di Fiore et al. reported that 3AB-OS, i.e., MG-63 treated with 3-aminobenzamide (3AB) for 100 days, possessed much greater capacity to form spheres, stronger self-renewal ability, and higher expression of genes associated with cell cycle (*ppRb, E2F2, cyclin A, B1, D1, E*), stemness (*Oct3/4, hTERT, nucleostemin, Nanog*), and inhibition of apoptosis (*HIF-1α, FLIP-L, Bcl-2, XIAP, IAPs, survivin*) than parental MG-63 (Di Fiore et al. 2009). They were also characterized by high expression of ABCG2 and CD133, indicating that CSCs were enriched by chemotherapeutics. Tang et al. found that MTX-resistant U2OS/MTX300 cells, cultured in DMEM with 300ng/ml MTX, showed CSC phenotypes (Tang et al. 2011). They were larger in size and showed a higher potential to form spheres *in vitro* and tumors *in vivo*. Flow cytometric analysis revealed that the percentage of SP cells was 0% for U2OS and 0.11% for U2OS/MTX300, whereas the percentage of $CD117^+/Stro\text{-}1^+$ cells was 0% for U2OS and 0.6% for U2OS/MTX300, concluding that chemotherapy enrichment was a feasible and practical way to enrich osteosarcoma CSCs.

CD133. Recent studies have demonstrated that CD133 (prominin-1) is a specific marker of CSCs in a wide spectrum of malignant tumors (Singh et al. 2003, Ricci-Vitiani et al. 2007, Maeda et al. 2008). CD133 was the first identified member of the prominin family of the five-transmembrane glycoprotein with an extracellular N-terminus, a cytoplasmic C-terminus, and two large extracellular loops with eight consensus sites for N-linked glycosylation (Mizrak et al. 2008, Chen et al. 2009). The characteristic feature of CD133 is its rapid downregulation during cell differentiation (Lee et al. 2005). This feature makes CD133 a unique cell surface marker for the identification and isolation of stem cells and progenitor cells in several tissues (Chen et al. 2009, Takahashi et al. 2011).

Tirino et al. were the first to identify that CD133$^+$ cells possessed many features of CSCs (Tirino et al. 2008). SaOS2, MG-63, and U2OS cell lines contained a small fraction of CD133$^+$ cells ranging from 3% to 5% and showed the following phenotype: high proliferation rate, cell cycle detection in a G2/M phase, positive for Ki-67, formation of spheres, and inclusion of a small subset of SP cells (0.97%). Their further analysis revealed that all of the 21 primary human sarcoma samples including osteosarcoma contained a small population of CD133$^+$ cells by flow cytometric analysis (Tirino et al. 2011). CD133$^+$ cells of 2 stabilized cell lines from osteosarcoma samples, 7.8% and 5.0%, respectively, also showed self-renewal, formation of spheres, adipogenic and osteogenic differentiation, high expression of stemness genes, and tumorigenicity *in vivo*, reinforcing the hypothesis that CD133 is a reliable marker. Veselska et al. identified the cells expressing nestin in all 18 of the osteosarcoma primary samples (Veselska et al. 2009). In 4 of the 18 stabilized cell lines, 3 contained nestin$^+$/CD133$^+$ cells at a frequency of 11–100% by immunohistochemistry.

The cell growth of CD133$^+$ cells increased using the Hydro Focusing Bioreactor (HBF), simulating aspects of hypogravity (Kelly et al. 2010). Interestingly, Kelly et al., using the SaOS2 cell line containing about 11% of CD133$^+$ cells, found that the hypogravity environment of the HFB greatly sensitized the CD133$^+$ cancer cells, which are normally resistant to chemotherapy, to become susceptible to various chemotherapeutic agents.

CD117 and Stro-1. The CD117/c-kit is a 145-kDa transmembrane glycoprotein and is expressed in both hematopoietic stem cells (HSCs) and MSCs (Adhikari et al. 2010). CD117 has been suggested as a prognostic marker for osteosarcoma because its higher expression is associated with a poorer outcome in patients, metastasis, and local recurrence (Entz-Werle et al. 2005, Sulzbacher et al. 2007, Wei et al. 2008). Stro-1, originally identified as an antigen expressed by stromal elements in human bone marrow, is exclusively expressed in MSCs (Gronthos et al. 1994).

CD117$^+$Stro-1$^+$ cells of osteosarcoma showed CSC phenotypes with high invasiveness and chemoresistance (Adhikari et al. 2010). Adhikari et al. found that spheres from murine osteosarcoma cell lines (318-1, P932, and K7M2) showed higher expression of*ABCG2, CXCR4, CD117,* and *Stro-1,* and higher tumorigenicity *in vivo* than those of the monolayer (Mo) cells. Furthermore, CD117$^+$Stro-1$^+$ (DP) cells had differentiation ability, enriched for CXCR4 and ABCG2, and showed higher resistance to DOX and higher tumorigenicity and metastatic ability than $CD117^-Stro\text{-}1^-$ (DN) cells and Mo cells. These properties of $CD117^+Stro\text{-}1^+$ (DP) cells were validated on human osteosarcoma cell lines, KHOS and MNNG/HOS, and a primary cell line, BCOS. Thus, both mouse and human $CD117^+Stro\text{-}1^+$ cells of osteosarcoma were shown to possess CSC properties, including self-renewal, drug resistance, tumorigenicity, and high metastatic potential. However, the reliability of $CD117^+Stro\text{-}1^+$ in clinical samples remains controversial. Tirino et al. reported that all 21 samples in their analysis were negative for CD117 (Tirino et al. 2011). Further investigation and discussion including clinical samples are necessary to confirm the possibility that they are appropriate markers of CSCs.

ALDH. ALDH is a detoxifying enzyme responsible for the oxidation of intracellular aldehydes (Sladek 2003). ALDH has been reported to play a role in the early differentiation of stem cells in the oxidation of retinol to retinoic acid (Ambroziak et al. 1999, Duester et al. 2003). High ALDH activity has been observed in murine and human hematopoietic and neural stem and progenitor cells (Armstrong et al. 2004, Hess et al. 2004, Matsui et al. 2004). ALDH activity may provide a common marker for both normal and malignant stem and progenitor cells. For example, an increase in ALDH activity has been found in stem cell populations in multiple myeloma and AML (Matsui et al. 2004, Pearce et al. 2005). The measurement of ALDH activity has been a useful approach in the identification and isolation of CSCs in several tumors (Awad et al. 2010, Charafe-Jauffret et al. 2010, Carpentino et al. 2009).

Wang et al. found that OS99-1, derived from a highly aggressive primary human osteosarcoma, contained cells with high ALDH activity ($ALDH^{br}$ cells) at a frequency of 45.07%, while 1.84% in HuO9, 1.56% in SaOS2, and 0.59% in MG63 (Wang et al. 2010). No statistically significant difference in tumor formation was found between $ALDH^{br}$ and $ALDH^{lo}$ cells sorted from the OS99-1 cell line. Interestingly, the percentage of $ALDH^{br}$ cells in the OS99-1 xenograft was decreased to 3.13%, which was around a 15-fold reduction, suggesting that ALDH activity might differ according to the environment of the tumor. $ALDH^{br}$ cells from the xenograft showed a higher proliferation rate, clone formation, and expression of *Oct3/4A*, *Nanog*, and *Sox2* than $ALDH^{lo}$ cells from the xenograft. In an additional animal experiment, $ALDH^{br}$ cells from tumors generated from a xenograft with as few as 100 cells maintained the phenotype after serial transplantation, whereas $ALDH^{lo}$ cells from the xenograft formed no tumors with at least $1x10^4$ cells. Thus, it seemed to be more reliable to use an *in vivo* model that provides a physiologic environment to develop the CSCs that can be isolated in human osteosarcoma (Wang et al. 2010). On the other hand, Honoki et al. demonstrated that MG63 contained 11% ALDH1 positive cells with the ability to form spheres, self-renew with secondary sphere formation, and increase expression level of *Oct3/4A, Nanog, Sox2, and Stat3*. Furthermore, these spheres had a higher level of chemoresistance to DOX and CDDP than monolayer adherent cells (Honoki et al. 2010).

Oct3/4. The transcription factor Oct3/4 (also known as POU5F1 and Oct-4) is regarded as one of the key regulators of pluripotency (Jong and Looijenga 2006). Expression in nonmalignant cells is restricted to the pluripotent cells in the embryo and the primordial germ cells that will pass pluripotency to future generations via the gametes. Aberrant expression of Oct3/4 has also been suggested to fulfill an oncogenic role in tumorigenesis (Gidekel et al. 2003, Tai et al. 2005, Monk and Holding 2001).

Levings et al. found that osteosarcoma cells derived from biopsies contained a small population of self-renewed spherical clones in an anchorage-independent, serum-starved environment in which the spheres showed a significant enhancement of *Oct-4* and *Nanog* (Levings et al. 2009). A tumorigenic osteosarcoma cell line from biopsy, OS521, was engineered to activate an Oct-4 promoter/GFP reporter (OS521Oct4-p). GFP^+ cells made up 23% of the culture and 67% of the xenograft. The mesenchymal stem cell-associated surface antigens, CD105 and ICAM-1, were positive in GFP^+ cells from the xenograft tumor. The GFP^+ cells were at least 100-fold more tumorigenic, capable of forming tumors at less than 300 cells, and formed metastases in the lung. On the other hand, in the GFP-depleted group, only 1 of 8 mice developed a tumor at 3,000 cells, and none of the mice developed tumors at 300 cells. Finally, OS521Oct4-$pGFP^+$ cells were capable of self-renewal in several passages, forming heterogeneous tumors for Oct-4/GFP expression.

Sox2. Sox2 is a transcription factor of the high mobility group (HMG) domain family that plays a critical role in embryonic development and in maintaining pluripotency and self renewal in embryonic stem cells (Yuan et al. 1995, Avilion et al. 2003, Masui 2007) and several cell lineages, such as neural (Pevny and Nicolis 2010) and tracheal cells (Que et al. 2009).

Basu-Roy et al. found that Sox2 mRNA and protein were highly expressed in 7 human osteosarcoma cell lines (MG63, HOS, U20S, OS187, SJSA, OS-99-1, SaOS2, and SaOS2-LM7) and 4 murine osteosarcoma cell lines (mOS-202M, mOS-379, m0S-482, and mOS-648). All 18 osteosarcoma tissues were also positive for Sox2 at variably high levels (Basu-Roy et al. 2011). Sox2 depletion by shRNA decreased colony formation in soft agar, migration, invasion, and tumorigenicity *in vivo*. Spheres of osteosarcoma were enriched for Sox2 and Sca-1, a stem cell antigen of the hematopoietic system. The Sca-1^{hi}Sox2^{hi} fraction was identified at frequencies of 45–74.5% in 3 murine cell lines and showed asymmetric division and adipogenic differentiation. Interestingly, high Sox2 expression was accompanied by reduced Wnt signaling, while activation of the Wnt signaling resulted in low Sox2 expression, suggesting that activation of Wnt signaling led to antagonize the effect of Sox2 in maintaining osteosarcoma cells.

Ewing sarcoma-Sarcosphere, CD133, and ALDH. Since its first description by James Ewing in 1921, Ewing sarcoma has been a cryptic malignancy (Jedlicka 2010). Ewing sarcoma is relatively uncommon: it accounts for 6–8% of primary malignant bone tumors and is less common than osteosarcoma and chondrosarcoma (Fletcher et al. 2002). The most common bony sites are the long bones of the extremities, pelvis, chest wall, and spine. Lesions of long bones typically involve the diaphysis. Biologically, Ewing sarcoma is a classic example of a malignancy driven by a fusion oncogene. Detection of EWS-/FLI-1 gene fusion transcripts by RT-PCR is widely used as a tool in the diagnosis of tumors of Ewing sarcoma (Downing et al. 1995). Since Ewing's sarcoma is relatively sensitive to chemotherapy and radiotherapy, the combination with surgery, chemotherapy, and radiotherapy is the standard treatment. Multi-disciplinary care incorporating advances in diagnosis, surgery, chemotherapy, and radiation has substantially improved the survival rate of patients with localized Ewing sarcoma to nearly 70% (Subbiah et al. 2009).

Fujii et al. demonstrated that the HTB166 cell line formed spheres at a frequency of 1/128 (466 colonies/60,000 cells), which expressed a higher level of Oct3/4, Nanog, STAT3, Sox2, Sox10, and EWS-FLI1 than the adherent cells (Fujii et al. 2009). The spheres also showed higher chemoresistance to DOX and CDDP and increased expression of DNA repair enzyme genes *MLH1 and MSH2*, similarly to osteosarcoma cells.

A subpopulation of ESFT cells expressing CD133 was demonstrated to show CSC phenotypes (Suva et al. 2009). Suva et al. isolated $CD133^+$ cells from 3 ESFT surgical samples using an autoMACS device and revealed that 4% to 8% of bulk ESFT cell populations expressed CD133. Both $CD133^+$ and $CD133^-$ cells were injected beneath the renal capsule of NOD-SCID mice, which displayed higher tumorigenicity in $CD133^+$ cells than in $CD133^-$ cells. Moreover, $CD133^+$ cells displayed spherogenic potential, differentiation along adipogenic, osteogenic, and chondrogenic lineages, and a high expression level of *Oct4* and *Nanog*. Jiang et al. also focused on CD133 as a CSC marker of ESFT (Jiang et al. 2010). Their analysis demonstrated heterogeneity in CD133 expression in ESFT tumors and cell lines. CD133 expression in 48 primary ESFT samples was absent or extremely low by qRT-PCR: < 1% in 37, 1–2% in 5, 2–3% in 2, > 3% in 4 tumors. Of the last 4 patients with $CD133^+$, two cases were drug-resistant, but the others were long-term, event-free survivors. Moreover, the frequency of $CD133^+$ cells ranged from 2–99%, and, with one exception, no differences were detected in the chemoresistance or tumorigenicity between $CD133^+$ and $CD133^-$ cells.

Briefly, only a few cases showed CSC phenotypes. Therefore, their study suggests that CD133 expression alone will be insufficient to isolate drug-resistant CSCs in ESFT.

Awad et al. investigated whether ALDH^{high} cells of ESFT possessed CSC phenotypes (Awad et al. 2010). They isolated the cells with the highest and lowest 2% of ALDH activity from 5 ESFT cell lines (TC-71, MHH-ES, SK-ES-1, A4573, RD-ES). ALDH^{high} cells were enriched for clonogenicity, sphere formation, expression of *Oct4*, *Bmi-1*, and *Nanog*, chemoresistance against DOX, and tumorigenicity *in vivo*. Interestingly, ALDH^{high} cells were resistant to DOX but sensitive to YK-4-279, a small molecule of the inhibitor of EWS-FLI1.

Chondrosarcoma-Sarcosphere, SP, CD133, and ALDH. Chondrosarcomas are a heterogeneous group of malignant tumors showing hyaline cartilage differentiation. Most cases of chondrosarcomas are slow-growing and rarely metastasize (Gelderblom et al. 2008). They are primarily bone tumors and are predominantly found in adults, among whom they are the third most common malignant bone tumor after myeloma and osteosarcoma (Fletcher et al. 2002). Prognosis depends largely on the histological grade. Conventional chondrosarcomas account for the vast majority, and approximately 90% of them are low- to intermediate-grade tumors characterized by indolent clinical behavior and low metastatic potential. Most of the others are high-grade lesions, which commonly metastasize. They are generally unresponsive to chemotherapy and radiotherapy. Treatment is usually limited to surgical resection; however, patient survival with high-grade chondrosarcoma is poor, even with wide surgical resection (Ozaki et al. 1997, Jamil et al. 2010).

Gibbs et al. found that chondrosarcoma cells that originated from patient biopsies formed spherical colonies in a serum-starved, anchorage-independent environment. These spheres, as well as tissue specimens, expressed activated *STAT3, Oct3/4,* and *Nanog* more than adherent cells. They also displayed Stro-1, CD105, and CD44 as well as the marker genes of mesodermal, ectodermal, and endodermal differentiation (Gibbs et al. 2005). Wu et al. identified SP cells in 2 of 4 primary tissue samples by flow cytometry (Wu et al. 2007); however, the frequencies were not proven. Tirino et al. identified CD133^{+} cells in all 6 clinical samples at a frequency of 0.39–3.5% by flow cytometry. Of 6 samples, 2 could be used to establish cell lines, which formed spheres and showed higher tumorigenicity in NOD-SCID mice than adherent cells (Tirino et al. 2011). To date, no functional analysis according to the markers has been conducted in chondrosarcoma.

Chordoma-CD133, and SP. Chordoma is a rare, low-grade, primary malignant bone tumor arising from primitive notochord remnants of the axial skeleton. The sacrum represents a more common anatomical site of origin accounting for 50–60% of all cases followed by the skull base region (spheno-occipital/nasal) (25–35% of cases), the cervical vertebrae (approximately 10% of cases), and the thoracolumbar vertebrae (approximately 5% of cases). Although it is considered to be of low metastatic potential, up to 40–60% of patients are reported to develop distant metastases over the course of their disease (Higinbotham et al. 1967, Chambers et al. 1979, Catton et al. 1996). Adequate wide surgery still remains the cornerstone of treatment for chordomas even though safe margins are often hard to obtain because of their anatomical sites of origin. Sensitivity to chemotherapy is very low and generally reported in a small subgroup of patients with high-grade dedifferentiated chordomas and agents active in high-grade sarcomas (Fleming et al. 1993). Proton radiotherapy may succeed in offering better tumor control and fewer side effect even if it is still not as readily available in comparison to external-beam radiotherapy. The 5- and 10-year survival rates of the entire series of patients were 76 and 59%, respectively (Ferraresi et al. 2010).

Only two studies have been reported on the markers of CSC in chordoma. Tirino et al. found that chordoma biopsy cells from the sacrum of a 39-year-old female contained CD133^{+} cells at a frequency of 0.8%. The cells were positive for CD90, CD44, and CD117 at 13.98%, 14.56%, and 0.01%, respectively (Tirino et al. 2011). Wu et al. reported that SP cells were found in a primary chordoma sample; however, the amount was not indicated (Wu et al. 2007). The functional analysis of CSC phenotypes would be the next step for chordoma CSC research.

Soft tissue sarcoma

Synovial sarcoma: CD133 and SP. Synovial sarcomas account for up to 10 percent of soft-tissue sarcomas and include two major histological subtypes, biphasic and monophasic, defined respectively by the presence and absence of glandular epithelial differentiation in a background of spindle tumor cells (Kawai et al. 1998). They can occur anywhere in the body and feature local invasiveness and a propensity to metastasize (Ferrari et al. 2008). A characteristic SYT–SSX fusion gene resulting from the chromosomal translocation t(X;18) (p11;q11) is detectable in almost all synovial sarcomas. Although treatment is predicated on surgery, adjuvant radiation and/or chemotherapy, such as DOX and IFO, may be beneficial, particularly in high-risk patients. Eilber et al. demonstrated that the 4-year disease-free survival of patients treated with ifosfamide was 88% compared with 67% for the patients who did not undergo chemotherapy (Eilber et al. 2007).

Terryet al. investigated 5 primary synovial sarcoma and 3 synovial sarcoma cell lines for the expression of CD133 by flow cytometry and immunohistochemistry. Subsequently, the Fuji, SYO-1, and KU-SS-1 cell lines contained 2.6–16.8% of CD133^{+} cells, and all 5 of the primary samples showed the presence of CD133 at a frequency of 1.5–20.5%. CD133$^+$ of three cell lines were dispersed and appeared to have dendritic processes (Terry et al. 2010). Tirino et al. also identified CD133$^+$ cells from the biopsies of 3 patients at a frequency of 0.84–7.23%. All samples were negative for CD117 (Tirino et al. 2011). No functional analysis on CSC phenotypes was conducted. Wu et al. found SP cells in 2/3 of cases, although the frequencies were not shown. One sample was sorted into both SP and non-SP cells, and SP cells formed tumors at a higher frequency than non-SP cells (Wu et al. 2007).

Leiomyosarcoma: SP and CD133. Leiomyosarcoma is a malignant tumor composed of cells showing distinct smooth muscle features. Soft-tissue leiomyosarcoma usually occurs in middle-aged or older individuals, although it may develop in young adults and even children. It arises in retroperitoneal lesions (40–45%), extremities (30–35%), skin (15–20%), and larger blood vessels (5%). Surgical wide resection is the most reliable treatment. Although the effectiveness of chemotherapy and radiotherapy is uncertain, a clear survival benefit is evident if surgical margins are narrow or not clear of tumor cells. The prognosis is poor, with survival rates among the lowest of all soft tissue sarcomas (Mankin et al. 2004). For patients with leiomyosarcomas in the extremities, Farshid et al. reported 10–25% as local recurrence rate, 64% as 5-year survival rate (Farshid et al. 2002).

SP cells were detected in 2 primary leiomyosarcoma samples, which were positive at about 3% in one sample (Wu et al. 2007). The other sample was grade 2 and did not contain SP cells. Tirino et al. reported that CD133$^+$ cells were present at 0.9% in a biopsy sample from the scapulohumeral region in a 39-year-old male (Tirino et al. 2011). The cells were positive at 78.45% for CD90 and 15.0% for CD44 and negative for CD117. Neither study included an *in vitro* and *in vivo* functional analysis, which should be considered in any further research.

Malignant fibrous histiocytoma (MFH): SP. In 2002, the World Health Organization (WHO) declassified MFH as a formal diagnostic entity and renamed it as an undifferentiated pleomorphic sarcoma not otherwise specified (NOS) (Fletcher et al. 2002). While it remains unclear how to most accurately organize these tumors, the term malignant fibrous histiocytoma represents the diagnosis for thousands of patients and is still commonly used by both patients and physicians. MFH of soft tissue typically presents in a patient that is approximately 50 to 70 years of age, although it can appear at any age. It can arise in any part of the body but most commonly in the lower extremities, especially the thigh. Since MFH tends to be resistant to chemotherapy and radiotherapy, surgical resection is the most reliable treatment. Chemotherapy has often been administered to patients who already have metastatic disease or who are at the highest risk for metastasis.

The overall 5-year survival rate of undifferentiated pleomorphic sarcomas/NOS of the extremities has been reported to be 60–80% (Lehnhardt et al. 2009).

Murase et al. demonstrated that the MFH2003 cell line contained 5.28% of SP cells. These cells showed a higher capacity for the formation of spherical colonies, self-renewal, differentiation into non-SP cells, and greater tumorigenic potential than non-SP cells *in vivo* (Murase et al. 2009). Wu et al. investigated SP cells in four primary leiomyosarcoma samples, three of which were positive. One sample of grade 3 contained SP cells at a frequency of 10%. Compared with other grade 1 and 2 samples, higher-grade tumors included a higher proportion of SP cells. SP cells could form tumors at 100 cells, whereas non-SP cells could not (Wu et al. 2008).

Liposarcoma: ALDH/CD133 and SP. Liposarcomas (LSs) are subdivided into four major types: atypical lipoma (ATL)/well-differentiated liposarcoma (WDLS), myxoidliposarcoma (MLS), pleomorphic liposarcoma (PLS), and mixed liposarcoma. Dedifferentiated liposarcoma (DDLS) is defined as a WDLS that shows abrupt transition to a non-lipogenic sarcoma at least several millimeters in diameter. In addition to distinctive morphology, each of the subgroups has a different prognosis and treatment strategy. Surgical resection is the mainstay of curative treatment, however, large, high-grade liposarcomas may benefit from multimodality treatment with chemotherapy and radiation. MLS is relatively chemosensitive in comparison to the other types (Jones et al. 2005). Although the prognosis of WDLS is good, the 5-year survival of DDLS is about 30% and that of liposarcoma of the retroperitoneum is much worse.

Stratford et al. demonstrated that $ALDH^{high}CD133^{high}$ cells of liposarcoma have CSC characteristics (Stratford et al. 2011). Immunohistochemical analyses of ALDH1 expression in liposarcoma patient samples including a range of subtypes of WDLS, DDLS, MLS, and PLS confirmed that all 10 samples expressed ALDH1. Furthermore, 8 of 10 samples expressed ALDH1 in more than 50% of the tumor cells. To determine whether $ALDH^{high}$ cells show CSC activity, they used a xenograft model of the SW872 cell line from a surgical specimen with histopathology of undifferentiated malignant liposarcoma. In their screening of xenograft derived cells for the presence of MSC markers and CSC markers, $ALDH^{high}CD133^{high}$ cells were observed in a small population (0.1–1.7%) in a stem cell medium but not a RPMI medium. These cells have an enhanced ability to form spheroids, to differentiate into adipocytes, and to form tumors more efficiently *in vivo* than $ALDH^{+}CD133^{-}$, $ALDH^{-}CD133^{+}$, and $ALDH^{-}CD133^{-}$ cells. Wu et al. investigated SP cells in one MLS and one PLS, both of which were positive for SP fractions, although the frequencies were not reported (Wu et al. 2008). Tirino et al. found $CD133^{+}$ cells in 3 patient biopsies at 0.31–5.20%.

All samples were negative for CD117 (Tirino et al. 2011). Further analysis of CSC phenotypes would be expected.

Rhabdomyosarcoma: SP, CD133 and FGFR3. Rhabdomyosarcoma (RMS) is the most common soft tissue sarcoma in childhood, representing 5 to 8% of all pediatric malignancies (De Giovannni et al. 2009). Histopathologically, RMS is classified into three types; embryonal (eRMS), alveolar (aRMS), and pleomorphic types. It occurs in most parts of the body but more frequently in spaces surrounding the brain, the trunk, and genitourinary tract (McDowell 2003). It has been suggested that mesenchymal stem cells (MSCs) might be the origin of rhabdomyosarcomas (Hirotsu et al. 2009, Merlino and Khanna 2007), whereas some reports indicate that also neuronal cells can transform into malignant myogenic cells after activation and a large number of neuronal genes are expressed in RMS. Hence, the origin of potential RMS stem cells remains to be determined (Galli et al. 2000, Wachtel et al. 2006). All cases with RMS are treated with chemotherapy. Depending upon the size and location of the primary tumor and how much of it can be surgically removed, most cases will also receive some combination of radiation therapy and surgery. Adult patients who had a complete response to chemotherapy had a 5-year survival rate of 57% compared with a rate of only 7% for poor responders (Esnaola et al. 2001).

Tsuchida et al. found an SP fraction in RH-4, a drug-resistant rhabdomyosarcoma cell line, at a frequency of 1.5–2.0% (Tsuchida et al. 2008). SP cells were resistant to CDDP treatment, and tumors formed after injection of 2.5×10^5 cells, whereas tumors did not form with non-SP cells. CDDP increased the SP fraction and enhanced the clonogenic activity and stemness gene expression of SP cells. "Rhabdospheres" formed from eRMS cells, which are enriched in a $CD133^+$ CSC population (Walter et al. 2011). Three eRMS cell lines (RD, Rh36, and Ruch2) formed spheres in a stem cell medium, which is more tumorigenic *in vivo* than their adherent cells. Stem cell genes, such as *Oct4*, *Nanog*, *c-Myc*, *Pax3*, and *Sox2*, are significantly upregulated in spheres that were differentiated into adipocytes, myocytes, and neuronal cells. Furthermore, CD133 was upregulated in rhabdospheres both on mRNA and protein levels. RD $CD133^+$ sorted cells were subsequently shown to be more tumorigenic and more resistant to cisplatin and chlorambucil. Finally, a tissue microarray of eRMS patients was performed and found that high expression of CD133 correlated with poor overall survival.

Hirotsu et al. screened many markers, including SP, CD9, CD10, CD13, CD29, CD31, CD34, CD44, CD117, CD113, FLT3, LNGFR, and FGFR3, for KYM-1, RD, and A204 cell lines. These markers contained a small population of $FGFR3^+$ cells at a frequency of 1.6–2.6%. Surprisingly, 2 of 6 mice inoculated with only single FGFR3$^+$ KYM-1 cells formed tumors *in vivo*, whereas those inoculated with FGFR3$^-$ KYM-1 cells did not. FGFR3$^+$ KYM-1 cells showed upregulation of undifferentiated markers, such as *CD34, PAX3, PAX7, MYF5, NANOG, OCT3*, and *SOX2*, and downregulation of differentiated muscle markers, such as *MYH1, desmin, myogenin*, and *dystrophin*. The expression of FGFR3 was validated in eRMS patient biopsy specimens to be positive or upregulated by immunohistochemistry and RT-PCR (Hirotsu et al. 2009).

Fibrosarcoma: Sarcosphere, CD133, and ALDH. Fibrosarcoma, in which histologically the predominant cells are fibroblasts that divide excessively without cellular control, can arise from both bone and soft tissue. Fibrosarcoma of bone can occur in patients of any age but are most common between the second and sixth decades of life. Fibrosarcoma of soft tissue is classified into infantile fibrosarcoma and adult fibrosarcoma. Surgical resection is the standard treatment, and some patients are candidates for chemotherapy, although the efficacy has not been proven. The 5-year-suvival rate of soft tissue fibrosarcoma is reported to be 58% in grade 1 and 2, 34% in grade 3, and 21% in grade 4 cases (Scott et al. 1989).

The HT1080 cell line was found to form spheres at a frequency of 1/180 (553 colonies/100,000 cells), which showed an increased level of *Oct3/4, Nanog, STAT3, Sox2, Sox10, MLH1*, and *MSH2* compared with adherent cells (Fujii et al. 2009). Honoki et al. found a population of cells expressing high ALDH activity with frequencies of 9% in an HT1080 cell line. ALDH1$^+$ cells showed high frequency of sphere formation in an anchorage-independent environment, in contrast to ALDH1$^-$ cells (Honoki et al. 2010). A small subset of CD133$^+$ cells were identified in 4 fibrosarcoma biopsies, ranging from 0.42% to 1.23% (Tirino et al. 2011). The HT1080 cell line expressed 3% of CD133$^+$ by flow cytometry. Spheres of HT1080 CD133$^+$ cells, showing increased expression of *Oct3/4, Nanog, Sox2*, and *nestin*, formed colonies in soft agar more efficiently and showed 2500-fold greater tumorigenicity than adherent culture cells.

Others: Hiwi, hTERT, Survivin, LGR5 and the other markers. Several reports have focused on the stem cell genes and cell surface markers of MSC and NCSC in various histological soft tissue sarcomas.

Because of difficulty in the detection and isolation of CSCs and controversy as to the presence of CSC, Taubert et al. approached this subject by analysis of stem cell-associated genes for stem cell identification (Taubert et al. 2007). The expression of three genes, *Hiwi, hTERT*, and *survivin*, was evaluated, which had previously been shown to be overexpressed in both stem cells and cancer cells. Hiwi belongs to the Piwi-domain proteins, which are components of ribonucleoprotein complexes. Taubert et al. had already reported an association between altered expression of *Hiwi* mRNA and a poor prognosis for soft tissue sarcoma patients (Qiao et al. 2002, Liu et al. 2006, Taubert et al. 2007). hTERT is the catalytic subunit of telomerase, which functions in the synthesis and extension of telomeric DNA. High telomerase activity is observed in more than 85% of human cancer cells, strongly suggesting a key role during tumorigenesis (Pendino et al. 2006). Survivin is a member of a family of apoptosis inhibitors (Altieri 2003) and is considered to be an interface molecule between the regulation of apoptosis and the control of cell proliferation (Altieri 2003). A cohort of 104 primary STS, including liposarcoma, MFH, neurogenic sarcoma, rhabdomyosarcoma, leiomyosarcoma, and other STS, was investigated by quantitative real-time PCR for their RNA coexpression of the three genes. Upon analysis of a single gene, only an increased expression of *hTERT* was associated with a poor outcome. Moreover, coexpression of all three genes showed a 74.6-fold increased risk of tumor-related death compared with patients with an intermediate level of *Hiwi*, low level of *survivin* and low level of *hTERT* mRNA expression. Additional analysis was performed with regard to the association among the expression of these three genes and the tumor stage and prognosis, and the findings show that the risk of tumor-related death was based on the coexpression of stem cell-associated genes rather than on tumor stage (Taubert et al. 2007).

The prognostic impact of the stem cell marker, cancer-associated gene, and Wnt/Tcf4 target gene LGR5/GPR49 in 77 frozen tumor samples from liposarcoma, MFH, neurogenic sarcoma, RMS, and the other STS was investigated by qRT-PCR (Rot et al. 2011). LGR5/GPR49 had been identified as a novel stem cell marker in intestinal epithelia, stomach, and hair follicles (Barker et al. 2010). Their analysis included a transcript variant of *LGR5/GPR49* lacking exon 5 (*LGR5/GPR49Δ5*). As a result, a low mRNA expression level of GPR49Δ5, a transcript variant of *LGR5/GPR49* lacking exon 5, but not of wild type LGR5/GPR49, was significantly correlated with a poor prognosis for the disease-associated survival of STS patients and associated with a shorter recurrence-free survival. Thus, transcript variant GPR49Δ5 was a negative prognostic marker for disease-associated and recurrence-free survival in STS patients (Rot et al. 2011).

Sadikovic et al. screened Ewing sarcoma, eRMS, aRMS, SS, MPNST, and the other undifferentiated soft tissue sarcomas (USTSs) for the presence of 3 mesenchymal stem cell markers (CD44, CD105, and CD166) and 5 neural stem cell markers (CD15, CD29, CD56, CD133, and nestin) by immunohistochemistry. Subsequently, CD56 could distinguish ES and USTS/eRMS, aRMS, MPNST, and SS, and CD166 could distinguish between ES and USTS by cluster analysis (Sadikovic et al. 2011). However, no discussion was presented with regard to the relationship with CSCs, which is a matter of interest.

Targeting CSCs in Sarcoma—Molecular Mechanisms Underlying the CSC Phenotype

Self-renewal pathways

For the principles of stem-cell biology to apply to tumorigenesis, cancers would have to be organized hierarchically into clonally derived populations of cells with different proliferative potentials, just like cells within normal tissues (Pardal et al. 2003). Considering the self-renewability of CSCs, there may be some kinds of self-renewal pathway in their biology. There are some reports on CSCs in carcinomas, but they are rare in sarcomas, which is the current matter of interest. Indeed, little is known about molecular mechanisms underlying the CSC phenotype of sarcomas. Multiple pathways including self-renewal pathway within sarcomas or MSCs might be clues to the mystery.

Wnt pathway. The Wnt signal transduction pathway coordinates myriad activities, from development and differentiation to proliferation and tumorigenesis (Matushansky et al. 2008). Aberrant Wnt signaling has been reported in tumors from patients with hepatocellular carcinoma, hepatoblastoma, colorectal cancer, AML, CML, multiple myeloma, gastric cancer, and Wilms tumor (reviewed in Takebe et al. 2011). In addition, deregulated Wnt signaling has been associated with CSC activity; specifically, cutaneous CSCs require β-catenin signaling to maintain their tumorigenic phenotype (Malanchi et al. 2008).

In striking contrast to carcinomas as described above, rarely has Wnt/β-catenin nuclear activity been noted in sarcomas (Matushansky et al. 2008). Ng et al. examined 549 cases of a diverse group of sarcomas. No high-level nuclear β-catenin expression was observed in any of 381 cases of fibrohistocytic, muscular, adipocytic, chondroid, or osseous tumors representing 42 diagnostic categories. However, there were several other categories that did show high levels of nuclear β-catenin, including desmoid-type fibromatosis (71% of cases), solitary fibrous tumors (40%), and synovial sarcoma (28%) (Ng et al. 2005). These data suggest that Wnt signaling may have a very different role in sarcomagenesis in comparison to the development of other malignancies (Matushansky et al. 2008). Matushansky et al. showed that the Wnt inhibitor Dickkopf-1 (Dkk1) is overexpressed in MFH and that it mediates the transformation of MSCs into MFH-like tumors *in vivo*. They also showed that activating Wnt signaling in MFH allows for their controlled differentiation toward mature connective tissue lineages (Matushansky et al. 2006).

Wnt signaling has also been well studied in MSCs. Gregory et al. demonstrated that Dkk1 promotes MSC self-renewal (Gregory et al. 2003). More importantly, Dkk1 expanded MSCs and maintained their pluripotentiality. In contrast to this work, Baksh et al. showed that Wnt3a (and resultant β-catenin activation) can similarly expand MSCs capable of maintaining their pluripotentiality (Baksh et al. 2007). Whether β-catenin promotes or inhibits MSC self-renewal remains unclear.

One study on regulating CSCs of sarcoma by inhibition of Wnt/β-catenin signaling was reported. Tang et al. investigated whether salinomycin acts against osteosarcoma stem cells (Tang et al. 2011). Salinomycin treatment was reported to overcome ABC transporter-mediated multi-drug resistance and regarded as a P-glycoprotein inhibitor. It was also recognized as a selective inhibitor of breast CSCs and leukemia stem cells (Gupta et al. 2009, Fuchs et al. 2010). Salinomycin was found to inhibit the sarcosphere formation of these cells and sensitize spheres to DOX *in vitro* and *in vivo.* Subsequently, salinomycin was shown to downregulate Wnt/β-catenin signaling by decreased β-catenin and cyclin D1 by immunohistochemistry and decreased protein level of phosphor-GSK3β and cyclin D1 by Western blot analysis (Tang et al. 2011).

Notch signaling. Notch signaling plays a key role in the normal development of many tissues and cell types through diverse effects on cell fate decision, stem cell renewal, differentiation, survival, and proliferation (Artavanis-Tsakonas et al. 1999). Notch itself is a transmembrane receptor. When Notch interacts with membrane-bound ligand Delta or Jagged on the surface of neighboring cells, the Notch intracellular domain (NICD) is proteolytically cleaved from the membrane by γ-secretase and translocates to the nucleus to complex with and to activate the transcription factor CSL (human CBF1, fly Suppressor of Hairless, worm Lag-1). CSL then recruits the co-activator Mastermind-like (MAML) and initiates transcription of target genes such as Hes and Hey (Bolós et al. 2007, Watt et al. 2008).

Notch signaling seems to function as an oncogene or a tumor suppressor, depending on the cellular context (Zhang et al. 2008). In addition to the reports on carcinomas and blood malignancies, the roles of Notch signaling in osteosarcoma, Ewing sarcoma, and rhabdomyosarcoma are demonstrated. Through pharmacological and direct retroviral modulation of the Notch pathway, Zhang et al. found that Notch signaling induces invasiveness and metastasis of osteosarcoma *in vitro* and *in vivo* but does not affect cell proliferation, survival, or tumorigenesis (Zhang et al. 2008). The Notch target gene *HES1* was sufficient to induce an invasive and metastatic phenotype in osteosarcoma. In ESFT, Bennani-Baiti et al. reported that Notch signaling is switched off despite the expression of Notch receptors and HES1 being uncoupled from Notch in ESFT (Bennani-Baiti et al. 2011). Furthermore, activated Notch proved to block ESFT cell proliferation and reduce the ESFT clonogenic potential, suggesting that Notch exerts tumor suppressor-like functions in ESFT. In rhabdomyosarcoma, the Notch target gene HES1 is upregulated and prevents tumor cell differentiation in a Notch-dependent manner (Raimondi et al. 2011). Raimondi et al. demonstrated that inhibition of Notch3 signaling induces rhabdomyosarcoma cell differentiation promoting p38 phosphorylation and p21Cip1 expression and hampers tumor cell growth *in vitro* and *in vivo*. However, the role of Notch signaling in sarcoma CSCs is unknown.

Notch signaling is also important for MSC differentiation into osteoblasts. Notch signaling improves osteoblastogenesis but does not necessarily enhance bone formation (Lin and Hankenson 2011). The osteoblast-specific gain of the Notch function in the Notch intracellular domain (NICD)-overexpressing transgenic mouse results in abnormally dense or osteosclerotic bone, whereas loss of Notch signaling via γ-secretase mutations leads to late-onset, age-related osteoporosis (Engin et al. 2008). The osteosclerotic phenotype is caused by enhanced proliferation of immature osteoblasts and not due to decreased osteoclastic activity. Conversely, the osteoporotic phenotype in loss-of-Notch-function mice displayed increased osteoclast numbers as a result of decreased Notch signaling in osteoblasts (Engin et al. 2008). Notch signaling also influences Runx2 function, a transcription factor required for osteoblastogenesis. Runx2 transcriptional activity is physically antagonized by the protein encoded by the Notch target gene *Hey1* (Zamurovic et al. 2004). Therefore, this signaling might have some role with regard to the CSC phenotype; however, none is known.

Hedgehog signaling. The hedgehog (Hh) pathway, which sets off a chain of events in target cells leading to the activation and repression of target genes by transcription factors in the GLI family, is implicated in the maintenance of stem cells in a variety of cancers, including multiple myeloma, myeloid leukemia, and colorectal cancer, therefore representing another potentially druggable target for cancer stem cell elimination (Alison et al. 2010). The Hh receptor Patched 1 (PTC1) inhibits signaling by the membrane protein Smoothened (Smo), and this inhibition is relieved by binding Sonic Hedgehog to PTC. Unrestricted Smo signaling can affect target gene transcription via the GLI family of transcription factors (GLI1, GLI2, and GLI3).

Reduction of human embryonal rhabdomyosarcoma tumor growth by inhibition of the Hh signaling pathway was reported (Tostar et al. 2010). Tostar et al. showed that Hh signaling is critical for the maintenance/survival of human eRMS tumor cell lines. In addition, inhibition of GLI1 by small molecule antagonists, such as GANT61, could be an effective therapeutic option in pediatric eRMS. On the other hand, Nagao et al. showed that GLI2 was aberrantly overexpressed in human osteosarcoma biopsy specimens. Inhibition of GLI2 prevented osteosarcoma growth *in vivo* and overexpression of GLI2 promoted MSC proliferation and accelerated their cell cycle progression (Nagao et al. 2011). No reports, however, has been reported on the role of the Hh pathway in CSCs of sarcomas.

Plaisant et al. showed that hMSC was endowed with a basal level of Hh signaling that decreased after differentiation of these cells (Plaisant et al. 2009). Inhibition of Hh signaling, using the classical inhibitor cyclopamine, or a siRNA directed against GLI2, leads to a decrease in hMSC proliferation (Plaisant et al. 2011). There seems to be some potential for regulating CSCs in sarcomas by controlling the Hh pathway.

Alternative molecular pathways

MAPK pathway. In addition to Wnt, Notch, and Hedgehog pathways playing a large role in many kinds of sarcomas and CSCs, the mitogen-activated protein kinase (MAPK) pathway has also been observed to play an important role in the pathogenesis of sarcoma. The MAPK signaling pathway occurs downstream of RAS, which is an intracellular central signal transduction mediator also activated by growth factors, such as the platelet-derived growth factor (PDGF) or the vascular endothelial growth factor (VEGF). This pathway consists of three kinases, including RAF1 (MAP kinase-kinase-kinase), MEK (MAP kinase-kinase), and ERK (extracellular signal-related kinase). This kinase cascade is closely related to tumor proliferation and metastasis in several types of cancer (Young et al. 2009, Wang et al. 2009, Ding et al. 2008).

Sasaki et al. demonstrated the role of the MAPK pathway in bone and soft tissue tumors. RAF1 and MEK 1/2 mRNA was detected in 6 human sarcoma cell lines (osteosarcoma and MFH) and all of the 7 surgically obtained OS specimens. Treatment with MEK inhibitor, U0126 [1,4-diamino-2,3-dicyano-1,4-bis (2-aminophynyltio) butadiene], resulted in dose- and time-dependent inhibition of cell proliferation and suppression of p-ERK expression, opposite to the promotion of p-MEK. In view of CSC regulation, Tsuchida et al. demonstrated that U0126 downregulated CDDP-induced VEGF and Flt1, whereas CDDP treatment could upregulate VEGF and Flt1 expression, leading to the survival and expansion of SP cells, as formerly described (Table 6).

miRNAs. MicroRNAs (miRNAs) have been focused on as a novel approach for regulating the phenotypes of CSCs. miRNAs are small regulatory RNA molecules that modulate the expression of their target genes and play important roles in a variety of physiological and pathological processes, such as development, differentiation, cell proliferation, apoptosis, and stress responses (Bartel 2004, Kosaka et al. 2010). Over the past several years, many miRNAs have been investigated in various human cancers (Croce

Table 6. Experimental trials for targeting CSCs or regulating the phenotypes of sarcoma CSCs.

Sarcoma	Agents	Target CSC	Mechanism	Year	Refs. (et al.)
Osteosarcoma	Salinomycin	Sarcosphere	Impairing of Wnt/β-catenin signaling by degradation of β-catenin	2011	Tang
Osteosarcoma	U0126 (MEK inhibitor)	Side population cells	Downregulation of CDDP-induced VEGF and Flt1 expression through MAPK/ERK1,2 signaling	2008	Tsuchida
Ewing sarcoma	miR-145	EWS-FLI-1-mediated hpMSC	Making a feedback loop with EWS-FLI-1 in reprogramming hpMSC into ESFT CSCs through repression of SOX2, a direct and an indirect EWS-FLI-1 target gene	2010	Riggi
Ewing sarcoma	YK-4-209 (EWS-FLI1 inhibitor)	$ALDH^{high}$ cells	Blocking RNA helicase A (RHA) binding to EWS-FLI1 (Not understood about the correlations with ALDH)	2010	Awad
MFH	Tc4C-6 (CTL clone)	Side population cells	Specific cytotoxicity against human leukocyte antigen (HLA) Class 1 molecules on the cell surface of the SP cells	2011	Tsukahara

2009). The deregulation of the expression of miRNAs has been shown to contribute to cancer development through various kinds of mechanisms, including deletions, amplifications, or mutations involving miRNA loci, epigenetic silencing, the dysregulation of transcription factors that target specific miRNAs, or the inhibition of processing (Kim et al. 2009).

The expression of microRNA-145 (miR-145), for example, is reported to be low in self-renewing human embryonic stem cells (hESCs) but highly upregulated during differentiation (Xu et al. 2009). Increased miR-145 expression inhibits hESC self-renewal, represses expression of pluripotency genes, and induces lineage-restricted differentiation. On the other hand, miR-145 is reported to be downregulated and to act as a tumor suppressor in various tumors (Chiyomaru et al. 2010). In the same chromosomal region of miR-145, miR-143 is found and reported to act as a tumor suppressor function in the lung metastasis of osteosarcoma (Osaki et al. 2011).

miRNA is also reported to regulate the differentiation of MSC. The role of miRNA in the osteogenic differentiation of MSCs has been indicated by several studies. MiR-125b, 206, and 26a acted as negative regulators of osteoblast differentiation of bone marrow-derived MSCs through negative regulation of ERBB2, Connexin 43, and SMAD1, respectively (reviewed in Guo et al. 2011). MiR-133 and -135 functionally inhibited the differentiation of osteoprogenitors by attenuating RUNX2 and SMAD5 pathways that synergistically contribute to bone formation. On the other hand, miR-196a, -210, and -2861 were shown to enhance osteogenic differentiation, possibly through its target genes *HOXC8, ACVR1b,* and *HDAC5,* respectively. Five miRNAs (miR-130b, -152, -28, -26b, and -193b) were found to be differentially expressed during chondrogenic differentiation of MSCs through microarray analysis, and four of them were confirmed by real-time polymerase chain reaction analysis. Furthermore, in the multipotent MSC line C2C12, miR-1, miR-206, and miR-133 exhibited opposing roles in modulating skeletal muscle proliferation and differentiation (reviewed in Guo et al. 2011). These miRNAs might regulate the phenotype of CSCs of sarcoma.

As for the regulation of the CSCs in sarcoma, several studies have emerged. High-throughput microRNA (miRNA) expression analysis revealed that the expression of miR-140 was associated with chemosensitivity in osteosarcoma tumor xenografts (Song et al. 2009). Tumor cells ectopically transfected with miR-140 were more resistant to MTX and 5-fluorouracil (5-FU). Overexpression of miR-140 inhibited cell proliferation in both osteosarcoma U-2 OS (wt-p53) and colon cancer HCT 116 (wt-p53) cell lines. Furthermore, the expression of endogenous miR-140 was significantly elevated in $CD133^+/CD44^+$ colon CSCs that exhibit a slow proliferation rate and strong chemoresistance. However, the effect of miR-140 on CSC of osteosarcoma has not been clarified.

Riggi et al. found that the EWS-FLI-1 fusion gene induces expression of stem cell genes *Oct-4, SOX2,* and *Nanog* in human pediatric MSCs (hpMSCs). hpMSCs expressing EWS-FLI-1 generate a cell subpopulation displaying ESFT CSC features *in vitro*. Their further analysis demonstrated that induction of the ESFT CSC phenotype is the result of the combined effect of EWS-FLI-1 on its target gene expression and repression of miR-145 promoter activity. Subsequently, EWS-FLI-1 and miR-145 proved to function in a mutually repressive feedback loop and identify their common target gene, SOX2, in addition to miR-145 itself, as key players in ESFT cell differentiation and tumorigenicity (Riggi et al. 2010).

Others. Li et al. demonstrated that diallyltrisulfide (DATS), a volatile organosulfur compound derived from garlic, could confer drug resistance of osteosarcoma by downregulation of P-glycoprotein. Their previous report had showed $CD133^+$ SaOS2 cells over expressed MDR1 (P-gp) in comparison with $CD133^-$ SaOS2 cells. The ratio of $CD133^+$ cells in flow cytometric analysis showed no significant difference (around 5.0%) under 0, 0.001, 0.01, 0.1, and 1 ppc of MTX without DATS. However, the ratio decreased (5.4, 3.2, 2.7, 1.4, and 0.92%, respectively) with the same concentration range of MTX in combination with 10μM DATS (Li et al. 2009).

Niche. The concept of "niche" defined as a specialized microenvironment that can promote the emergence of tumor stem cells and provide all the required factors for their development has recently emerged in the literature. Whereas the role of the niche in maintaining stem cells has been shown in several systems, the sarcoma CSC niche has not been identified. However, if sarcoma CSCs arise from MSCs or NCSCs, it is presumable that they might reside within the proposed niche. Several studies suggest that MSCs reside in a perivascular niche in almost all adult tissues, where they are in close contact with blood vessels (Shi and Gronthos 2003, da Silva Meirelles et al. 2008). The location of MSCs within perivascular niches is proposed to support the migration of MSCs in response to injury or disease (Kuhn and Tuan 2010). Similarly, location within a perivascular niche may support the metastasis of sarcomas.

There are some implications that niche plays a certain role in sarcoma CSCs (Siclari and Qin 2010). First, the differences in behavior of sarcoma CSCs grown *in vitro* compared to *in vivo* have been observed. For example, whereas the CSCs are characterized by being quiescent *in vivo*, they are more proliferative than the non-CSCs *in vitro* (Gibbs et al. 2005, Iwasaki et al. 2009). Secondly, especially with relation to bone sarcomas, the bone tissue is a hypoxic environment (Heddeleston et al. 2010, Siclari and Qin 2010). Activation of the hypoxia signaling pathway activates many pathways important for stem cell and CSC maintenance, such as brain CSC, which includes HIF-1α and HIF-2α (Heddeleston et al. 2010). In osteosarcoma, studies have revealed that the expression of hypoxia-relevant genes occurs mainly in necrotic areas, which suggests that there is an *in vivo* relationship between HIF-1α and apoptosis (Ishenko et al. 2008). In the cases of chondrosarcoma, HIF-1α protein might be a useful prognostic marker due to its important role in tumor angiogenesis and cellular proliferation (Kubo et al. 2008). In Ewing sarcoma, Aryee et al. found that the EWS-FLI1 protein was upregulated in a HIF-1α-dependent manner and that HIF-1α induced EWS-FLI1 accumulation in a time-dependent dynamic study (Aryee et al. 2010). Thirdly, there are some reports on morphological implications. David et al. presented "bone niche" in chondrosarcoma. Pathological findings of chondrosarcoma include infiltration of chondrosarcoma cells into the bone tissue (permeation) and chondrosarcoma development associated with bone resorption foci. They assume that chondrosarcoma growth is strongly linked to bone tissue (David et al. 2011). Indeed, permeation of bony trabeculae is considered an absolute sign of chondrosarcoma and is useful in differentiating Grade 1 chondrosarcoma (malignant) from enchondroma (benign).

These findings support the concept of a niche; however, there is no evidence identifying the role of a niche in sarcoma CSC biology, which had been anticipated.

Immunotherapy

Many clinical trials of CTL-based immunotherapy using peptide vaccination have demonstrated the potency of this new therapeutic modality for various cancers that are resistant to standard chemotherapy (Tsukahara et al. 2004). However, it remains unknown whether CTL-based immunotherapy can kill CSC. Kano et al. found that the SP cells showed the expression of the human leukocyte antigen (HLA) Class I molecules on the cell surface. The CTL clone Tc4C-6, induced by mixed lymphocyte tumor cell culture using autologous peripheral blood mononuclear cells and freshly isolated SP cells, showed specific cytotoxicity against the SP cells. Moreover, the cytotoxicity against SP cells was blocked by the anti-HLA Class I antibody W6/32. These results indicate that CTL-recognizing CSCs certainly exist in the tumor microenvironment and circulating peripheral blood and that SP cells can be killed by CTL. Therefore, CTL-based immunotherapy could be one of the options for targeting sarcoma CSCs.

Conclusions and Perspectives

A great deal of effort has been made in the research on not only the origin of sarcomas but also cancer stem cells of sarcomas. Considering the characteristics of sarcoma CSCs, such as their self-renewal, resistance to systemic therapy, and high migratory ability, sarcoma CSCs seem to have a lot of similarities with normal tissue stem cells, such as MSCs and NCSCs. However, the relationship between tissue stem cells and cancer stem cells remains unclear. This might be because of the problems categorized below.

The first problem is on the ambiguity of the definitions of cancer stem cells and mesenchymal stem cells. These precise definitions remain to be developed. Clarke et al. reported in 2006 that the consensus definition of a cancer stem cell that was developed for the American Association for Cancer Research (AACR) Workshop was "a cell within a tumor that possesses the capacity to self-renew and to cause the heterogeneous lineages of cancer cells that comprise the tumor." In addition, cancer stem cells could "only be defined experimentally by their ability to recapitulate the generation of a continuously growing tumor" (Clarke et al. 2006). On the other hand, MSC is also defined only by its pluripotent "ability" as described above. The next step would be the identification of the common and different features between the cells that meet the "concept" of these cells. In agreement with the theory that tumorigenesis could be seen as aberrant organogenesis (Reya et al. 2001), mutated MSCs have all that they need to source and maintain sarcoma (Honoki 2010) as recently exemplified by the effects of HMGA2 alteration in mesenchymal stem-like cells (Henriksen et al. 2010). However, it is of interest that mutated MSC or NCSC possesses malignant phenotypes, such as chemoresistance and metastasis.

The second problem is that any reliable markers for the sarcoma CSC have not been specified. Although sarcoma stem cells have been isolated and identified by various methods, no one has been able to completely eliminate non-CSCs. The final achievement in detecting CSC markers would be identifying the makers specific to sarcoma CSCs; however, only one or two markers seem too simple to represent sarcoma CSCs, which possess the multiple phenotype described above. Thus, further markers adding to current ones should be discussed to purify sarcoma CSCs more exclusively. The ideal goal would be to apply these CSC markers to practical pathological use to help make decisions regarding treatment strategies, including the current molecular therapeutics of CSCs.

For the achievement of this purpose, more evidence on clinical biology is also needed. The essence of identifying CSCs in sarcoma lies in their clinical implications. Therefore, the third problem is that there is no evidence of the correlation between sarcoma CSCs and clinical behaviors. A major goal in order to discover the best therapies against tumor progression and to address the question of CSC hypothesis would require analyzing patient samples at different stages of disease and, in particular, a follow-up of CSC numbers during and after treatment. Little evidence is available that deals exclusively with sarcoma biopsy specimens; moreover, there is still no study in which the specimens of post-chemotherapy and metastasis have been analyzed. This is an important aspect remaining to be addressed to confirm the clinical significance of this research.

Despite the inconsistency in the definition and the CSC markers, new approaches have to be developed to overcome the most lethal characteristics, such as chemoresistance and metastasis. The final problem deals with targeting and killing these cells. Several suggestions and trials demonstrated in this chapter are summarized below.

1) Inhibition of the self-renewal pathway: The major focus has been on developmental signaling pathways, including Hedgehog, Notch, and Wnt, and novel agents inhibiting these pathways that have been used to target CSCs in multiple diseases. Clinical trials using inhibitors of the Hedgehog pathway (GDC-0449, PF-04449913, BMS-833923, IPI-926, and TAK-441), Notch pathway (RO4929097, BMS-906024, and MK0752), Wnt pathway (PRI-724 and telomerase [GRN163L]) have begun to emerge, but their efficacy against CSC function remains to be determined (Rasheed et al. 2011). Since little has been tried on sarcomas, novel trials against sarcomas would be expected. 2) Inhibition of the other molecular pathways: Although not reached to clinical trials, MEK inhibitor has been proposed for osteosarcoma CSC treatment. The evidences on the activation of MAPK pathways have emerged also in the other cancers. Then, multiple pathways including self-renewal or the other molecular pathways might be associated with the CSC phenotypes. Therefore, the most influential pathways should be discussed in each type of sarcomas. However, we have no idea how many pathways are enough to be inhibited for regulation of CSC phenotypes. Further analysis above single molecular pathways would be required. 3) Antibody drugs: Targeting antibodies specific to CSCs would be the ideal approach for the elimination of CSCs. However, a commonly used marker in CSC studies, CD133 (also known as prominin) for example, has been implied as a normal stem cell marker. Thus, targeting CD133 has the potential for significant side effects. This makes it difficult to target on the reported markers of CSCs, so that the novel markers that specify only cancer cells would be required. 4) microRNA: The use of a natural miRNA to suppress the CSC phenotypes is also a promising possibility as a new therapeutic strategy for the treatment of CSCs. miRNA also possesses the potential for use as a biomarker. The development of biomarker strategies that quantify CSCs in a serial fashion may provide novel endpoints to monitor CSC-based clinical trials. 5) Immunotherapy: The CTL-based immunotherapy targeting on SP cells has been recently proposed. Further analysis including in vivo assay or the strategy targeting on the other CSC markers would raise expectations for sarcoma treatments. 6) Others: The other alternatives include oncolytic viruses that seem to be well suited to eliminate CSCs because the viruses are

cytotoxic and not subject to the drug efflux, such as ABC transporters and defective apoptotic signaling (Cripe et al. 2009). Virotherapy could also be used to sensitize tumor cells to radiation and as tools for immunotherapy (Thorne et al. 2010). Furthermore, the stem cell niche should be analyzed and targeted because of the implications of their essential role in deciding the stem cell fate and resistance to therapeutic drugs.

CSC studies on sarcoma are in their infancy compared to studies of carcinomas and hematopoietic malignancy. In this regard, we hope for the increasing preclinical and clinical trials of targeting on sarcoma CSCs. Researchers would have to continue their efforts to provide better prognosis for many sarcoma patients all over the world.

Acknowledgments

This work was supported in part by a grant-in-aid for the Third-Term Comprehensive 10-Year Strategy for Cancer Control of Japan; a grant-in-aid for Scientific Research on Priority Areas Cancer from the Japanese Ministry of Education, Culture, Sports, Science, and Technology; and the Program for Promotion of Fundamental Studies in Health Sciences of the National Institute of Biomedical Innovation of Japan.

References

Adhikari, A.S., N. Agarwal, B.M. Wood, C. Porretta, B. Ruiz, R.R. Pochampally, and T. Iwakuma. 2010. CD117 and Stro-1 identify osteosarcoma tumor-initiating cells associated with metastasis and drug resistance. Cancer Res 70: 4602–4612.

Al-Hajj, M., M.S. Wicha, A. Benito-Hernandez, S.J. Morrison, and M.F. Clarke. 2003. Prospective identification of tumorigenic breast cancer cells. Proc Natl Acad Sci USA 100: 3983–3988.

Alison, M.R., S.M. Lim, and L.J. Nicholson. 2010. Cancer stem cells: problems for therapy? J Pathol 27.

Altieri, D.C. 2003. Validating survivin as a cancer therapeutic target. Nat Rev Cancer 3: 46–54.

Ambroziak, W., G. Izaguirre, and R. 1999. Pietruszko. Metabolism of retinaldehyde and other aldehydes in soluble extracts of human liver and kidney. J Biol Chem 274: 33366–33373.

Aogi, K., A. Woodman, V. Urquidi, D.C. Mangham, D.T. arin, and S. Goodison. 2000. Telomerase activity in soft-tissue and bone sarcomas. Clin Cancer Res. 6: 4776–4781.

Aoyama, T., T. Okamoto, S. Nagayama, K. Nishijo, T. Ishibe, K. Yasura, T. Nakayama, T. Nakamura, and J. Toguchida. 2004. Methylation in the core-promoter region of the chondromodulin-I gene determines the cell-specific expression by regulating the binding of transcriptional activator Sp3. J Biol Chem 279: 28789–28797.

Armstrong, L., M. Stojkovic, I. Dimmick, S. Ahmad, P. Stojkovic, N. Hole, and M. Lako. 2004. Phenotypic characterization of murine primitive hematopoietic progenitor cells isolated on basis of aldehyde dehydrogenase activity. Stem Cells 22: 1142–1151.

Artavanis-Tsakonas, S., M.D. Rand, and R.J. Lake. 1999. Notch signaling: cell fate control and signal integration in development. Science 284: 770–776.

Aryee, D.N., S. Niedan, M. Kauer, R. Schwentner, I.M. Bennani-Baiti, J. Ban, K. Muehlbacher, M. Kreppel, R.L. Walker, P. Meltzer, C. Poremba, R. Kofler, and H. Kovar. 2010. Hypoxia modulates EWS-FLI1 transcriptional signatureand enhances the malignant properties of Ewing's sarcoma cells *in vitro*, Cancer Res 70: 4015–4023.

Avilion, A.A., S.K. Nicolis, L.H. Pevny, L. Perez, N. Vivian, and R. Lovell-Badge. 2003. Multipotent cell lineages in early mouse development depend on SOX2 function. Genes Dev 17: 126–140.

Awad, O., J.T. Yustein, P. Shah, N. Gul, V. Katuri, A. O'Neill, Y. Kong, M.L. Brown, J.A. Toretsky, and D.M. Loeb. 2010. High aldh activity identifies chemotherapy-resistant Ewing's sarcoma stem cells that retain sensitivity to ews-fli1 inhibition. PLoS One 5: e13943.

Bacci, G., S. Ferrari, F. Bertoni, P. Ruggieri, P. Picci, A. Longhi, R. Casadei, N. Fabbri, C. Forni, M. Versari, and M. Campanacci. 2000. Long-term outcome for patientswith nonmetastatic osteosarcoma of the extremity treated at theIstituto Ortopedico Rizzoli according to the Istituto Ortopedico Rizzoli/osteosarcoma-2 protocol: an updated report. J Clin Oncol 18: 4016–4027.

Baksh, D., and R.S. Tuan. 2007. Canonical and non-canonical Wnts differentially affect the developmentpotential of primary isolate of human bone marrow mesenchymal stem cells. J Cell Physiol 212: 817–826.

Banas, A., T. Teratani, Y. Yamamoto, M. Tokuhara, F. Takeshita, M. Osaki, M. Kawamata, T. Kato, H. Okochi, and T. Ochiya. 2008. IFATS collection: *in vivo* therapeutic potential of human adipose tissue mesenchymal stem cells after transplantation into mice with liver injury. Stem Cells 26: 2705–2712.

Barker, N., M. Huch, P. Kujala, M. van de Wetering, H.J. Snippert, J.H. van Es, T. Sato, D.E. Stange, H. Begthel, M. van den Born, E. Danenberg, S. van den Brink, J. Korving, A. Abo, P.J. Peters, N. Wright, R. Poulsom, and H. Clevers. 2010. Lgr5(+ve) stem cells drive self-renewal in the stomach and build long-lived gastric units *in vitro*. Cell Stem Cell 6: 25–36.

Bartel, D.P. 2004. MicroRNAs: genomics, biogenesis, mechanism, and function. Cell 116: 281–297.

Basu-Roy, U., E. Seo, L. Ramanathapuram, T.B. Rapp, J.A. Perry, S.H. Orkin, A. Mansukhani, and C. Basilico. 2011. Sox2 maintains self renewal of tumor-initiating cells in osteosarcomas. Oncogene. doi: 10.1038/onc.2011.405

Bennani-Baiti, I.M., D.N. Aryee, J. Ban, I. Machado, M. Kauer, K. Mühlbacher, G. Amann, A. Llombart-Bosch, and H. Kovar. 2011. Notch signalling is off and is uncoupled from HES1 expression in Ewing's sarcoma. J Pathol 225: 353–363.

Bielack, S., B. Kempf-Bielack, D. Schwenzer, T. Birkfellner, G. Delling, V. Ewerbeck, G.U. Exner, N. Fuchs, U. Göbel, N. Graf, U. Heise, K. Helmke, A.R. von Hochstetter, H. Jürgens, R. Maas, N. Münchow, M. Salzer-Kuntschik, J. Treuner, U. Veltmann, M. Werner, W. Winkelmann, A. Zoubek, and R. Kotz. 1999. Neoadjuvanttherapy for localized osteosarcoma of extremities: results from the Cooperative osteosarcoma study group COSS of 925 patients. Klin Padiatr 211: 260–270.

Bolós, V., J. Grego-Bessa, and J.L. de la Pompa. 2007. Notch signaling in development and cancer. Endocr Rev 28: 339–363.

Brady, M.S., G. Perino, G. Tallini, P. Russo, and J.M. Woodruff. 1996. Malignant mesenchymoma. Cancer 77: 467–473.

Carpentino, J.E., M.J. Hynes, H.D. Appelman, T. Zheng, D.A. Steindler, E.W. Scott, and E.H. Huang. 2009. Aldehyde dehydrogenase-expressing colon stem cells contribute to tumorigenesis in the transition from colitis to cancer. Cancer Res 69: 8208–8215.

Catton, C., B. O'Sullivan, R. Bell, N. Laperriere, B. Cummings, V. Fornasier, and J. Wunder. 1996. Chordoma: Long-term follow-up after radical photonirradiation. Radiother Oncol 41: 67–72.

Cha, E.J. 2011. Dedifferentiated liposarcoma of the small bowel mesentery presenting as a submucosal mass. World J Gastrointest Oncol 3: 116–118.

Chambers, P.W., and C.P. Schwinn. 1979. Chordoma. A clinicopathologic study of metastasis. Am J Clin Pathol 72: 765–776.

Charafe-Jauffret, E., C. Ginestier, F. Iovino, C. Tarpin, M. Diebel, B. Esterni, G. Houvenaeghel, J.M. Extra, F. Bertucci, J. Jacquemier, L. Xerri, G. Dontu, G. Stassi, Y. Xiao, S.H. Barsky, D. Birnbaum, P. Viens, and M.S. Wicha. 2010. Aldehyde dehydrogenase 1-positive cancer stem cells mediate metastasis and poor clinical outcome in inflammatory breast cancer. Clin Cancer Res. 16: 45–55.

Chiba, T., S. Miyagi, A. Saraya, R. Aoki, A. Seki, Y. Morita, Y. Yonemitsu, O. Yokosuka, H. Taniguchi, H. Nakauchi, and A. Iwama. 2008. The polycomb gene product bmi1 contributes to the maintenance of tumor-initiating side population cells in hepatocellular carcinoma. Cancer Res 68: 7742–7749.

Cheng, J.X., B.L. Liu, and X. Zhang. 2009. How powerful is CD133 as a cancer stem cell marker in brain tumors? Cancer Treat Rev 35: 403–408.

Chiyomaru, T., H. Enokida, S. Tatarano, K. Kawahara, Y. Uchida, K. Nishiyama, L. Fujimura, N. Kikkawa, N. Seki, and M. Nakagawa. 2010. miR-145 and miR-133a function as tumor suppressors and directly regulate FSCN1 expression in bladder cancer. Br J Cancer 102: 883–891.

Chugh, R., H. Tawbi, D.R. Lucas, J.S. Biermann, S.M. Schuetze, and L.H. Baker. 2007. Chordoma: the nonsarcoma primary bone tumor. Oncologist 12: 1344–1350.

Clarke, M.F., J.E. Dick, P.B. Dirks, C.J. Eaves, C.H. Jamieson, D.L. Jones, J. Visvader, I.L. Weissman, and G.M. Wahl. 2006. Cancer stem cells—perspectives on current status and future directions: AACR Workshop on cancer stem cells. Cancer Res 66: 9339–9344.

Clevers, H. 2011. The cancer stem cell: premises, promises and challenges. Nat Med 17: 313–319.

Collins, A.T., P.A. Berry, C. Hyde, M.J. Stower, and N.J. Maitland. 2005. Prospective identification of tumorigenic prostate cancer stem cells. Cancer Res 65: 10946–10951.

Cripe, T.P., P.Y. Wang, P. Marcato, Y.Y. Mahller, and P.W. Lee. 2009. Targeting cancer-initiating cells with oncolytic viruses. Mol Ther 17: 1677–1682.

Croce, C.M. 2009. Causes and consequences of microRNA dysregulation in cancer. Nat Rev Genet 10: 704–714.

Dahlin, D.C., and J.W. Beabout. 1971. Dedifferentiation of lowgrade chondrosarcomas. Cancer 28: 461–466.

Da Silva Meirelles, L., T.T. Sand, R.J. Harman, D.P. Lennon, and A.I. Caplan. 2009. MSC frequency correlates with blood vessel density in equine adipose tissue. Tissue Eng Part A 15: 221–229.

David, E., F. Blanchard, M.F. Heymann, G. De Pinieux, F. Gouin, F. Rédini, and D. Heymann. 2011. The Bone Niche of Chondrosarcoma: A Sanctuary for Drug Resistance, Tumor Growth and also a Source of New Therapeutic Targets. Sarcoma 2011: 932451.

De Giovanni, C., L. Landuzzi, G. Nicoletti, P.L. Lollini, and P. Nanni. 2009. Molecularand cellular biology of rhabdomyosarcoma. Future Oncol 5: 1449–1475.

De Jong, J., and L.H. Looijenga. 2006. Stem cell marker OCT3/4 in tumor biology and germ cell tumor diagnostics: history and future. Crit Rev Oncog 12: 171–203.

Di Fiore, R., A. Santulli, R.D. Ferrante, M. Giuliano, A. De Blasio, C. Messina, G. Pirozzi, V. Tirino, G. Tesoriere, and R. Vento. 2009. Identification and expansion of human osteosarcoma-cancer-stem cells by long-term 3-aminobenzamide treatment. J Cell Physiol 219: 301–313.

Ding, Y., E.A. Boguslawski, B.D. Berghuis, J.J. Young, Z. Zhang, K. Hardy, K. Furge, E. Kort, A.E. Frankel, R.V. Hay, J.H. Resau, and N.S. Duesbery. 2008. Mitogen-activated protein kinase kinase signaling promotes growth and vascularization of fibrosarcoma. Mol Cancer Ther 7: 648–658.

Dorfman, H.D., and B. Czerniak. 1995. Bone cancers. Cancer 75: 203–210.

Downing, J.R., A. Khandekar, S.A. Shurtleff, D.R. Head, D.M. Parham, B.L. Webber, A.S. Pappo, M.G. Hulshof, W.P. Conn, and D.N. Shapiro. 1995. Multiplex RT-PCR assay for the differential diagnosis of alveolar rhabdomyosarcoma and Ewing's sarcoma. Am J Pathol 146: 626–634.

Duester, G., F.A. Mic, and A. Molotkov. 2003. Cytosolic retinoid dehydrogenases govern ubiquitous metabolism of retinol to retinaldehyde followed by tissue-specific metabolism to retinoic acid. Chem Biol Interact 143–144: 201–210.

Eilber, F.C., M.F. Brennan, F.R. Eilber, J.J Eckardt, S.R. Grobmyer, E. Riedel, C. Forscher, R.G. Maki, and S. Singer. 2007. Chemotherapy is associated with improved survival in adult patients with primary extremity synovial sarcoma. Ann Surg 246: 105–113.

Engin, F., Z. Yao, T. Yang, G. Zhou, T. Bertin, M.M. Jiang, Y. Chen, L. Wang, H. Zheng, R.E. Sutton, B.F. Boyce, and B. Lee. 2008. Dimorphic effects of Notch signaling in bone homeostasis. Nat Med 14: 299–305.

Entz-Werle, N., L. Marcellin, M.P. Gaub, E. Guerin, A. Schneider, P. Berard-Marec, C. Kalifa, L. Brugiere, H. Pacquement, C. Schmitt, M.D. Tabone, C. Jeanne-Pasquier, P. Terrier, F. Dijoud, P. Oudet, P. Lutz, and A. Babin-Boilletot. 2005. Prognostic significance of allelic imbalance at the c-kit gene locus and c-kit overexpression by immunohistochemistry in pediatric osteosarcomas. J Clin Oncol 23: 2248–2255.

Entz-Werle, N., M.P. Gaub, T. Lavaux, L. Marcellin, N. Metzger, P. Marec-Berard, C. Schmitt, L. Brugiere, C. Kalifa, M.D. Tabone, H. Pacquement, P. Gentet, P. Lutz, P. Oudet, and A. Babin. 2007. KIT gene in pediatric osteosarcomas: could it be a new therapeutic target? Int J Cancer 120: 2510–2516.

Esiashvili, N., M. Goodman, and R.B. Marcus Jr. 2008. Changes in incidence and survival of Ewing sarcoma patients over the past 3 decades: Surveillance Epidemiology and End Results data. J Pediatr Hematol Oncol 30: 425–430.

Esnaola, N.F., B.P. Rubin, E.H. Baldini, N. Vasudevan, G.D. Demetri, C.D. Fletcher, and S. Singer. 2001. Response to chemotherapy and predictors of survival in adult rhabdomyosarcoma. Ann Surg 234: 215–223.

Evans, H.L., A.G. Ayala, and M.M. Romsdahl. 1977. Prognostic factors in chondrosarcoma of bone: a clinicopathologic analysis with emphasis on histologic grading. Cancer 40: 818–831.

Evans, H.L. 1979. Liposarcoma: a study of 55 cases with areassessment of its classification. Am J Surg Pathol 3: 507–523.

Farshid, G., M. Pradhan, J. Goldblum, and S.W. Weiss. 2002. Leiomyosarcoma of somatic soft tissues: a tumor of vascular origin with multivariate analysis of outcome in 42 cases. Am J Surg Pathol 26: 14–24.

Ferguson, W., and A. Goorin. 2001. Current treatment of osteosarcoma. Cancer Invest 19: 292–315.

Ferraresi, V., C. Nuzzo, C. Zoccali, F. Marandino, A. Vidiri, N. Salducca, M. Zeuli, D. Giannarelli, F. Cognetti, and R. Biagini. 2010. Chordoma: clinical characteristics, management and prognosis of a case series of 25 patients. BMC Cancer 10: 22.

Ferrari, A., G. Bisogno, R. Alaggio, G. Cecchetto, P. Collini, A. Rosolen, C. Meazza, P. Indolfi, A. Garaventa, L. De Sio, P. D'Angelo, P. Tamaro, M. Casanova, and M. Carli. 2008. Synovial sarcoma of children and adolescents: the prognostic role of axial sites. Eur J Cancer 44: 1202–1209.

Ferrari, S., E. Palmerini, E.L. Staals, M. Mercuri, B. Franco, P. Picci, and G. Bacci. 2009. The treatment of nonmetastatic high grade osteosarcoma of the extremity: review of the Italian Rizzoli experience. Impact on the future. Cancer Treat Res 152: 275–287.

Fibbe, W.E. 2002. Mesenchymal stem cells. A potential source for skeletal repair. Ann Rheum Dis 61: ii29–31.

Fletcher, C.D.M., K.K. Unni, and F. Mertens. 2002. World Health Organisation classification of tumors. Pathology and genetics of tumors of soft tissue and bone. IARC Press, Lyon, France.

Fleming, G.F., P.S. Heimann, J.K. Stephens, M.A. Simon, M.K. Ferguson, R.S. Benjamin, and B.L. Samuels. 1993. Dedifferentiated chordoma. Response toaggressive chemotherapy in two cases. Cancer 72: 714–718.

Friedenstein, A.J., J.F. Gorskaja, and N.N. Kulagina. 1976. Fibroblast precursors in normal and irradiated mouse hematopoietic organs. ExpHematol 4: 267–274.

Fuchs, D., V. Daniel, M. Sadeghi, G. Opelz, and C. Naujokat. 2010. Salinomycin overcomes ABC transporter-mediated multidrug and apoptosis resistance in human leukemia stem cell-like KG-1a cells. Biochem Biophys Res Commun 394: 1098–1104.

Fujii, H., K. Honoki, T. Tsujiuchi, A. Kido, K. Yoshitani, and Y. Takakura. 2009. Sphere-forming stem-like cell populations with drug resistance in human sarcoma cell lines. Int J Oncol 34: 1381–1386.

Furlong, M.A., T. Mentzel, and J.C. Fanburg-Smith. 2001. Pleomorphic rhabdomyosarcoma in adults: a clinicopathologic study of 38 cases with emphasis on morphologic variants and recent skeletal muscle-specific markers. Mod Pathol 14: 595–603.

Galli, R., U. Borello, A. Gritti, M.G. Minasi, C. Bjornson, M. Coletta, M. Mora, M.G. De Angelis, R. Fiocco, G. Cossu, and A.L. Vescovi. 2000. Skeletal myogenic potential of human and mouse neural stem cells. Nat Neurosci 3: 986–991.

Geller, D.S., and R. Gorlick. 2010. Osteosarcoma: a review of diagnosis, management, and treatment strategies. Clin Adv Hematol Oncol 8: 705–718.

Gelderblom, H., P.C. Hogendoorn, S.D. Dijkstra, C.S. van Rijswijk, A.D. Krol, A.H. Taminiau, and J.V. Bovée. 2008. The clinical approach towards chondrosarcoma. Oncologist 13: 320–329.

Gibbs, C.P., V.G. Kukekov, J.D. Reith, O. Tchigrinova, O.N. Suslov, E.W. Scott, S.C. Ghivizzani, T.N. Ignatova, and D.A. Steindler. 2005. Stem-like cells in bone sarcomas: implications for tumorigenesis. Neoplasia 7: 967–976.

Gidekel, S., G. Pizov, Y. Bergman, and E. Pikarsky. 2003. Oct-3/4 is a dose-dependent oncogenic fate determinant. Cancer Cell 4: 361–370.

Goodell, M.A., M. Rosenzweig, H. Kim, D.F. Marks, M. DeMaria, G. Paradis, S.A. Grupp, C.A. Sieff, R.C. Mulligan, and R.P. Johnson. 1997. Dye efflux studies suggest that hematopoietic stem cells expressing low or undetectable levels of CD34 antigen exist in multiple species. Nat Med 3: 1337–1345.

Gregory, C.A., H. Singh, A.S. Perry, and D.J. Prockop. 2003. The Wnt signaling inhibitor dickkopf-1 is required for reentry into the cell cycle of human adult stem cells from bone marrow. J Biol Chem 278: 28067–28078.

Gronthos, S., S.E. Graves, S. Ohta, and P.J. Simmons. 1994. The STRO-1+ fractionof adult human bone marrow contains the osteogenic precursors. Blood 84: 4164–4173.

Guo, L., R.C. Zhao, and Y. Wu. 2011. The role of microRNAs in self-renewal and differentiation of mesenchymal stem cells. Exp Hematol 39: 608–616.

Gupta, P.B., T.T. Onder, G. Jiang, K. Tao, C. Kuperwasser, R.A. Weinberg, and E.S. Lander. 2009. Identification of selective inhibitors ofcancer stem cells by high-throughput screening. Cell 138: 645–659.

Halldorsson, A., S. Brooks, S. Montgomery, and S. Graham. 2009. Lung metastasis 21 years after initial diagnosis of osteosarcoma: a case report. J Med Case Reports 3: 9298.

Heddleston, J.M., Z. Li, J.D. Lathia, S. Bao, A.B. Hjelmeland, and J.N. Rich. 2010. Hypoxia inducible factors in cancer stem cells. Br J Cancer 102: 789–795.

Henriksen, J., M. Stabell, L.A. Meza-Zepeda, S.A. Lauvrak, M. Kassem, and O. Myklebost. 2010. Identification of targetgenes for wild type and truncated HMGA2 in mesenchymal stem-likecells. BMC Cancer 10: 329.

Hermann, P.C., S.L. Huber, T. Herrler, A. Aicher, J.W. Ellwart, M. Guba, C.J. Bruns, and C. Heeschen. 2007. Distinct populations of cancer stem cells determine tumor growth and metastatic activity in human pancreatic cancer. Cell Stem Cell 1: 313–323

Hess, D.A., T.E. Meyerrose, L. Wirthlin, T.P. Craft, P.E. Herrbrich, M.H. Creer, and J.A. Nolta. 2004. Functional characterization of highly purified human hematopoietic repopulating cells isolated according to aldehyde dehydrogenase activity. Blood 104: 1648–1655.

Higinbotham, N.L., R.F. Phillips, H.W. Farr, and H.O. Hustu. 1967. Chordoma. Thirty-five year study at Memorial Hospital. Cancer 20: 1841–1850.

Hirotsu, M., T. Setoguchi, Y. Matsunoshita, H. Sasaki, H. Nagao, et al. 2009. Tumor formation by single fibroblast growth factor receptor 3-positiverhabdomyosarcoma-initiating cells. Br J Cancer 101: 2030–2037.

Honoki, K., H. Fujii, A. Kubo, A. Kido, T. Mori, Y. Tanaka, and T. Tsujiuchi. 2010. Possible involvement of stem-like populations with elevated ALDH1 in sarcomas for chemotherapeutic drug resistance. Oncol Rep 24: 501–505.

Honoki, K. 2010. Do stem-like cells play a role in drug resistance of sarcomas? Expert Rev Anticancer Ther 10: 261–270.

Huang, X., and J. Saint-Jeannet. 2004. Induction of the neural crest and the opportunities of life on the edge. Developmental Biology 275: 1–11.

Huber, M.A., N. Kraut, and H. Beug. 2005. Molecular requirements for epithelial-mesenchymal transition during tumor progression. Curr Opin Cell Biol 17: 548–558.

Ibrahim, S.F., A.H. Diercks, T.W. Petersen, and G. van den Engh. 2007. Kinetic analyses as a critical parameter in defining the side population (SP) phenotype. Exp Cell Res 313: 1921–1926.

Ischenko, I., H. Seeliger, M. Schaffer, K.W. Jauch, and C.J. Bruns. 2008. Cancer stem cells: how can we target them? Curr Med Chem 15: 3171–3184.

Iwasaki, H., and T. Suda. 2009. Cancer stem cells and their niche. Cancer Sci 100: 1166–1172.

James, G., M. Crocker, A. King, I. Bodi, A. Ibrahim, and B.P. Chitnavis. 2008. Malignant triton tumors of the spine. J Neurosurg Spine 8: 567–573.

Jamil, N., S. Howie, and D.M. Salter. 2010. Therapeutic molecular targets in human chondrosarcoma. Int J Exp Pathol 91: 387–393.

Jedlicka, P. 2010. Ewing Sarcoma, an enigmatic malignancy of likely progenitor cell origin, driven by transcription factor oncogenic fusions. Int J Clin Exp Pathol 19 3: 338–347.

Jones, R.L., C. Fisher, O. Al-Muderis, and I.R. Judson. 2005. Differential sensitivity of liposarcoma subtypes to chemotherapy. Eur J Cancer 41: 2853–2860.

Jiang, X., Y. Gwye, D. Russell, C. Cao, D. Douglas, L. Hung, H. Kovar, T.J. Triche, E.R. Lawlor. 2010. CD133 expression in chemo-resistant Ewing sarcoma cells. BMC Cancer 26 10: 116.

Katoh, M., C. Shaw, Q. Xu, N. Van Driessche, T. Morio, H. Kuwayama, S. Obara, H. Urushihara, Y. Tanaka, and G. Shaulsky. 2004. An orderly retreat: Dedifferentiation is a regulated process. Proc Natl Acad Sci USA 101: 7005–7010.

Kawamoto, E.H., N. Weidner, R.M. Agostini Jr, and R. Jaffe. 1987. Malignant ectomesenchymoma of soft tissue report of two cases and review of the literature. Cancer 59: 1791–1802.

Kawai, A., J. Woodruff, J.H. Healey, M.F. Brennan, C.R. Antonescu, and M. Ladanyi. 1998. SYT-SSX gene fusion as a determinant of morphology and prognosis in synovial sarcoma. N Engl J Med 338: 153–160.

Kelly, S.E., A. Di Benedetto, A. Greco, C.M. Howard, V.E. Sollars, D.A. Primerano, J.V. Valluri, and P.P. Claudio. 2010. Rapid selection and proliferation of CD133+ cells from cancer cell lines: chemotherapeutic implications. PLoS One 5: e10035.

Kim, V.N., J. Han, and M.C. Siomi. 2009. Biogenesis of small RNAs in animals. Nat Rev Mol Cell Biol 10: 126–139.

Kosaka, N., H. Iguchi, and T. Ochiya. 2010. Circulating microRNA in body fluid: a new potential biomarker for cancer diagnosis and prognosis. Cancer Sci 101: 2087–2092.

Kubo, T., T. Sugita, S. Shimose, T. Matsuo, K. Arihiro, and M. Ochi. 2008. Expression of hypoxia-inducible factor-1alpha and its relationship to tumor angiogenesis and cell proliferation in cartilage tumors. J Bone Joint Surg Br 90: 364–370.

Kuhn, N.Z., and R.S. Tuan. 2010. Regulation of stemness and stem cell niche ofmesenchymal stem cells: implications in tumorigenesis and metastasis. J Cell Physiol 222: 268–277.

Lapidot, T., C. Sirard, J. Vormoor, B. Murdoch, T. Hoang, J. Caceres-Cortes, M. Minden, B. Paterson, M.A. Caligiuri, and J.E. Dick. 1994. A cell initiating human acute myeloid leukemia after transplantation into SCID mice. Nature 367: 645–648.

Le Douarin, N.M., G.W. Calloni, and E. Dupin. 2008. The stem cells of the neural crest. Cell Cycle 7: 1013–1019.

Lee, A., J.D. Kessler, T.A. Read, C. Kaiser, D. Corbeil, W.B. Huttner, J.E. Johnson, and R.J. Wechsler-Reya. 2005. Isolation of neural stem cells from the postnatal cerebellum. Nat Neurosci 8: 723–729.

Lehnhardt, M., A. Daigeler, H.H. Homann, V. Schwaiberger, O. Goertz, C. Kuhnen, H.U. Steinau. 2009. MFH revisited: outcome after surgical treatment of undifferentiated pleomorphic or not otherwise specified (NOS) sarcomas of the extremities—an analysis of 140 patients. Langenbecks Arch Surg 394: 313–320.

Levings, P.P., S.V. McGarry, T.P. Currie, D.M. Nickerson, S. McClellan, S.C. Ghivizzani, D.A. Steindler, and C.P. Gibbs. 2009. Expression of an exogenous human Oct-4 promoter identifies tumor-initiating cells in osteosarcoma. Cancer Res 69: 5648–5655.

Li, C., D.G. Heidt, P. Dalerba, C.F. Burant, L. Zhang, V. Adsay, M. Wicha, M.F. Clarke, and D.M. Simeone. 2007. Identification of pancreatic cancer stem cells. Cancer Res 67: 1030–1037.

Li, J., W. Liu, K. Zhao, Y. Zhang, X. Li, Q. Yang, Z. Li, and J. Li. 2009. Diallyl trisulfide reverses drug resistance and lowers the ratio of CD133+ cells in conjunction with methotrexate in a human osteosarcoma drug-resistant cell subline. Mol Med Report 2: 245–252.

Lin, G.L., and K.D. Hankenson. 2011. Integration of BMP, Wnt, and notch signaling pathways in osteoblast differentiation. J Cell Biochem 112: 3491–3501.

Lindner, N.J., O. Ramm, A. Hillmann, R. Roedl, G. Gosheger, C. Brinkschmidt, H. Juergens, and W. Winkelmann. 1999. Limb salvage and outcome of osteosarcoma. The University of Muenster experience. Clin Orthop Relat Res 358: 83–89.

Liu, X., Y. Sun, J. Guo, H. Ma, J. Li, B. Dong, G. Jin, J. Zhang, J. Wu, L. Meng, and C. Shou. 2006. Expression of hiwi gene in human gastric cancer wasassociated with proliferation of cancer cells. Int J Cancer 118: 1922–1929.

Longhi, A., C. Errani, M. De Paolis, M. Mercuri, and G. Bacci. 2006. Primary boneosteosarcoma in the pediatric age: state of the art. Cancer Treat Rev 32: 423–436.

Maeda, S., H. Shinchi, H. Kurahara, Y. Mataki, K. Maemura, M. Sato, S. Natsugoe, T. Aikou, and S. Takao. 2008. CD133 expression is correlated with lymph node metastasis and vascular endothelial growth factor-c expression in pancreatic cancer. Br J Cancer 98: 1389–1397.

Malanchi, I., H. Peinado, D. Kassen, T. Hussenet, D. Metzger, P. Chambon, M. Huber, D. Hohl, A. Cano, W. Birchmeier, and J. Huelsken. 2008. Cutaneous cancer stem cell maintenance is dependent on beta-catenin signalling. Nature 452: 650–653.

Mankin, H.J., J. Casas-Ganem, J.I. Kim, M.C. Gebhardt, F.J. Hornicek, and E.N. Zeegen. 2004. Leiomyosarcoma of somatic soft tissues. Clin Orthop Relat Res 421: 225–231.

Marina, N., M. Gebhardt, L. Teot, and R. Gorlick. 2004. Biology and therapeutic advances for pediatric osteosarcoma. Oncologist 9: 422–441.

Masui, S., Y. Nakatake, Y. Toyooka, D. Shimosato, R. Yagi, K. Takahashi, H. Okochi, A. Okuda, R. Matoba, A.A. Sharov, M.S. Ko, and H. Niwa. 2007. Pluripotency governed by Sox2 via regulation of Oct3/4 expression in mouse embryonic stem cells. Nat Cell Biol 9: 625–635.

Matsui, W., C.A. Huff, Q. Wang, M.T. Malehorn, J. Barber, Y. Tanhehco, B.D. Smith, C.I. Civin, and R.J. Jones. 2004. Characterization of clonogenic multiple myeloma cells. Blood 103: 2332–2336.

Matushansky, I., R.G. Maki, and C. Cordon-Cardo. 2008. A context dependent role for Wnt signaling in tumorigenesis and stem cells. Cell Cycle 7: 720–724.

Matushansky, I., E. Hernando, N.D. Socci, J.E. Mills, T.A. Matos, M.A. Edgar, S. Singer, R.G. Maki, and C. Cordon-Cardo. 2007. Derivation of sarcomas from mesenchymal stem cells via inactivation of the Wnt pathway. J Clin Invest 117: 3248–3257.

McDowell, H.P. 2003. Update on childhood rhabdomyosarcoma. Arch Dis Child 88: 354–357.

Merlino, G., and C. Khanna. 2007. Fishing for the origins of cancer. Genes Dev 21: 1275–1279.

Meyers, P.A., R. Gorlick, G. Heller, E. Casper, J. Lane, A.G. Huvos, and J.H. Healey. 1998. Intensification of preoperativechemotherapy for osteogenic sarcoma: results of the Memorial Sloan-Kettering (T12) protocol. J Clin Oncol 16: 2452–2458.

Meyers, P.A., C.L. Schwartz, M.D. Krailo, J.H. Healey, M.L. Bernstein, D. Betcher, W.S. Ferguson, M.C. Gebhardt, A.M. Goorin, M. Harris, E. Kleinerman, M.P. Link, H. Nadel, M. Nieder, G.P. Siegal, M.A. Weiner, R.J. Wells, R.B. Womer, and H.E. Grier. 2008. Children's Oncology Group. Osteosarcoma: the addition of muramyl tripeptide to chemotherapy improves overall survival—a report from the Children's Oncology Group. J Clin Oncol 26: 633–638.

Misra, A., N. Mistry, R. Grimer, and F. Peart. 2009. The management of soft tissue sarcoma. J Plast Reconstr Aesthet Surg 62: 161–174.

Mizrak, D., M. Brittan, and M.R. Alison. 2008. CD133: Molecule of the moment. J Pathol 214: 3–9.

Mohseny, A.B., K. Szuhai, S. Romeo, E.P. Buddingh, I. Briaire-de Bruijn, D. de Jong, M. van Pel, A.M. Cleton-Jansen, and P.C. Hogendoorn. 2009. Osteosarcoma originates from mesenchymal stem cells in consequence of aneuploidization and genomic loss of Cdkn2. J Pathol 219: 294–305.

Monk, M., and C. Holding. 2001. Human embryonic genes re-expressed in cancer cells. Oncogene 20: 8085–8091.

Morrison, S.J., P.M. White, C. Zock, and D.J. Anderson. 1999. Prospective identification, isolation by flow cytometry, and *in vivo* self-renewal of multipotent mammalian neural crest stem cells. Cell 96: 737–749.

Murase, M., M. Kano, T. Tsukahara, A. Takahashi, T. Torigoe, S. Kawaguchi, S. Kimura, T. Wada, Y. Uchihashi, T. Kondo, T. Yamashita, and N. Sato. 2009. Side population cells have the characteristics of cancer stem-like cells/cancer-initiating cells in bone sarcomas. Br J Cancer 101: 1425–1432.

Nagao, H., K. Ijiri, M. Hirotsu, Y. Ishidou, T. Yamamoto, S. Nagano, T. Takizawa, K. Nakashima, S. Komiya, T. Setoguchi. 2011. Role of GLI2 in the growth of human osteosarcoma. J Pathol 224: 169–179.

Nagoshi, N., S. Shibata, M. Nakamura, Y. Matsuzaki, Y. Toyama, and H. Okano. 2009. Neural crest-derived stem cells display a wide variety of characteristics. J Cell Biochem 107: 1046–1052.

Nagoshi, N., S. Shibata, Y. Kubota, M. Nakamura, Y. Nagai, E. Satoh, S. Morikawa, Y. Okada, Y. Mabuchi, H. Katoh, S. Okada, K. Fukuda, T. Suda, Y. Matsuzaki, Y. Toyama, and H. Okano. 2008. Ontogeny and multipotency of neural crest-derived stem cells in mouse bone marrow, dorsal root ganglia, and whisker pad. Cell Stem Cell 2: 392–403.

Naka, N., S. Takenaka, N. Araki, T. Miwa, N. Hashimoto, K. Yoshioka, S. Joyama, K. Hamada, Y. Tsukamoto, Y. Tomita, T. Ueda, H. Yoshikawa, and K. Itoh. 2010. Synovial sarcoma is a stem cell malignancy. Stem Cells 28: 1119–1131.

Ng, T.L., A.M. Gown, T.S. Barry, M.C. Cheang, A.K. Chan, D.A. Turbin, F.D. Hsu, R.B. West, and T.O. Nielsen. 2005. Nuclear beta-catenin in mesenchymal tumors. Mod Pathol 18: 68–74.

Osaki, M., F. Takeshita, Y. Sugimoto, N. Kosaka, Y. Yamamoto, Y. Yoshioka, E. Kobayashi, T. Yamada, A. Kawai, T. Inoue, H. Ito, M. Oshimura, and T. Ochiya. 2011. MicroRNA-143 regulates human osteosarcoma metastasis by regulating matrix metalloprotease-13 expression. Mol Ther 19: 1123–1130.

Ozaki, T., A. Hillmann, N. Lindner, S. Blasius, and W. Winkelmann. 1997. Chondrosarcoma of the pelvis. Clin Orthop Relat Res 337: 226–239.

Pardal, R., M.F. Clarke, and S.J. Morrison. 2003. Applying the principles of stem-cell biology to cancer. Nat Rev Cancer 3: 895–902.

Patrawala, L., T. Calhoun, R. Schneider-Broussard, J. Zhou, K. Claypool, and D.G. Tang. 2005. Side population is enriched in tumorigenic, stem-like cancer cells, whereas ABCG2+ and ABCG2- cancer cells are similarly tumorigenic. Cancer Res 65: 6207–6219.

Pearce, D.J., D. Taussig, C. Simpson, K. Allen, A.Z. Rohatiner, T.A. Lister, and D. Bonnet. 2005. Characterization of cells with a high aldehyde dehydrogenase activity from cord blood and acute myeloid leukemia samples. Stem Cells 23: 752–760.

Pendino, F., I. Tarkanyi, C. Dudognon, J. Hillion, M. Lanotte, J. Aradi, and E. Ségal-Bendirdjian. 2006. Telomeres and telomerase: Pharmacological targets for new anticancer strategies? Curr Cancer Drug Targets 6: 147–180.

Pevny, L.H., and S.K. Nicolis. 2010. Sox2 roles in neural stem cells. Int J Biochem Cell Biol 42: 421–424.

Plaisant, M., S. Giorgetti-Peraldi, M. Gabrielson, A. Loubat, C. Dani, and P. Peraldi. Inhibition of hedgehog signaling decreases proliferation and clonogenicity of human mesenchymal stem cells. PLoS One 6: e16798.

Provisor, A., L. Ettinger, J. Nachman, M.D. Krailo, J.T. Makley, and E.J. Yunis. 1997. Treatment of nonmetastatic osteosarcoma of the extremity with preoperative andpostoperative chemotherapy: a report from the Children's Cancer Group. J Clin Oncol 15: 76–84.

Qiao, D., A.M. Zeeman, W. Deng, L.H. Looijenga, and H. Lin. 2002. Molecular characterization of hiwi, a human member of thepiwi gene family whose overexpression is correlated toseminomas. Oncogene 21: 3988–3999.

Que, J., X. Luo, R.J. Schwartz, and B.L. Hogan. 2009. Multiple roles for Sox2 in the developing and adult mouse trachea. Development 136: 1899–1907.

Raimondi, L., R. Ciarapica, M. De Salvo, F. Verginelli, M. Gueguen, C. Martini, L. De Sio, G. Cortese, M. Locatelli, T.P. Dang, N. Carlesso, L. Miele, S. Stifani, I. Limon, F. Locatelli, and R. Rota. 2011. Inhibition of Notch3 signalling induces rhabdomyosarcoma cell differentiation promoting p38 phosphorylation and p21(Cip1) expression and hampers tumor cell growth *in vitro* and *in vivo*. Cell Death Differ 25.

Rasheed, Z.A., J. Yang, Q. Wang, J. Kowalski, I. Freed, C. Murter, S.M. Hong, J.B. Koorstra, N.V. Rajeshkumar, X. He, M. Goggins, C. Iacobuzio-Donahue, D.M. Berman, D. Laheru, A. Jimeno, M. Hidalgo, A. Maitra, and W. Matsui. 2010. Prognostic significance of tumorigenic cells with mesenchymal features in pancreatic adenocarcinoma. J Natl Cancer Inst 102: 340–351.

Rasheed, Z.A., J. Kowalski, B.D. Smith, and W. Matsui. 2011. Concise review: Emerging concepts in clinical targeting of cancer stem cells. Stem Cells 29: 883–887.

Reya, T., S.J. Morrison, M.F. Clarke, and I.L. Weissman. 2001. Stem cells, cancer, and cancerstem cells. Nature 414: 105–111.

Reynolds, B.A., and S. Weiss. 1992. Generation of neurons and astrocytes from isolated cells of the adult mammalian central nervous system. Science 255: 1707–1710.

Ricci-Vitiani, L., D.G. Lombardi, E. Pilozzi, M. Biffoni, M. Todaro, C. Peschle, and R. De Maria. 2007. Identification and expansion of human colon-cancer-initiating cells. Nature 445: 111–115.

Riggi, N., M.L. Suvà, C. De Vito, P. Provero, J.C. Stehle, K. Baumer, L. Cironi, M. Janiszewska, T. Petricevic, D. Suvà, S. Tercier, J.M. Joseph, L. Guillou, and I. Stamenkovic. EWS-FLI-1 modulates miRNA145 and SOX2 expression to initiate mesenchymal stem cell reprogrammingtoward Ewing sarcoma cancer stem cells. Genes Dev 24: 916–932.

Rot, S., H. Taubert, M. Bache, T. Greither, P. Wurl, A.W. Eckert, J. Schubert, D. Vordermark, and M. Kappler M. 2011. A novel splice variant of the stem cell marker LGR5/GPR49 is correlated with the risk of tumor-related death in soft-tissue sarcoma patients. BMC Cancer 11: 429.

Sadikovic, B., C. Graham, M. Ho, M. Zielenska, and G.R. Somers. 2011. Immunohistochemical expression and cluster analysis of mesenchymal and neural stem cell-associated proteins in pediatric soft tissue sarcomas. Pediatr Dev Pathol. 14: 259–272.

Sasaki, K., T. Hitora, O. Nakamura, R. Kono, and T. Yamamoto. 2011. The role of MAPK pathway in bone and soft tissue tumors. Anticancer Res 31: 549–553.

Scott, S.M., H.M. Reiman, D.J. Pritchard, and D.M. Ilstrup. 1989. Soft tissue fibrosarcoma. A clinicopathologic study of 132 cases. Cancer 64: 925–931.

Shi, S., and S. Gronthos. 2003. Perivascular niche of postnatal mesenchymal stem cells in human bone marrow and dental pulp. J Bone Miner Res 18: 696–704.

Shima, Y., T. Okamoto, T. Aoyama, K. Yasura, T. Ishibe, K. Nishijo, K.R. Shibata, Y. Kohno, K. Fukiage, S. Otsuka, D. Uejima, T. Nakayama, T. Nakamura, T. Kiyono, and J. Toguchida. 2007. *In vitro* transformation of mesenchymal stem cells by oncogenic H-rasVal12. Biochem Biophys Res Commun 353: 60–66.

Shimono, Y., M. Zabala, R.W. Cho, N. Lobo, P. Dalerba, D. Qian, M. Diehn, H. Liu, S.P. Panula, E. Chiao, F.M. Dirbas, G. Somlo, R.A. Pera, K. Lao, and M.F. Clarke. 2009. Downregulation of miRNA-200c links breast cancer stem cells with normal stem cells. Cell 138: 592–603.

Siclari, V.A., and L. Qin. 2010. Targeting the osteosarcoma cancer stem cell. J Orthop Surg Res 5: 78.

Singh, S.K., I.D.Clarke, M. Terasaki, V.E. Bonn, C. Hawkins, J. Squire, and P.B. Dirks. 2003. Identification of a cancer stem cell in human brain tumors. Cancer Res 63: 5821–5828.

Singh, S.K., I.D. Clarke, T. Hide, and P.B. Dirks. 2004. Cancer stem cells in nervous system tumors. Oncogene. 23: 7267–7273.

Sladek, N.E. 2003. Human aldehyde dehydrogenases: Potential pathological, pharmacological, and toxicological impact. J Biochem Mol Toxicol 17: 7–23.

Song, B., Y. Wang, Y. Xi, K. Kudo, S. Bruheim, G.I. Botchkina, E. Gavin, Y. Wan, A. Formentini, M. Kornmann, O. Fodstad, and J. Ju. 2009. Mechanism of chemoresistance mediated by miR-140 in human osteosarcoma and colon cancer cells. Oncogene 28: 4065–4074.

Stratford, E.W., R. Castro, A. Wennerstrøm, R. Holm, M. Else, L. Silj, B. Bodil, and M. Ola. 2011. Liposarcoma cells with aldefluor and CD133 activity have a cancer stem cell potential Clinical Sarcoma Research 1: 8.

Subbiah, V., P. Anderson, A.J. Lazar, E. Burdett, K. Raymond, and J.A. Ludwig. 2009. Ewing's sarcoma: standard and experimental treatment options. Curr Treat Options Oncol 10: 126–140.

Sulzbacher, I., P. Birner, C. Toma, N. Wick, and P.R. Mazal. 2007. Expression of c-kit in human osteosarcoma and its relevance as a prognostic marker. J Clin Pathol 60: 804–807.

Suvà, M.L., N. Riggi, J.C. Stehle, K. Baumer, S. Tercier, J.M. Joseph, D. Suvà, V. Clément, P. Provero, L. Cironi, M.C. Osterheld, L. Guillou, and I. Stamenkovic. 2009. Identification of cancer stem cells in Ewing's sarcoma. Cancer Res 69: 1776–1781.

Ta, H.T., C.R. Dass, P.F. Choong, and D.E. Dunstan: Osteosarcoma treatment: stateof the art. Cancer Metastasis Rev 28: 247–263.

Tai, M.H., C.C. Chang, L.K. Olson, and J.E. Trosko. 2005. Oct4 expression in adult human stem cells: evidence in support of the stem cell theory of carcinogenesis. Carcinogenesis 26: 495–502.

Takahashi, R., T. Fumitaka, T. Fujiwara, O. Makiko, and T. Ochiya. 2011. Review: Cancer Stem Cells in Breast Cancer. Cancers 3: 1311–1328.

Takebe, N., P.J. Harris, R.Q. Warren, and S.P. Ivy. 2011. Targeting cancer stem cells by inhibiting Wnt, Notch, and Hedgehog pathways. Nat Rev Clin Oncol 8: 97–106.

Tang, Q.L., Y. Liang, X.B. Xie, J.Q. Yin, C.Y. Zou, Z.Q. Zhao, J.N. Shen, and J. Wang. 2011. Enrichment of osteosarcoma stem cells by chemotherapy. Chin J Cancer 30: 426–432.

Tang, Q.L., Z.Q. Zhao, J.C. Li, Y. Liang, J.Q. Yin, C.Y. Zou, X.B. Xie, Y.X. Zeng, J.N. Shen, T. Kang, and J. Wang. 2011. Salinomycin inhibits osteosarcoma by targeting its tumor stem cells. Cancer Lett 311: 113–121.

Taubert, H., T. Greither, D. Kaushal, P. Wurl, M. Bache, F. Bartel, A. Kehlen, C. Lautenschläger, L. Harris, K. Kraemer, A. Meye, M. Kappler, H. Schmidt, H.J. Holzhausen, and S. Hauptmann. 2007. Expression of the stem cell self-renewal geneHiwi and risk of tumor-related death in patients with soft tissuesarcoma. Oncogene 26: 1098–1100.

Taubert, H., P. Wurl, T. Greither, M. Kappler, M. Bache, F. Bartel, A. Kehlen, C. Lautenschlager, L.C. Harris, D. Kaushal, S. Fussel, A. Meye, A. Bohnke, H. Schmidt, H.J. Holzhausen, and S. Hauptmann. 2007. Stem cell-associated genes are extremely poor prognostic factors for soft-tissue sarcoma patients. Oncogene 26: 7170–7174.

Terry, J., and T. Nielsen. 2010. Expression of CD133 in synovial sarcoma. Appl Immunohistochem Mol Morphol 18: 159–165.

Thorne, S.H., W. Liang, P. Sampath, T. Schmidt, R. Sikorski, A. Beilhack, and C.H. Contag. 2010. Targeting localized immune suppression within the tumor through repeat cycles of immune cell-oncolytic virus combination therapy. Mol Ther 18: 1698–1705.

Tirino, V., V. Desiderio, R. d'Aquino, F. De Francesco, G. Pirozzi, A. Graziano, U. Galderisi, C. Cavaliere, A. De Rosa, G. Papaccio, and A. Giordano. 2008. Detection and characterization of CD133+ cancer stem cells in human solid tumors. PLoS One 3: e3469.

Tirino, V., V. Desiderio, F. Paino, A. De Rosa, F. Papaccio, F. Fazioli, G. Pirozzi, and G. Papaccio. 2011. Human primary bone sarcomas contain CD133+ cancer stem cells displaying high tumorigenicity *in vivo*. FASEB J 25: 2022–2030.

Toguchida J. Sarcoma-initiating cells and tissue stem cells. pp. 65–71. In: H. Esumi, N. Takakura, K. Miyazono, M. Mori. [eds.]. 2009. Experimental Medicine. 27. Yodosha, Tokyo, Japan.

Toro, J.R., L.B. Travis, H.J. Wu, K. Zhu, C.D. Fletcher, and S.S. Devesa. 2006. Incidence patterns of soft tissue sarcomas, regardless of primary site, in the surveillance, epidemiology and end results program, 1978–2001: An analysis of 26,758 cases. Int J Cancer 119: 2922–2930.

Tostar, U., R. Toftgård, P.G. Zaphiropoulos, and T. Shimokawa. 2010. Reduction of human embryonal rhabdomyosarcoma tumor growth by inhibition of the hedgehog signaling pathway. Genes Cancer 1: 941–951.

Tsuchida, R., B. Das, H. Yeger, G. Koren, M. Shibuya, P.S. Thorner, S. Baruchel, and D. Malkin. 2008. Cisplatin treatment increases survival and expansion of a highly tumorigenic side-population fraction by upregulating VEGF/Flt1 autocrine signaling. Oncogene 27: 3923–3934.

Tsukahara, T., Y. Nabeta, S. Kawaguchi, H. Ikeda, Y. Sato, K. Shimozawa, K. Ida, H. Asanuma, Y. Hirohashi, T. Torigoe, H. Hiraga, S. Nagoya, T. Wada, T. Yamashita, and N. Sato. 2004. Identification of human autologous cytotoxic T-lymphocyte-defined osteosarcoma gene that encodes a transcriptional regulator, papillomavirus binding factor. Cancer Res 64: 5442–5448.

Tsukahara, T., T. Torigoe, Y. Tamura, T. Wada, S. Kawaguchi, T. Tsuruma, K. Hirata, T. Yamashita, and N. Sato. 2008. Antigenic peptide vaccination: Provoking immune response and clinical benefit for cancer. Curr Immunol Rev 4: 235–241.

Tuan, R.S., G. Boland, and R. Tuli. 2003. Adult mesenchymal stem cells and cell-based tissue engineering. Arthritis Research and Therapy 5: 32–45.

Veselska, R., M. Hermanova, T. Loja, P. Chlapek, I. Zambo, K. Vesely, K. Zitterbart, and J. Sterba. 2008. Nestin expression in osteosarcomas and derivation of nestin/CD133 positive osteosarcoma cell lines. BMC Cancer 8: 300.

Visvader, J.E. 2011. Cells of origin in cancer. Nature 469: 314–322.

Visvader, J.E., and G.J. Lindeman. 2008. Cancer stem cells in solid tumors: accumulating evidence and unresolved questions. Nat Rev Cancer 8: 755–768.

Von Levetzow, C., X. Jiang, Y. Gwye, G. von Levetzow, L. Hung, A. Cooper, J.H. Hsu, and E.R. Lawlor. 2011. Modeling initiation of Ewing sarcoma in human neural crest cells. PLoS One 6: e19305.

Wachtel, M., T. Runge, I. Leuschner, S. Stegmaier, E. Koscielniak, J. Treuner, B. Odermatt, S. Behnke, F.K. Niggli, and B.W. Schäfer. 2006. Subtype and prognostic classification of rhabdomyosarcoma by immunohistochemistry. J Clin Oncol 24: 816–822.

Walter, D., S. Satheesha, P. Albrecht, B.C. Bornhauser, V. D'Alessandro, S.M. Oesch, H. Rehrauer, I. Leuschner, E. Koscielniak, C. Gengler, H. Moch, M. Bernasconi, F.K. Niggli, and B.W. Schäfer. 2011. CD133 positive embryonal rhabdomyosarcoma stem-like cell population is enriched in rhabdospheres. PLoS One 6: e19506.

Wang, C., T. Maass, M. Krupp, F. Thieringer, S. Strand, M.A. Wörns, A.P. Barreiros, P.R. Galle, and A. Teufel. 2009. A systems biology perspective on cholangiocellular carcinoma development: focus on MAPK-signaling and the extracellular environment. J Hepatol 50: 1122–1131.

Wang, L., P. Park, and C.Y. Lin. 2009. Characterization of stem cell attributes in human osteosarcoma cell lines. Cancer Biol Ther 8.

Wang, L., P. Park, H. Zhang, F. La Marca, and C.Y. Lin. 2010. Prospective identification of tumorigenic osteosarcoma cancer stem cells in OS99-1 cells based on high aldehyde dehydrogenase activity. Int J Cancer 128: 294–303.

Warner, J.K., J.C. Wang, K.J. Hope, L. Jin, and J.E. Dick. 2004. Concepts of human leukemic development. Oncogene 23: 7164–7177.

Watt, F.M., S. Estrach, and C.A. Ambler. 2008. Epidermal Notch signalling: differentiation, cancer and adhesion. Curr Opin Cell Biol 20: 171–179.

Wei, H., M.Q. Zhao, W. Dong, Y. Yang, and J.S. Li. 2008. Expression of c-kit proteinand mutational status of the c-kit gene in osteosarcoma and theirclinicopathological significance. J Int Med Res 36: 1008–1014.

Weis, L.D. 1999. The success of limb-salvage surgery in the adolescentpatient with osteogenic sarcoma. Adolesc Med 10: 451–458.

Wesolowski, R., and G.T. Budd. 2010. Use of chemotherapy for patients with bone and soft-tissue sarcomas. Cleve Clin J Med 77 Suppl 1: S23–26.

Wilson, H., M. Huelsmeyer, R. Chun, K.M. Young, K. Friedrichs, and D.J. Argyle. 2008. Isolation and characterization of cancer stem cells from canine osteosarcoma. Vet J 175: 69–75.

Wick, M.R., G.P. Siegal, S.E. Mills, R.C.Thompson, D. Sawhney, and R.E. Fechner. 1987. Dedifferentiated chondrosarcoma of bone. An immunohistochemical and lectin-histochemical study. Virchows Arch A Pathol Anat Histopathol 411: 23–32.

Wu, C., and B.A. Alman. 2008. Side population cells in human cancers. Cancer Lett 268: 1–9.

Wu, C., Q. Wei, V. Utomo, P. Nadesan, H. Whetstone, R. Kandel, J.S. Wunder, and B.A. Alman. 2007. Side population cells isolated from mesenchymal neoplasms have tumor initiating potential. Cancer Res 67: 8216–8222.

Xu, N., T. Papagiannakopoulos, G. Pan, J.A. Thomson, and K.S. Kosik. 2009. MicroRNA-145 regulates OCT4, SOX2, and KLF4 and represses pluripotency in human embryonic stem cells. Cell. 137: 647–658.

Yin, S., J. Li, C. Hu, X. Chen, M. Yao, M. Yan, G. Jiang, C. Ge, H. Xie, D. Wan, S. Yang, S. Zheng, and J. Gu. 2007. CD133 positive hepatocellular carcinoma cells possess high capacity for tumorigenicity. Int J Cancer 120: 1444–1450.

Young, A., J. Lyons, A.L. Miller, V.T. Phan, I.R. Alarcón, and F. McCormick. 2009. Ras signaling and therapies. Adv Cancer Res 102: 1–17.

Yuan, H., N. Corbi, C. Basilico, and L. Dailey. 1995. Developmental-specific activity of the FGF-4 enhancer requires the synergistic action of Sox2 and Oct-3. Genes Dev 9: 2635–2645.

Zamurovic, N., D. Cappellen, D. Rohner, and M. Susa. 2004. Coordinated activation of notch, Wnt, and transforming growth factor-beta signaling pathways in bone morphogenic protein 2-induced osteogenesis. Notch target gene Hey1 inhibits mineralization and Runx2 transcriptional activity. J Biol Chem 279: 37704–37715.

Zeng, W., R. Wan, Y. Zheng, S.R. Singh, and Y. Wei. 2011. Hypoxia, stem cells and bone tumor. Cancer Lett 313: 129–136.

Zhang, P., Y. Yang, P.A. Zweidler-McKay, and D.P. Hughes. 2008. Critical role of notch signaling in osteosarcoma invasion and metastasis. Clin Cancer Res 14: 2962–2969.

Zhou, S., J.D. Schuetz, K.D. Bunting, A.M. Colapietro, J. Sampath, J.J. Morris, I. Lagutina, G.C. Grosveld, M. Osawa, H. Nakauchi, and B.P. Sorrentino. 2001. The ABC transporter Bcrp1/ABCG2 is expressed in a wide variety of stem cells and is a molecular determinant of the side-population phenotype. Nat Med 7: 1028–1034.

Chapter 3

Role of Hematopoietic Stem and Progenitor Cells in the Origination of Mature B-cell Lymphomas

Jose A. Martinez-Climent

Introduction

Most human cancers are organized as a cellular hierarchy initiated and sustained by a self-renewing cell population of *cancer stem cells* that give rise to the full repertoire of cells composing the tumor and are responsible for their continuous growth and propagation (Reya et al. 2001, Pardal et al. 2003, Scadden 2004, Clarke and Fuller 2006, Polyak and Hahn 2006, Visvader 2011, Nguyen et al. 2012). Although the cancer stem cell (CSC) concept was postulated in early reports (Bruce and Van Der Gaag 1963, Park et al. 1971, Hamburger and Salmon 1977, Sabbath et al. 1985), definite proof of their existence was shown in leukemias, where cells with a $CD34^{+}CD38^{-}$ phenotype could initiate, regenerate and maintain the tumor entirely after transplantation into immunocompromised mice (Lapidot et al. 1994, Blair et al. 1997, Bonnet and Dick 1997). A number of phenotypically and functionally different CSCs have been subsequently identified in epithelial cancers, including among others breast, head and neck, and prostate carcinomas, some of which can be detected by expressing the cell-surface glycoprotein CD44 (Al-Hajj et al. 2003, Al-Hajj and Clarke 2004, Collins et al. 2005, Li et al. 2007, Prince et al. 2007, Stingl and Caldas 2007, Hermann et

Division of Oncology, Center for Applied Medical Research, University of Navarra, Avda. Pio XII, 55, 31008 Pamplona, Spain.
E-mail: jamcliment@unav.es

al. 2010). Notably, a $CD44^+$ cell-specific gene expression signature containing a large number of stem-cell genes predicted the outcome of patients with breast cancer (Liu et al. 2007, Shipitsin et al. 2007). The CD133 glycoprotein is another cell surface marker expressed by neural and hematopoietic stem cells that defines the CSCs of brain, liver and colon carcinomas (Singh et al. 2004, Ma et al. 2007, O'Brien et al. 2007, Ricci-Vitiani et al. 2007, Hermann et al. 2010). Remarkably, glioblastoma $CD133^+$ CSCs were more resistant to chemo and radiotherapy than the $CD133^-$ cell population (Bao et al. 2006). Indeed, current therapies that successfully target the differentiated cancer cells may be ineffective against the CSCs, suggesting that novel drugs aiming to eliminate this cell population are required (Martinez-Climent et al. 2006, Zhang and Rosen 2006, Apperley 2007, Hermann et al. 2010, Visvader 2011, Nguyen et al. 2012). Thus, the CSC concept is not only changing our current understanding of cancer biology but is also having a hearing on cancer therapeutics.

B-cell lymphomas are clonal tumors of malignant B lymphocytes that recapitulate the various stages of normal B-cell differentiation (Kuppers 2005). Despite early studies demonstrating the presence of clonogenic lymphoma cell subpopulations with stem cell-like properties in patients with mature B-cell lymphomas (Jones et al. 1979, Bobak and Whisler 1980, Tweeddale et al. 1987), it is widely accepted that these tumors derive from germinal or post-germinal center B lymphocytes (Kuppers et al. 1999, Shaffer et al. 2002, Kuppers 2005, Klein and Dalla-Favera 2008). However, recent reports have suggested that early hematopoietic progenitor cells may have a role in the origination of different lymphoma subgroups. Thus, rare cell subpopulations with stem cell features that are responsible for the generation and maintenance of the predominant mature B-cell tumor burden have been found in patients with mantle cell lymphoma, follicular lymphoma and Hodgkin disease (Jones et al. 2009, Chen et al. 2010, Lee et al. 2012). More importantly, genetic and functional studies in human and mice have demonstrated that hematopoietic stem and progenitor cells can be involved in the pathogenesis of some mature lymphoid tumors (Kikushige et al. 2011, Quivoron et al. 2011, Vicente-Dueñas et al. 2012). In this chapter we discuss the role of the diverse hematopoietic and lymphoid cell subpopulations as the putative cells of origin in the various mature B-cell lymphomas.

Lymphoma-Initiating Cells and Lymphoma-Stem Cells

Like other tumors, lymphomas derive from the multistep accumulation of genetic and epigenetic mutations in cells that progressively acquire neoplastic features such as increased proliferation, evasion of apoptosis and blockade of differentiation, eventually driving their malignant transformation into cancer cells (Kuppers 2005, Hanahan and Weinberg 2011). By definition, the cell that harbors the first mutation is the *lymphoma-initiating cell*, which is generally unable to drive tumor development and need additional mutations for doing so. Like in most human epithelial tumors and myeloid leukemias, these lymphoma-initiating require the properties of self-renewal and multi-differentiation in order to generate and sustain the heterogeneous lymphoma cell population (Reya et al. 2001, Clarke and Fuller 2006, Visvader 2011). By definition, the cell able to generate and maintain the entire B-cell lymphoma is the *lymphoma-stem cell* (or lymphoma-propagating cell). But how can a pre-malignant lymphoid cell become a fully malignant stem cell-like cancer cell? A simple explanation is that committed early lymphocytes represent the lymphoma-initiating cells, which sequentially accumulate mutations that target reprogramming pathways and reactivate stemness properties, thus becoming the lymphoma-stem cells. Another possibility is that the first mutation occurs in cells with inherent stem-cell properties, such as bone marrow hematopoietic stem/progenitor cells, germinal-center B centroblasts or long-lived memory B lymphocytes, thus allowing lymphoma-initiating cell to directly behave as lymphoma-stem cells (Fearon et al. 2001, Lanzavecchia and Sallusto 2002). The identification of the cellular nature of the lymphoma-initiating and lymphoma-stem cells in the different B-cell lymphoam subgroups will be critical to fully understand the lymphomagenesis process.

Chromosomal Translocations as Time-Course Markers of Lymphoma Initiation

Balanced chromosomal translocations that involve immunoglobulin (*IG*) genes and oncogenic partner genes are genetic hallmarks of mature B-cell lymphomas, being considered the first (or one of the earliest) genetic abnormalities occurring in the lymphomagenesis process (Willis and Dyer 2000, Chen et al. 2010, Nussenzweig and Nussenzweig 2010, Gostissa et al. 2011). Studies in mice have shown that these *IG*-related chromosomal fusions occurring in B lymphocytes are consequences of the physiological DNA damage induced during the antigenic B-cell receptor (BCR) gene diversification (Nussenzweig and Nussenzweig 2010, Gostissa et al. 2011). This process initiates when variable, diversity and joining (V(D)J) gene segment recombination assembles IG heavy or light chain (*IGH* or *IGL*) variable region exons in early B-cell precursors in the bone marrow, which is mediated by the RAG1/2 recombinase (Busslinger 2004, Matthias and Rolink 2005, Medina and Singh 2005). Once a functional surface BCR is expressed by these immature B cells, they leave the bone marrow to populate peripheral lymphoid organs, entering in contact with the antigen and forming specialized dynamic microenvironmental structures termed the germinal centers (GCs) (MacLennan 1994, Matthias and Rolink 2005, Allen et al. 2007). There, B lymphocytes continue BCR remodeling by using the cytidine deaminase (AID)-mediated class switch recombination (CSR) process, a mechanism that exchanges *IGH* constant region exons to generate different antibody isotypes (MacLennan 1994, Muramatsu et al. 2000). Also within the GC, somatic hypermutation (SHM) introduces nucleotide changes, deletions and duplications at a very high frequency in both productive and non-productive variable regions of rearranged *IG* genes (IGVH), resulting in the generation of antibody variants with high affinity for the antigen (Muramatsu et al. 2000, Kuppers 2003). Finally, GC cells expressing favorable antibody mutants are positively selected, vigorously expanded and released into the periphery as immunoglobulin-secreting plasma cells or long-lived memory B lymphocytes.

Errors in either V(D)J recombination in bone marrow B cells or in CSR or SHM remodeling in GC B lymphocytes can result in *IG*-related chromosomal translocations (Zhu et al. 2002, Ramiro et al. 2004, Robbiani et al. 2008, Robbiani et al. 2009, Nussenzweig and Nussenzweig 2010, Gostissa et al. 2011). Functionally, these genetic rearrangements induce deregulated expression of selected oncogenes that accelerate cell proliferation, decrease cell survival or block lymphocyte differentiation (Willis and Dyer 2000, Chen et al. 2010, Nussenzweig and Nussenzweig 2010). The different mature B-cell lymphoma subgroups are characterized by specific chromosomal fusions of *IG* genes with selected oncogenes, among others *BCL2* in follicular lymphoma (FCL), *CCND1* in mantle cell lymphoma (MCL), *MALT1* in mucosa-associated lymphoid tissue (MALT) lymphoma, *MYC* in Burkitt lymphoma and *BCL6* in diffuse large cell lymphoma (DLBCL) (Dalla-Favera et al. 1982, Willis and Dyer 2000, Sanchez-Izquierdo et al. 2003). The analysis of the molecular structure of these genetic rearrangements can assign each translocation to a specific developmental cell stage, potentially defining the lymphoma-initiating cell in the diverse lymphoma subgroups (Rolink et al. 2000, Kuppers and Dalla-Favera 2001, Busslinger 2004, Matthias and Rolink 2005, Medina and Singh 2005).

Bone Marrow Precursor B Lymphocytes as Lymphoma-Initiating Cells

Molecular analysis of the common translocation breakpoints has provided insights into the cellular stage at which they occurred. The majority of chromosomal translocations involving *BCL2* and *CCND1* genes usually target the non-functional *IG* gene at the 5′ end of J heavy-chain gene segments, indicating that were caused by V(D)J recombination mistakes at the D to

J_H stage in bone marrow lymphoid progenitors (namely, early lymphoid progenitors and common lymphoid progenitos) or in B-cell precursors (pro-B and pre-B cells) (Willis and Dyer 2000, Busslinger 2004). However, at least one-third of FCLs carry *BCL2-IGH* gene rearrangements arised at later V(D)J recombination stages (primarily V to DJ_H) involving pre-B and immature B cells (Cotter et al. 1990, Stamatopoulos et al. 1997, Jager et al. 2000). Thus, early B-cell precursors carrying V(D)J-associated translocations of *BCL2* and *CCND1* genes can be considered the lymphoma-initiating cells in FCL and MCL, respectively (Martinez-Climent et al. 2010). Nevertheless, mice with ectopic expression of the *CCND1* or *BCL2* oncogenes in early B-cell precursors did not develop malignancy spontaneously (McDonnell et al. 1989, McDonnell and Korsmeyer 1991, Bodrug et al. 1994, Lovec et al. 1994, Beltran et al. 2011). In addition, circulating B lymphocytes carrying the *BCL2-IGH* or *CCND1-IGH* gene fusions can be detected in the peripheral blood of healthy individuals, only a minority of which will eventually develop malignancy (Schuler et al. 2003, Roulland et al. 2006, Lecluse et al. 2009). These data suggest that human and mouse B lymphocytes carrying these genetic rearrangements need additional secondary mutations to fully drive tumor development.

But what is the nature of these secondary mutations and where do they take place? The pre-malignant lymphoma-initiating cell must acquire self-renewal and multidifferentiation properties to be capable of generating and maintaining tumor development. Accordingly, the secondary mutations can either block differentiation and freeze cells in a stage with inherent stem-cell properties, induce cell reprogramming that dedifferentiate the cells back to stem-cell like cells, or target stem-cell genes that directly activate aberrant stem-cell functions (Reya et al. 2001, Pardal et al. 2005, Visvader 2011, Nguyen et al. 2012). Although these secondary genetic rearrangements can occur throughout the normal stages of differentiation, most of them are induced within the GCs. The molecular analysis of the SHM and CSR-induced mutation patterns in the mature B-cell lymphoma cells can trace the cellular stage at which the secondary changes took place. FCL and germinal center-derived DLBCL (GBC-DLBCL) show intraclonal *IGH* nucleotide variation consisting in heterogeneous *IVHG* sequences and multiple isotype transcripts (Ottensmeier et al. 1998, Lossos et al. 2000, Ottensmeier and Stevenson 2000). The presence of such "ongoing SHM indicates that the lymphoma-initiating cell carrying the t(14;18) (q32;q21) acquired a second mutation at pre-germinal or early GC B-cell stages, which was followed by clonal expansion of the B lymphocytes that suffered divergences in the SHM process. Conversely, ongoing SHM is rarely observed in DLBCL with activated B-cell phenotype (ABC-DLBCL), suggesting that the secondary mutations occurred in lymphocytes about to exit or already out of the GCs (Lenz et al. 2007).

What are the genes commonly mutated in the GCs? The transcriptional repressor BCL6 is highly expressed in GC cells, facilitating IG affinity maturation by repressing target genes involved in DNA damage responses and cell cycle, and by blocking differentiation of GC B cells into plasma cells, in part by directly or indirectly repressing the *PRDM1* gene (Crotty et al. 2010). Once the GC B lymphocytes express a BCR with high affinity for the antigen, BCL6 is down-regulated to allow these cells to exit the GC and differentiate into IG-secreting plasma cells or long-lived memory B lymphocytes. In most GC-derived B-cell lymphomas, BCL6 is constitutively expressed due to promoter chromosomal translocations or point mutations of negative regulatory elements (Ci et al. 2009, Basso and Dalla-Favera 2010). Therefore, these BCL6-expressing cells are frozen into the GC stage, retaining the inherent stem-cell features of centroblasts that may allow full malignant transformation. A similar mechanism that can block B-cell differentiation at the GC exit is caused by genetic or epigenetic inactivation of *PRDM1*, which are frequently observed in ABC-DLBCL (Tam et al. 2006, Mandelbaum et al. 2010). These data pinpoint *BCL6* and *PRDM1* as critical secondary genes that may transform the lymphoma-initiating cell into a GC-derived lymphoma-stem cell.

Composite Lymphomas and Transplant-Related Lymphomas Can Provide Information to Define the Cells of Origin

The molecular analysis of composite B-cell lymphomas, defined as the simultaneous presence of two distinct lymphoma subtypes in the same biopsy, has also provided valuable information on the cellular origins of the tumors. Most patients carrying composite lymphomas (preferentially including FCL or MCL) show identical clonal V(D)J gene rearrangements and the same *IGH*-related chromosomal translocations in the two phenotypically different tumors, but they differ in the pattern of SHM or CSR-induced mutations (Cleary et al. 1988, Brauninger et al. 1999, Schmitz et al. 2005). These data further support that chromosomal translocations generally arise in B-cell precursor-derived lymphoma-initiating cells that acquire divergent secondary mutations during the lymphocyte differentiation pathway able to induce the development of two distinct lymphomas. Interestingly, the molecular analysis of two donor-recipient pairs, who both developed lymphoma several years after allogeneic bone marrow transplantation, allowed to characterizing the timing of genetic alterations contributing to lymphoma transformation. Both tumor pairs were classified as FCL with similar histopathologically features, presenting an identical *BCL2-IGH* gene fusion and showing the same V(D)J rearrangement, which underwent SHM before and after clonal divergence (Hart et al. 2007, Weigert et al. 2012).

The remarkable genetic similarities between the lymphomas in donors and recipients pointed to a B-cell precursor cell carrying the *BCL2-IGH* rearrangement as the lymphoma-initiating cell, which evolved to a GC-derived lymphoma-stem cell capable of generating similar tumors after transplantation.

Germinal-Center B Lymphocytes and Post-Germinal Memory B cells: The Usual Suspects Capable of Lymphoma Development

A different molecular class of chromosomal translocations occurs in *IGH* switch sequences, being consequence of illegitimate CSR in the GC. These include translocations of *MYC* gene in some Burkitt lymphomas, of *BCL6* gene in DLBCL, and of *IRF4, c-MAF, FGFR3, CCND1* and *MMSET* genes in multiple myeloma (Willis and Dyer 2000, Kuppers and Dalla-Favera 2001, Kuehl and Bergsagel 2002). Very rarely, FCLs display CSR-mediated *BCL2-IGH* rearrangements (Fenton et al. 2002). DNA breaks generated by SHM may also result in aberrant recombination and thus promote chromosomal translocations. Indeed, in many Burkitt lymphoma translocations involving the *MYC* oncogene, the breakpoint in *IGH* often lies within the region targeted by SHM, suggesting that these rearrangements may be consequence of errors during this process (Haluska et al. 1986, Neri et al. 1988). SHM may also target non-*IG* genes, for instance in the 5′ site of the *BCL6* oncogene, therefore promoting chromosomal translocations of *BCL6* with different partners (Lo Coco et al. 1994, Pasqualucci et al. 2001). Theoretically, because GC B lymphocytes experience asymmetric cell division during the GC reaction and are capable of self-renewal and clonal expansion in order to generate identical long-lived plasmatic or memory cells, chromosomal rearrangements arising in GC cells may induce the development of lymphoid tumors without requiring further genetic lesions (Allen et al. 2007, Schroeder 2007, Barnett et al. 2012, Duffy et al. 2012). Likewise, long-lived memory B cells show stem cell-like capacities for self-renewal and plasticity, which are the basis for the continual generation of effector lymphocytes from the memory pool (Schittek and Rajewsky 1990, McHeyzer-Williams et al. 2000, Fearon et al. 2001, Lanzavecchia and Sallusto 2002, Good-Jacobson and Shlomchik 2010). Moreover, these memory B cells share a transcriptional program of self-renewal with hematopoietic stem cells which is mediated by the polycomb complex and STAT5 (Scheeren et al. 2005, Luckey et al. 2006). Similarly to GC B cells, mutations occurring in long-lived memory B lymphocytes may transform them into lymphoma-stem cells capable of generating and maintaining the heterogeneous lymphoma cell populations. This scenario is compatible with the fact that the majority of human B-cell lymphomas show GC or post-GC phenotypes (Jaffe et al. 2008).

Evidences Supporting Hematopoietic Stem/Progenitor Cells as Lymphoma-Stem Cells in Mature B-cell Lymphomas

Hematopoietic stem cells (HSCs) show self-renewal and multi-differentiation features, which are partially retained in bone marrow multipotent and early progenitor cells (Reya et al. 2001, Pardal et al. 2003, Scadden 2004, Dick 2008). Relative quiescence is another defining characteristic of HSCs, while their progeny have increased proliferative ability and inexorably move toward terminal differentiation (Dick 2008, Mercier et al. 2011). Accordingly, genetic rearrangements occurring in hematopoietic and progenitor cell populations may theoretically induce their transformation in lymphoma-stem cells able to drive lymphoma development and maintenance (Castellanos et al. 2010). Obviously, these long-lived HSC populations have more time to accumulate genetic abnormalities than short-lived cells. However, because hematopoietic stem and progenitor cells do not normally have an active V(D)J recombination program, it seems unlikely that the *IG*-related chromosomal translocations may arise in these cells (Martinez-Climent et al. 2010). Alternatively, they may acquire other type of mutations and drive B-cell lymphoma development. In line with this hypothesis, and in contrast to the widely assumed concept supporting that quiescence is a protective mechanism for stem cells to minimize endogenous stress, HSC quiescence can promote error-prone DNA repair that makes these cells intrinsically vulnerable to mutagenesis following DNA damage (Mohrin et al. 2010). A recent report showed that some patients with mature B-cell lymphomas harbored inactivating mutations of *TET2* gene in hematopoietic stem/progenitor cells that promoted lymphoma development (Quivoron et al. 2011). Importantly, this is the first study that provides genetic evidence for the implication of hematopoietic progenitor cells in the origination of human B-cell lymphomas.

Hematopoietic Stem Cells and B-cell Chronic Lymphocytic Leukemia

B-cell chronic lymphocytic leukemia (B-CLL) is classified among the mature B-cell lymphoid malignancies lymphoma according to the WHO classification (Jaffe et al. 2008). A remarkable study has revealed that in B-CLL the propensity to generate clonal B cells is already acquired at the HSC stage (Kikushige et al. 2011). Thus, HSCs purified from patients with B-CLL displayed lymphoid-lineage gene priming and produced a high number of polyclonal B-cell progenitors. Strikingly, their maturation into B cells was restricted to mono- or oligo-clones with B-CLL-like phenotype in xenogeneic recipients. These B-cell clones were independent of the original

B-CLL clones because they had different V(D)J gene rearrangements. Furthermore, they used preferentially *IGVH* genes frequently used in human B-CLL, presumably reflecting the role of BCR signalling in clonal selection. Notably, these HSCs did not show the typical chromosomal abnormalities of B-CLL but rather these genetic changes were acquired at later stages of cell differentiation (Kikushige et al. 2011). These results show that HSCs are implicated in the pathogenesis of a B-CLL, functioning as lymphoma-initiating cells, and opens the possibility that similar cell populations may be the source of other mature B-cell malignancies (Alizadeh and Majeti 2011).

TET2 Gene Mutations, Epigenetic Reprogramming and Lymphoma-Stem Cells

But what are the mutations that can transform a normal HSC into a lymphoma-prone HSC? As outlined above, inactivating mutations of *TET2* gene have been detected in hematopoietic and progenitor cells from patients with B and T-cell lymphomas (Quivoron et al. 2011). TET2 catalyzes the conversion of 5-methyl-cytosine (mC) to 5-hydroxymethyl-cytosine (5hmC) and participates in the epigenetic regulation of gene expression during embryogenesis and cancer (Ko et al. 2010). Loss-of-function mutations of TET2 are common in patients with myeloid malignancies (Delhommeau et al. 2009, Langemeijer et al. 2009). Tet2 is a critical regulator of self-renewal and differentiation of HSCs. Indeed, Tet2 deficiency in mice led to enhanced hematopoiesis of lymphoid and myeloid lineages and delayed HSC differentiation, presumably by controlling 5hmC levels at genes important for the self-renewal, proliferation, and differentiation of HSCs (Ko et al. 2011, Moran-Crusio et al. 2011). Therefore, TET2 inactivation could induce an abnormal epigenetic program in human HSCs that may initiate the lymphomagenesis process. In line with this hypothesis, a different study reported that mature B-cell lymphomas with different genetic backgrounds show a common set of genes silenced by hypermethylation that was significantly enriched for polycomb targets in embryonic stem cells. These findings indicate that B-cell lymphomas share a similar stem cell-like epigenetic pattern, suggesting that they may derive from lymphoma-initiating cells with stem cell features (Martin-Subero et al. 2009). Whether TET2 inactivation may be responsible for this abnormal epigenetic pattern is unknown, but due to the rarity of TET2 mutations detected in patients with B-cell lymphomas these could only explain a minority cases, suggesting that other genes may be involved in this process (Quivoron et al. 2011).

Expression of *MALT1* Oncogene in Mouse Hematopoietic/ Progenitor Cells Generates Human-Like Mature B-cell Lymphomas

Chromosomal translocations involving *MALT1* gene are hallmarks of mucosa-associated lymphoid tissue (MALT) lymphoma (Dierlamm et al. 1999, Sanchez-Izquierdo et al. 2003). However, targeting these translocations to mouse B-cells has failed to reproduce human disease (Isaacson and Du 2004, Farinha and Gascoyne 2005). Using a recently developed technology, ectopic expression of *MALT1* expression in mouse $Sca1^{+}Lin^{-}$ hematopoietic stem/progenitor cells led to NF-κB activation and early lymphoid priming (Perez-Caro et al. 2009, Vicente-Dueñas et al. 2012). These cells were selectively skewed towards B-cell differentiation, accumulated in extranodal tissues and gave rise to clonal tumors recapitulating the principal clinical and biological features of human MALT lymphoma. Deletion of *p53* gene accelerated tumor onset and induced the transformation of MALT lymphoma to human-like ABC-DLBCL. Treatment of MALT1-induced lymphomas with a specific inhibitor of MALT1 proteolytic activity decreased cell viability, indicating that Malt1 signaling is required for tumor cell survival (Vicente-Dueñas et al. 2012). To evaluate whether these findings could be relevant to human disease, bone marrow hematopoietic $CD34^{+}$ cells isolated from patients with MALT lymphoma were analyzed and found to display an aberrant transcriptional signature that included activation of NF-κB signaling and early B-cell priming (Vicente-Dueñas et al. 2012). This study shows that human-like lymphomas can be modeled in mice by targeting MALT1 expression to hematopoietic stem/progenitor cells, suggesting that these cells may be involved in the pathogenesis of human mature B-cell lymphomas.

Plasticity of B lymphocytes Contribute to Malignant Transformation

An exceptional plasticity is a common feature of normal and tumoral B lymphocytes. Xie et al. demonstrated that enforced expression of CEBPα/β transcription factors in mature B-cells can trans-differentiate them into macrophages by inhibiting Pax5 and synergizing with PU.1 (Xie et al. 2004). Another study reported an extensive reprogramming occurring in B cells to become Hodgkin and Reed-Sternberg lymphoma cells upon mutation (Mathas et al. 2006). Cobaleda et al. showed that a conditional *Pax5* deletion in mice allowed mature B cells from peripheral lymphoid organs to dedifferentiate back to early uncommitted progenitors *in vivo*, giving rise to progenitor cell-derived lymphomas (Cobaleda et al. 2007).

Moreover, direct nuclear reprogramming of terminally differentiated mouse mature B lymphocytes to pluripotent stem cells was achieved by expressing four transcription factors (Oct4, Sox2, Klf4, and c-Myc) in addition to either CEBPα/β expression or PAX5 knockdown (Hanna et al. 2008). These results point to CEBPα/β and PAX5 as key reprogramming factors of the normal and tumoral differentiation of lymphoid cells. In a different report, immunofluoresence and fluorescence *in situ* hybridization analyses of human lymphoma biopsies detected the typical *IG*-related chromosomal translocations involving *BCL2*, *CCND1*, *MYC* and *MALT1* genes in neoplastic B cells but also in over 30% of the surrounding microvascular endothelial cells. These findings suggest that the tumor-related endothelial cells derived from mature B lymphocytes carrying the translocation that were reprogrammed to form tumor vessels in response to hypoxia (Streubel et al. 2004). Together, these results underscore that the inherent plasticity potential of the lymphoid cells may contribute to reprogram the pre-malignant lymphoma-initiating cells into lymphoma-stem cells with full tumorigenic potential.

Lymphoma-Stem Cells and Therapeutic Resistance

Clonogenic peripheral blood $CD45^{+}CD19^{-}$ cells isolated from patients with leukemic MCL were able to reconstitute the tumor after transplantation into immunodeficient mice (Chen et al. 2010). These potential MCL-stem cells showed self-renewal and differentiation properties. Similarly, a rare fraction of FCL cells with stem-cell features isolated from patients with FCL was found to interact with follicular dendritic cells to maintain tumourigenic potential, being able to generate the lymphoma after xenotransplantation in immunodeficient mice (Lee et al. 2012). While it is unclear whether these cell populations represent true cancer stem cells, both studies show that these cells are more resistant to several chemotherapeutic agents compared to the mature B-cell population composing the entire lymphoma (Chen et al. 2010, Lee et al. 2012). Interestingly, MCL and FCL-stem cells exhibited high expression of ABC transporters, which are known regulators of multi-drug resistance (Chen et al. 2010, Jung et al. 2011, Lee et al. 2012). Subsequent studies revealed that the $CD45^{+}CD19^{-}$ MCL-stem cells are also resistant to the proteasome inhibitor bortezomib, despite constitutive NF-κB signaling (Jung et al. 2012). These results might explain at least in part the inherent therapeutic resistance of MCL, which still remains an incurable disease with currently available therapies (Jares et al. 2007, Perez-Galan et al. 2011). Likewise, most patients with FCL initially respond to chemotherapy, but they typically experience a continuous pattern of relapses and eventually transform into high-grade therapy-resistant lymphoma (Lossos and Levy 2003). It is thus conceivable that the MCL and FCL-stem cells expressing high levels of ABC transporters stand resistant to chemotherapy, whereas most other lymphoma cells die upon treatment. When therapy ceases, these lymphoma-stem cells regenerate the lymphoma completely.

In summary, detailed isolation and molecular characterization of the specific cells that originate and maintain the diverse mature B-cell lymphomas will lead to a better understanding of the lymphomagenesis process. This knowledge will contribute to design novel therapies targeting molecules or pathways damaged in the lymphoma-stem cells but not in normal cells that may be more effective and eventually increase the survival of patients with B-cell lymphoma.

References

Al-Hajj, M., and M.F. Clarke. 2004. Self-renewal and solid tumor stem cells. Oncogene 23: 7274–7282.

Al-Hajj, M., M.S. Wicha, A. Benito-Hernandez, S.J. Morrison, and M.F. Clarke. 2003. Prospective identification of tumorigenic breast cancer cells. Proc Natl Acad Sci USA 100: 3983–3988.

Alizadeh, A.A., and R. Majeti. 2011. Surprise! HSC are aberrant in chronic lymphocytic leukemia. Cancer Cell 20: 135–136.

Allen, C.D., T. Okada, and J.G. Cyster. 2007. Germinal-center organization and cellular dynamics. Immunity 27: 190–202.

Apperley, J.F. 2007. Part I: Mechanisms of resistance to imatinib in chronic myeloid leukaemia. The Lancet Oncology 8: 1018–1029.

Bao, S., Q. Wu, R.E. McLendon, Y. Hao, Q. Shi, A.B. Hjelmeland, M.W. Dewhirst, D.D. Bigner, and J.N. Rich. 2006. Glioma stem cells promote radioresistance by preferential activation of the DNA damage response. Nature 444: 756–760.

Barnett, B.E., M.L. Ciocca, R. Goenka, L.G. Barnett, J. Wu, T.M. Laufer, J.K. Burkhardt, M.P. Cancro, and S.L. Reiner. 2012. Asymmetric B cell division in the germinal center reaction. Science 335: 342–344.

Basso, K., and R. Dalla-Favera. 2010. BCL6: master regulator of the germinal center reaction and key oncogene in B cell lymphomagenesis. Adv Immunol 105: 193–210.

Beltran et al. PNAS 2011 Jul 26; 108(30): 12461–6; PMID: 21746927.

Blair, A., D.E. Hogge, L.E. Ailles, P.M. Lansdorp, and H.J. Sutherland. 1997. Lack of expression of Thy-1 (CD90) on acute myeloid leukemia cells with long-term proliferative ability *in vitro* and *in vivo*. Blood 89: 3104–3112.

Bobak, D., and R. Whisler. 1980. Human B lymphocyte colony responses. I. General characteristics and modulation by monocytes. J Immunol 125: 2764–2769.

Bodrug, S.E., B.J. Warner, M.L. Bath, G.J. Lindeman, A.W. Harris, and J.M. Adams. 1994. Cyclin D1 transgene impedes lymphocyte maturation and collaborates in lymphomagenesis with the myc gene. Embo J 13: 2124–2130.

Bonnet, D., and J.E. Dick. 1997. Human acute myeloid leukemia is organized as a hierarchy that originates from a primitive hematopoietic cell. Nat Med 3: 730–737.

Brauninger, A., M.L. Hansmann, J.G. Strickler, R. Dummer, G. Burg, K. Rajewsky, and R. Kuppers. 1999. Identification of common germinal-center B-cell precursors in two patients with both Hodgkin's disease and non-Hodgkin's lymphoma. N Engl J Med 340: 1239–1247.

Bruce, W.R., and H. Van Der Gaag. 1963. A Quantitative Assay For The Number Of Murine Lymphoma Cells Capable Of Proliferation *in vivo*. Nature 199: 79–80.

Busslinger, M. 2004. Transcriptional control of early B cell development. Annu Rev Immunol 22: 55–79.
Castellanos, A., C. Vicente-Duenas, E. Campos-Sanchez, J.J. Cruz, F.J. Garcia-Criado, M.B. Garcia-Cenador, P.A. Lazo, J. Perez-Losada, and I. Sanchez-Garcia. 2010. Cancer as a reprogramming-like disease: implications in tumor development and treatment. Semin Cancer Biol 20: 93–97.
Chen, J., O. Odenike, and J.D. Rowley. 2010. Leukaemogenesis: more than mutant genes. Nat Rev Cancer 10: 23–36.
Chen, Z., P. Ayala, M. Wang, L. Fayad, R.L. Katz, J. Romaguera, N. Caraway, S.S. Neelapu, L.W. Kwak, P.J. Simmons, and N. McCarty. 2010. Prospective isolation of clonogenic mantle cell lymphoma-initiating cells. Stem Cell Res 5: 212–225.
Ci, W., J.M. Polo, L. Cerchietti, R. Shaknovich, L. Wang, S.N. Yang, K. Ye, P. Farinha, D.E. Horsman, R.D. Gascoyne, O. Elemento, and A. Melnick. 2009. The BCL6 transcriptional program features repression of multiple oncogenes in primary B cells and is deregulated in DLBCL. Blood 113: 5536–5548.
Clarke, M.F., and M. Fuller. 2006. Stem cells and cancer: two faces of eve. Cell 124: 1111–1115.
Cleary, M.L., N. Galili, M. Trela, R. Levy, and J. Sklar. 1988. Single cell origin of bigenotypic and biphenotypic B cell proliferations in human follicular lymphomas. J Exp Med 167: 582–597.
Cobaleda, C., W. Jochum, and M. Busslinger. 2007. Conversion of mature B cells into T cells by dedifferentiation to uncommitted progenitors. Nature 449: 473–477.
Collins, A.T., P.A. Berry, C. Hyde, M.J. Stower, and N.J. Maitland. 2005. Prospective identification of tumorigenic prostate cancer stem cells. Cancer Res 65: 10946–10951.
Cotter, F., C. Price, E. Zucca, and B.D. Young. 1990. Direct sequence analysis of the 14q+ and 18q-chromosome junctions in follicular lymphoma. Blood 76: 131–135.
Crotty, S., R.J. Johnston, and S.P. Schoenberger. 2010. Effectors and memories: Bcl-6 and Blimp-1 in T and B lymphocyte differentiation. Nat Immunol 11: 114–120.
Dalla-Favera, R., M. Bregni, J. Erikson, D. Patterson, R.C. Gallo, and C.M. Croce. 1982. Human c-myc onc gene is located on the region of chromosome 8 that is translocated in Burkitt lymphoma cells. Proc Natl Acad Sci USA 79: 7824–7827.
Delhommeau, F., S. Dupont, V. Della Valle, C. James, S. Trannoy, A. Masse, O. Kosmider, J.P. Le Couedic, F. Robert, A. Alberdi, Y. Lecluse, I. Plo, F.J. Dreyfus, C. Marzac, N. Casadevall, C. Lacombe, S.P. Romana, P. Dessen, J. Soulier, F. Viguie, M. Fontenay, W. Vainchenker, and O.A. Bernard. 2009. Mutation in TET2 in myeloid cancers. N Engl J Med 360: 2289–2301.
Dick, J.E. 2008. Stem cell concepts renew cancer research. Blood 112: 4793–4807.
Dierlamm, J., M. Baens, I. Wlodarska, M. Stefanova-Ouzounova, J.M. Hernandez, D.K. Hossfeld, C. De Wolf-Peeters, A. Hagemeijer, H. Van den Berghe, and P. Marynen. 1999. The apoptosis inhibitor gene API2 and a novel 18q gene, MLT, are recurrently rearranged in the t(11;18)(q21;q21)p6ssociated with mucosa-associated lymphoid tissue lymphomas. Blood 93: 3601–3609.
Duffy, K.R., C.J. Wellard, J.F. Markham, J.H. Zhou, R. Holmberg, E.D. Hawkins, J. Hasbold, M.R. Dowling, and P.D. Hodgkin. 2012. Activation-induced B cell fates are selected by intracellular stochastic competition. Science 335: 338–341.
Farinha, P., and R.D. Gascoyne. 2005. Molecular pathogenesis of mucosa-associated lymphoid tissue lymphoma. J Clin Oncol 23: 6370–6378.
Fearon, D.T., P. Manders, and S.D. Wagner. 2001. Arrested differentiation, the self-renewing memory lymphocyte, and vaccination. Science 293: 248–250.
Fenton, J.A., J.W. Vaandrager, W.M. Aarts, R.J. Bende, K. Heering, M. van Dijk, G. Morgan, C.J. van Noesel, E. Schuuring, and P.M. Kluin. 2002. Follicular lymphoma with a novel t(14;18) breakpoint involving the immunoglobulin heavy chain switch mu region indicates an origin from germinal center B cells. Blood 99: 716–718.

Good-Jacobson, K.L., and M.J. Shlomchik. 2010. Plasticity and heterogeneity in the generation of memory B cells and long-lived plasma cells: the influence of germinal center interactions and dynamics. J Immunol 185: 3117–3125.

Gostissa, M., F.W. Alt, and R. Chiarle. 2011. Mechanisms that promote and suppress chromosomal translocations in lymphocytes. Annu Rev Immunol 29: 319–350.

Haluska, F.G., S. Finver, Y. Tsujimoto, and C.M. Croce. 1986. The t(8; 14) chromosomal translocation occurring in B-cell malignancies results from mistakes in V-D-J joining. Nature 324: 158–161.

Hamburger, A.W., and S.E. Salmon. 1977. Primary bioassay of human tumor stem cells. Science 197: 461–463.

Hanahan, D., and R.A. Weinberg. 2011. Hallmarks of cancer: the next generation. Cell 144: 646–674.

Hanna, J., S. Markoulaki, P. Schorderet, B.W. Carey, C. Beard, M. Wernig, M.P. Creyghton, E.J. Steine, J.P. Cassady, R. Foreman, C.J. Lengner, J.A. Dausman, and R. Jaenisch. 2008. Direct reprogramming of terminally differentiated mature B lymphocytes to pluripotency. Cell 133: 250–264.

Hart, J., A.R. Turner, L. Larratt, J. Russell, B. Franko, C. Frantz, T. Paonessa, A. Mansoor, and R. Lai. 2007. Transmission of a follicular lymphoma by allogeneic bone marrow transplantation--evidence to support the existence of lymphoma progenitor cells. Br J Haematol 136: 166–167.

Hermann, P.C., S. Bhaskar, M. Cioffi, and C. Heeschen. 2010. Cancer stem cells in solid tumors. Semin Cancer Biol 20: 77–84.

Isaacson, P.G., and M.Q. Du. 2004. MALT lymphoma: from morphology to molecules. Nat Rev Cancer 4: 644–653.

Jaffe, E.S., N.L. Harris, H. Stein, and P.G. Isaacson. 2008. Classification of lymphoid neoplasms: the microscope as a tool for disease discovery. Blood 112: 4384–4399.

Jager, U., S. Bocskor, T. Le, G. Mitterbauer, I. Bolz, A. Chott, M. Kneba, C. Mannhalter, and B. Nadel. 2000. Follicular lymphomas' BCL-2/IgH junctions contain templated nucleotide insertions: novel insights into the mechanism of t(14;18) translocation. Blood 95: 3520–3529.

Jares, P., D. Colomer, and E. Campo. 2007. Genetic and molecular pathogenesis of mantle cell lymphoma: perspectives for new targeted therapeutics. Nat Rev Cancer 7: 750–762.

Jones, R.J., C.D. Gocke, Y.L. Kasamon, C.B. Miller, B. Perkins, J.P. Barber, M.S. Vala, J.M. Gerber, L.L. Gellert, M. Siedner, M.V. Lemas, S. Brennan, R.F. Ambinder, and W. Matsui. 2009. Circulating clonotypic B cells in classic Hodgkin lymphoma. Blood 113: 5920–5926.

Jones, S.E., A.W. Hamburger, M.B. Kim, and S.E. Salmon. 1979. Development of a bioassay for putative human lymphoma stem cells. Blood 53: 294–303.

Jung, H.J., Z. Chen, L. Fayad, M. Wang, J. Romaguera, L.W. Kwak, and N. McCarty. 2012. Bortezomib-resistant nuclear factor kappaB expression in stem-like cells in mantle cell lymphoma. Exp Hematol 40: 107–118 e102.

Jung, H.J., Z. Chen, and N. McCarty. 2011. Stem-like tumor cells confer drug resistant properties to mantle cell lymphoma. Leuk Lymphoma 52: 1066–1079.

Kikushige, Y., F. Ishikawa, T. Miyamoto, T. Shima, S. Urata, G. Yoshimoto, Y. Mori, T. Iino, T. Yamauchi, T. Eto, H. Niiro, H. Iwasaki, K. Takenaka, and K. Akashi. 2011. Self-renewing hematopoietic stem cell is the primary target in pathogenesis of human chronic lymphocytic leukemia. Cancer Cell 20: 246–259.

Klein, U., and R. Dalla-Favera. 2008. Germinal centres: role in B-cell physiology and malignancy. Nat Rev Immunol 8: 22–33.

Ko, M., H.S. Bandukwala, J. An, E.D. Lamperti, E.C. Thompson, R. Hastie, A. Tsangaratou, K. Rajewsky, S.B. Koralov, and A. Rao. 2011. Ten-Eleven-Translocation 2 (TET2) negatively regulates homeostasis and differentiation of hematopoietic stem cells in mice. Proc Natl Acad Sci USA 108: 14566–14571.

Ko, M., Y. Huang, A.M. Jankowska, U.J. Pape, M. Tahiliani, H.S. Bandukwala, J. An, E.D. Lamperti, K.P. Koh, R. Ganetzky, X.S. Liu, L. Aravind, S. Agarwal, J.P. Maciejewski, and A. Rao. 2010. Impaired hydroxylation of 5-methylcytosine in myeloid cancers with mutant TET2. Nature 468: 839–843.

Kuehl, W.M., and P.L. Bergsagel. 2002. Multiple myeloma: evolving genetic events and host interactions. Nat Rev Cancer 2: 175–187.

Kuppers, R. 2003. Somatic hypermutation and B cell receptor selection in normal and transformed human B cells. Ann N Y Acad Sci 987: 173–179.

Kuppers, R. 2005. Mechanisms of B-cell lymphoma pathogenesis. Nat Rev Cancer 5: 251–262.

Kuppers, R., and R. Dalla-Favera. 2001. Mechanisms of chromosomal translocations in B cell lymphomas. Oncogene 20: 5580–5594.

Kuppers, R., U. Klein, M.L. Hansmann, and K. Rajewsky. 1999. Cellular origin of human B-cell lymphomas. N Engl J Med 341: 1520–1529.

Langemeijer, S.M., R.P. Kuiper, M. Berends, R. Knops, M.G. Aslanyan, M. Massop, E. Stevens-Linders, P. van Hoogen, A.G. van Kessel, R.A. Raymakers, E.J. Kamping, G.E. Verhoef, E. Verburgh, A. Hagemeijer, P. Vandenberghe, T. de Witte, B.A. van der Reijden, and J.H. Jansen. 2009. Acquired mutations in TET2 are common in myelodysplastic syndromes. Nat Genet 41: 838–842.

Lanzavecchia, A., and F. Sallusto. 2002. Progressive differentiation and selection of the fittest in the immune response. Nat Rev Immunol 2: 982–987.

Lapidot, T., C. Sirard, J. Vormoor, B. Murdoch, T. Hoang, J. Caceres-Cortes, M. Minden, B. Paterson, M.A. Caligiuri, and J.E. Dick. 1994. A cell initiating human acute myeloid leukaemia after transplantation into SCID mice. Nature 367: 645–648.

Lecluse, Y., P. Lebailly, S. Roulland, A.C. Gac, B. Nadel, and P. Gauduchon. 2009. t(11;14)-positive clones can persist over a long period of time in the peripheral blood of healthy individuals. Leukemia 23: 1190–1193.

Lee, C.G., B. Das, T.L. Lin, C. Grimes, X. Zhang, T. Lavezzi, L. Huang, J. Cole, L. Yau, and L. Li. 2012. A rare fraction of drug-resistant follicular lymphoma cancer stem cells interacts with follicular dendritic cells to maintain tumourigenic potential. Br J Haematol.

Lenz, G., I. Nagel, R. Siebert, A.V. Roschke, W. Sanger, G.W. Wright, S.S. Dave, B. Tan, H. Zhao, A. Rosenwald, H.K. Muller-Hermelink, R.D. Gascoyne, E. Campo, E.S. Jaffe, E.B. Smeland, R.I. Fisher, W.M. Kuehl, W.C. Chan, and L.M. Staudt. 2007. Aberrant immunoglobulin class switch recombination and switch translocations in activated B cell-like diffuse large B cell lymphoma. J Exp Med 204: 633–643.

Li, C., D.G. Heidt, P. Dalerba, C.F. Burant, L. Zhang, V. Adsay, M. Wicha, M.F. Clarke, and D.M. Simeone. 2007. Identification of pancreatic cancer stem cells. Cancer Res 67: 1030–1037.

Liu, R., X. Wang, G.Y. Chen, P. Dalerba, A. Gurney, T. Hoey, G. Sherlock, J. Lewicki, K. Shedden, and M.F. Clarke. 2007. The prognostic role of a gene signature from tumorigenic breast-cancer cells. N Engl J Med 356: 217–226.

Lo Coco, F., B.H. Ye, F. Lista, P. Corradini, K. Offit, D.M. Knowles, R.S. Chaganti, and R. Dalla-Favera. 1994. Rearrangements of the BCL6 gene in diffuse large cell non-Hodgkin's lymphoma. Blood 83: 1757–1759.

Lossos, I.S., A.A. Alizadeh, M.B. Eisen, W.C. Chan, P.O. Brown, D. Botstein, L.M. Staudt, and R. Levy. 2000. Ongoing immunoglobulin somatic mutation in germinal center B cell-like but not in activated B cell-like diffuse large cell lymphomas. Proc Natl Acad Sci USA 97: 10209–10213.

Lossos, I.S., and R. Levy. 2003. Higher grade transformation of follicular lymphoma: phenotypic tumor progression associated with diverse genetic lesions. Semin Cancer Biol 13: 191–202.

Lovec, H., A. Grzeschiczek, M.B. Kowalski, and T. Moroy. 1994. Cyclin D1/bcl-1 cooperates with myc genes in the generation of B-cell lymphoma in transgenic mice. Embo J 13: 3487–3495.

Luckey, C.J., D. Bhattacharya, A.W. Goldrath, I.L. Weissman, C. Benoist, and D. Mathis. 2006. Memory T and memory B cells share a transcriptional program of self-renewal with long-term hematopoietic stem cells. Proc Natl Acad Sci USA 103: 3304–3309.

Ma, S., K.W. Chan, L. Hu, T.K. Lee, J.Y. Wo, I.O. Ng, B.J. Zheng, and X.Y. Guan. 2007. Identification and characterization of tumorigenic liver cancer stem/progenitor cells. Gastroenterology 132: 2542–2556.

MacLennan, I.C. 1994. Germinal centers. Annu Rev Immunol 12: 117–139.

Mandelbaum, J., G. Bhagat, H. Tang, T. Mo, M. Brahmachary, Q. Shen, A. Chadburn, K. Rajewsky, A. Tarakhovsky, L. Pasqualucci, and R. Dalla-Favera. 2010. BLIMP1 is a tumor suppressor gene frequently disrupted in activated B cell-like diffuse large B cell lymphoma. Cancer Cell 18: 568–579.

Martin-Subero, J.I., M. Kreuz, M. Bibikova, S. Bentink, O. Ammerpohl, E. Wickham-Garcia, M. Rosolowski, J. Richter, L. Lopez-Serra, E. Ballestar, H. Berger, X. Agirre, H.W. Bernd, V. Calvanese, S.B. Cogliatti, H.G. Drexler, J.B. Fan, M.F. Fraga, M.L. Hansmann, M. Hummel, W. Klapper, B. Korn, R. Kuppers, R.A. Macleod, P. Moller, G. Ott, C. Pott, F. Prosper, A. Rosenwald, C. Schwaenen, D. Schubeler, M. Seifert, B. Sturzenhofecker, M. Weber, S. Wessendorf, M. Loeffler, L. Trumper, H. Stein, R. Spang, M. Esteller, D. Barker, D. Hasenclever, and R. Siebert. 2009. New insights into the biology and origin of mature aggressive B-cell lymphomas by combined epigenomic, genomic, and transcriptional profiling. Blood 113: 2488–2497.

Martinez-Climent, J.A., E.J. Andreu, and F. Prosper. 2006. Somatic stem cells and the origin of cancer. Clin Transl Oncol 8: 647–663.

Martinez-Climent, J.A., L. Fontan, R.D. Gascoyne, R. Siebert, and F. Prosper. 2010. Lymphoma stem cells: enough evidence to support their existence? Haematologica 95: 293–302.

Mathas, S., M. Janz, F. Hummel, M. Hummel, B. Wollert-Wulf, S. Lusatis, I. Anagnostopoulos, A. Lietz, M. Sigvardsson, F. Jundt, K. Johrens, K. Bommert, H. Stein, and B. Dorken. 2006. Intrinsic inhibition of transcription factor E2A by HLH proteins ABF-1 and Id2 mediates reprogramming of neoplastic B cells in Hodgkin lymphoma. Nat Immunol 7: 207–215.

Matthias, P., and A.G. Rolink. 2005. Transcriptional networks in developing and mature B cells. Nat Rev Immunol 5: 497–508.

McDonnell, T.J., N. Deane, F.M. Platt, G. Nunez, U. Jaeger, J.P. McKearn, and S.J. Korsmeyer. 1989. bcl-2-immunoglobulin transgenic mice demonstrate extended B cell survival and follicular lymphoproliferation. Cell 57: 79–88.

McDonnell, T.J., and S.J. Korsmeyer. 1991. Progression from lymphoid hyperplasia to high-grade malignant lymphoma in mice transgenic for the t(14; 18). Nature 349: 254–256.

McHeyzer-Williams, L.J., M. Cool, and M.G. McHeyzer-Williams. 2000. Antigen-specific B cell memory: expression and replenishment of a novel b220(-) memory b cell compartment. J Exp Med 191: 1149–1166.

Medina, K.L., and H. Singh. 2005. Genetic networks that regulate B lymphopoiesis. Curr Opin Hematol 12: 203–209.

Mercier, F.E., C. Ragu, and D.T. Scadden. 2011. The bone marrow at the crossroads of blood and immunity. Nat Rev Immunol 12: 49–60.

Mohrin, M., E. Bourke, D. Alexander, M.R. Warr, K. Barry-Holson, M.M. Le Beau, C.G. Morrison, and E. Passegue. 2010. Hematopoietic stem cell quiescence promotes error-prone DNA repair and mutagenesis. Cell Stem Cell 7: 174–185.

Moran-Crusio, K., L. Reavie, A. Shih, O. Abdel-Wahab, D. Ndiaye-Lobry, C. Lobry, M.E. Figueroa, A. Vasanthakumar, J. Patel, X. Zhao, F. Perna, S. Pandey, J. Madzo, C. Song, Q. Dai, C. He, S. Ibrahim, M. Beran, J. Zavadil, S.D. Nimer, A. Melnick, L.A. Godley, I. Aifantis, and R.L. Levine. 2011. Tet2 loss leads to increased hematopoietic stem cell self-renewal and myeloid transformation. Cancer Cell 20: 11–24.

Muramatsu, M., K. Kinoshita, S. Fagarasan, S. Yamada, Y. Shinkai, and T. Honjo. 2000. Class switch recombination and hypermutation require activation-induced cytidine deaminase (AID), a potential RNA editing enzyme. Cell 102: 553–563.

Neri, A., F. Barriga, D.M. Knowles, I.T. Magrath, and R. Dalla-Favera. 1988. Different regions of the immunoglobulin heavy-chain locus are involved in chromosomal translocations in distinct pathogenetic forms of Burkitt lymphoma. Proc Natl Acad Sci USA 85: 2748–2752.

Nguyen, L.V., R. Vanner, P. Dirks, and C.J. Eaves. 2012. Cancer stem cells: an evolving concept. Nat Rev Cancer 12: 133–143.

Nussenzweig, A., and M.C. Nussenzweig. 2010. Origin of chromosomal translocations in lymphoid cancer. Cell 141: 27–38.

O'Brien, C.A., A. Pollett, S. Gallinger, and J.E. Dick. 2007. A human colon cancer cell capable of initiating tumour growth in immunodeficient mice. Nature 445: 106–110.

Ottensmeier, C.H., and F.K. Stevenson. 2000. Isotype switch variants reveal clonally related subpopulations in diffuse large B-cell lymphoma. Blood 96: 2550–2556.

Ottensmeier, C.H., A.R. Thompsett, D. Zhu, B.S. Wilkins, J.W. Sweetenham, and F.K. Stevenson. 1998. Analysis of VH genes in follicular and diffuse lymphoma shows ongoing somatic mutation and multiple isotype transcripts in early disease with changes during disease progression. Blood 91: 4292–4299.

Pardal, R., M.F. Clarke, and S.J. Morrison. 2003. Applying the principles of stem-cell biology to cancer. Nat Rev Cancer 3: 895–902.

Pardal, R., A.V. Molofsky, S. He, and S.J. Morrison. 2005. Stem cell self-renewal and cancer cell proliferation are regulated by common networks that balance the activation of proto-oncogenes and tumor suppressors. Cold Spring Harb Symp Quant Biol 70: 177–185.

Park, C.H., D.E. Bergsagel, and E.A. McCulloch. 1971. Mouse myeloma tumor stem cells: a primary cell culture assay. J Natl Cancer Inst 46: 411–422.

Pasqualucci, L., P. Neumeister, T. Goossens, G. Nanjangud, R.S. Chaganti, R. Kuppers, and R. Dalla-Favera. 2001. Hypermutation of multiple proto-oncogenes in B-cell diffuse large-cell lymphomas. Nature 412: 341–346.

Perez-Caro, M., C. Cobaleda, I. Gonzalez-Herrero, C. Vicente-Duenas, C. Bermejo-Rodriguez, M. Sanchez-Beato, A. Orfao, B. Pintado, T. Flores, M. Sanchez-Martin, R. Jimenez, M.A. Piris, and I. Sanchez-Garcia. 2009. Cancer induction by restriction of oncogene expression to the stem cell compartment. Embo J 28: 8–20.

Perez-Galan, P., M. Dreyling, and A. Wiestner. 2011. Mantle cell lymphoma: biology, pathogenesis, and the molecular basis of treatment in the genomic era. Blood 117: 26–38.

Polyak, K., and W.C. Hahn. 2006. Roots and stems: stem cells in cancer. Nat Med 12: 296–300.

Prince, M.E., R. Sivanandan, A. Kaczorowski, G.T. Wolf, M.J. Kaplan, P. Dalerba, I.L. Weissman, M.F. Clarke, and L.E. Ailles. 2007. Identification of a subpopulation of cells with cancer stem cell properties in head and neck squamous cell carcinoma. Proc Natl Acad Sci USA 104: 973–978.

Quivoron, C., L. Couronne, V. Della Valle, C.K. Lopez, I. Plo, O. Wagner-Ballon, M. Do Cruzeiro, F. Delhommeau, B. Arnulf, M.H. Stern, L. Godley, P. Opolon, H. Tilly, E. Solary, Y. Duffourd, P. Dessen, H. Merle-Beral, F. Nguyen-Khac, M. Fontenay, W. Vainchenker, C. Bastard, T. Mercher, and O.A. Bernard. 2011. TET2 inactivation results in pleiotropic hematopoietic abnormalities in mouse and is a recurrent event during human lymphomagenesis. Cancer Cell 20: 25–38.

Ramiro, A.R., M. Jankovic, T. Eisenreich, S. Difilippantonio, S. Chen-Kiang, M. Muramatsu, T. Honjo, A. Nussenzweig, and M.C. Nussenzweig. 2004. AID is required for c-myc/IgH chromosome translocations *in vivo*. Cell 118: 431–438.

Reya, T., S.J. Morrison, M.F. Clarke, and I.L. Weissman. 2001. Stem cells, cancer, and cancer stem cells. Nature 414: 105–111.

Ricci-Vitiani, L., D.G. Lombardi, E. Pilozzi, M. Biffoni, M. Todaro, C. Peschle, and R. De Maria. 2007. Identification and expansion of human colon-cancer-initiating cells. Nature 445: 111–115.

Robbiani, D.F., A. Bothmer, E. Callen, B. Reina-San-Martin, Y. Dorsett, S. Difilippantonio, D.J. Bolland, H.T. Chen, A.E. Corcoran, A. Nussenzweig, and M.C. Nussenzweig 2008. AID is required for the chromosomal breaks in c-myc that lead to c-myc/IgH translocations. Cell 135: 1028–1038.

Robbiani, D.F., S. Bunting, N. Feldhahn, A. Bothmer, J. Camps, S. Deroubaix, K.M. McBride, I.A. Klein, G. Stone, T.R. Eisenreich, T. Ried, A. Nussenzweig, and M.C. Nussenzweig. 2009. AID produces DNA double-strand breaks in non-Ig genes and mature B cell lymphomas with reciprocal chromosome translocations. Mol Cell 36: 631–641.

Rolink, A.G., C. Schaniel, M. Busslinger, S.L. Nutt, and F. Melchers. 2000. Fidelity and infidelity in commitment to B-lymphocyte lineage development. Immunol Rev 175: 104–111.

Roulland, S., J.M. Navarro, P. Grenot, M. Milili, J. Agopian, B. Montpellier, P. Gauduchon, P. Lebailly, C. Schiff, and B. Nadel. 2006. Follicular lymphoma-like B cells in healthy individuals: a novel intermediate step in early lymphomagenesis. J Exp Med 203: 2425–2431.

Sabbath, K.D., E.D. Ball, P. Larcom, R.B. Davis, and J.D. Griffin. 1985. Heterogeneity of clonogenic cells in acute myeloblastic leukemia. J Clin Invest 75: 746–753.

Sanchez-Izquierdo, D., G. Buchonnet, R. Siebert, R.D. Gascoyne, J. Climent, L. Karran, M. Marin, D. Blesa, D. Horsman, A. Rosenwald, L.M. Staudt, D.G. Albertson, M.Q. Du, H. Ye, P. Marynen, J. Garcia-Conde, D. Pinkel, M.J. Dyer, and J.A. Martinez-Climent 2003. MALT1 is deregulated by both chromosomal translocation and amplification in B-cell non-Hodgkin lymphoma. Blood 101: 4539–4546.

Scadden, D.T. 2004. Cancer stem cells refined. Nat Immunol 5: 701–703.

Scheeren, F.A., M. Naspetti, S. Diehl, R. Schotte, M. Nagasawa, E. Wijnands, R. Gimeno, F.A. Vyth-Dreese, B. Blom, and H. Spits. 2005. STAT5 regulates the self-renewal capacity and differentiation of human memory B cells and controls Bcl-6 expression. Nat Immunol 6: 303–313.

Schittek, B., and K. Rajewsky. 1990. Maintenance of B-cell memory by long-lived cells generated from proliferating precursors. Nature 346: 749–751.

Schmitz, R., C. Renne, R. Rosenquist, M. Tinguely, V. Distler, F. Menestrina, M. Lestani, T. Stankovic, B. Austen, A. Brauninger, M.L. Hansmann, and R. Kuppers. 2005. Insights into the multistep transformation process of lymphomas: IgH-associated translocations and tumor suppressor gene mutations in clonally related composite Hodgkin's and non-Hodgkin's lymphomas. Leukemia 19: 1452–1458.

Schroeder, T. 2007. Asymmetric cell division in normal and malignant hematopoietic precursor cells. Cell Stem Cell 1: 479–481.

Schuler, F., C. Hirt, and G. Dolken. 2003. Chromosomal translocation t(14;18) in healthy individuals. Semin Cancer Biol 13: 203–209.

Shaffer, A.L., A. Rosenwald, and L.M. Staudt. 2002. Lymphoid malignancies: the dark side of B-cell differentiation. Nat Rev Immunol 2: 920–932.

Shipitsin, M., L.L. Campbell, P. Argani, S. Weremowicz, N. Bloushtain-Qimron, J. Yao, T. Nikolskaya, T. Serebryiskaya, R. Beroukhim, M. Hu, M.K. Halushka, S. Sukumar, L.M. Parker, K.S. Anderson, L.N. Harris, J.E. Garber, A.L. Richardson, S.J. Schnitt, Y. Nikolsky, R.S. Gelman, and K. Polyak. 2007. Molecular definition of breast tumor heterogeneity. Cancer Cell 11: 259–273.

Singh, S.K., C. Hawkins, I.D. Clarke, J.A. Squire, J. Bayani, T. Hide, R.M. Henkelman, M.D. Cusimano, and P.B. Dirks. 2004. Identification of human brain tumour initiating cells. Nature 432: 396–401.

Stamatopoulos, K., C. Kosmas, C. Belessi, T. Papadaki, S. Afendaki, D. Anagnostou, and D. Loukopoulos. 1997. t(14;18) chromosomal translocation in follicular lymphoma: an event occurring with almost equal frequency both at the D to J(H) and at later stages in the rearrangement process of the immunoglobulin heavy chain gene locus. Br J Haematol 99: 866–872.

Stingl, J., and C. Caldas. 2007. Molecular heterogeneity of breast carcinomas and the cancer stem cell hypothesis. Nat Rev Cancer 7: 791–799.

Streubel, B., A. Chott, D. Huber, M. Exner, U. Jager, O. Wagner, and I. Schwarzinger. 2004. Lymphoma-specific genetic aberrations in microvascular endothelial cells in B-cell lymphomas. N Engl J Med 351: 250–259.

Tam, W., M. Gomez, A. Chadburn, J.W. Lee, W.C. Chan, and D.M. Knowles. 2006. Mutational analysis of PRDM1 indicates a tumor-suppressor role in diffuse large B-cell lymphomas. Blood 107: 4090–4100.

Tweeddale, M.E., B. Lim, N. Jamal, J. Robinson, J. Zalcberg, G. Lockwood, M.D. Minden, and H.A. Messner. 1987. The presence of clonogenic cells in high-grade malignant lymphoma: a prognostic factor. Blood 69: 1307–1314.

Vicente-Dueñas, C., L. Fontan, I. Gonzalez-Herrero, I. Romero-Camarero, V. Segura, M., A. Aznar, A. Sagardoy, C. Bertolo, X. Agirre, I.S. Lossos, J.M. Hernandez-Rivas, M. Gonzalez, A. Salar, B. Bellosillo, E. Conde, R. Siebert, F. Prosper, X. Sagaert, C. Cobaleda, I. Sanchez-Garcia, and J.A. Martinez-Climent. 2012. Expression of *MALT1* oncogene in mouse hematopoietic stem/progenitor cells recapitulates the pathogenesis of human B-cell lymphoma. Proceedings of the National Academy of Sciences of the United States of America.

Visvader, J.E. 2011. Cells of origin in cancer. Nature 469: 314–322.

Weigert, O., N. Kopp, A.A. Lane, A. Yoda, S.E. Dahlberg, D. Neuberg, A.Y. Bahar, B. Chapuy, J. Lutok, M. Salois, D.C. Fisher, E.A. Fox, S.J. Rodig, J.H. Antin, and D.M. Weinstock. 2012. Molecular ontogeny of donor-derived follicular lymphomas ocurring after hematopoietic cell transplantation. Cancer Discovery 2012 Jan; 2(1): 47–55.

Willis, T.G., and M.J. Dyer. 2000. The role of immunoglobulin translocations in the pathogenesis of B-cell malignancies. Blood 96: 808–822.

Xie, H., M. Ye, R. Feng, and T. Graf. 2004. Stepwise reprogramming of B cells into macrophages. Cell 117: 663–676.

Zhang, M., and J.M. Rosen. 2006. Stem cells in the etiology and treatment of cancer. Curr Opin Genet Dev 16: 60–64.

Zhu, C., K.D. Mills, D.O. Ferguson, C. Lee, J. Manis, J. Fleming, Y. Gao, C.C. Morton, and F.W. Alt. 2002. Unrepaired DNA breaks in p53-deficient cells lead to oncogenic gene amplification subsequent to translocations. Cell 109: 811–821.

Chapter 4

The Leukemia Stem Cell

Uri Rozovski[a] and Zeev Estrov[b,*]

Introduction

The cancer stem cell hypothesis dates back to the middle of the 19th century. In 1858, Rudolf Virchow proposed that cancer arises from embryonic tissue. This idea was later extended by Cohnheim and Durante, who suggested that adult tissue contains remnant dormant embryonic tissue and that when these embryonic cells are activated, they become cancerous (Hendrix et al. 2007, Virchow 1858).

At the turn of the 20th century, European researchers suggested that all circulating blood cells arise from a single cell, later called the hematopoietic stem cell (HSC) (Maximow 1909, Pappenheim 1917). Six decades later, Till and McCulloch provided experimental evidence that HSCs exist (Till and Mc 1961), and in the mid-1990s, xenotransplantation mouse models suggested that leukemia stem cells (LSCs) arising from the HSCs are responsible for the initiation and progression of human acute leukemia, in particular, acute myeloid leukemia (Lapidot et al. 1994). Since then, the cancer stem cell hypothesis has been exploited in diverse malignancies such as brain tumors, colon cancer, prostate cancer, and malignant melanoma.

In this chapter, we critically review the LSC hypothesis and the LSC research conducted in recent years. We describe normal HSC characteristics and how they were used to introduce the LSC model. We also review the data that led to the conclusion that LSCs exist in different types of leukemia and the implications of this conclusion for current and future therapies.

The Department of Leukemia, The University of Texas MD Anderson Cancer Center, 1515 Holcombe Blvd Houston Tx 77030.
[a]E-mail: Urozovski@mdadnerson.org
[b]E-mail: Zestrov@mdadnerson.org
*Corresponding author

Hematopoietic Stem Cells

In 1961, Till and McCulloch demonstrated that intravenous injection of bone marrow cells into lethally irradiated mice resulted in the formation of splenic myeloid colonies. The subpopulation of bone marrow cells that formed spleen colonies was termed *colony-forming unit–spleen* (CFU-S). A rare subpopulation of these spleen colony cells possess clonogenic potential and are capable of further repopulating a secondary irradiated host. Although it was previously known that spleen colonies are formed in irradiated mice, Till and McCulloch's pioneering studies established the first experimental evidence that a distinct type of rare hematopoietic cells are capable of generating blood elements. These studies formed the basis for the HSC hypothesis (Till and Mc 1961).

Morphologically, HSCs resemble small lymphocytes and have been detected in the bone marrow, peripheral blood, umbilical cord blood, and fetal liver. The HSC is considered the prototype of a pluripotent stem cell because it possesses the three necessary features characterizing a somatic stem cell, namely, self-renewal capacity, clonogenic capacity, and the capability to differentiate into multiple lineages. On the basis of transplantation experiments, HSCs were further classified into long-term repopulating HSCs (LT-HSCs), which have an indefinite self-renewal capacity, and short-term HSCs (ST-HSCs), which repopulate the bone marrow of a lethally irradiated mouse short-term, thus having no self-renewal capacity. In the mouse bone marrow, an estimated 1 in every 5,000 cells is an LT-HSC, and 1 in every 1,000 cells is an ST-HSC (Passegue et al. 2003). *In vitro* studies have demonstrated that HSCs differentiate in a hierarchical manner into immature progenitors that further differentiate into mature progenitors and eventually into terminally differentiated mature cells (Ogawa et al. 1983).

Using high-speed multiparameter cell sorting, cell surface expression markers have been used to obtain a cell population that is highly enriched in HSC. For example, the combination of $Lin^{-}Thy1^{+}CD34^{+}CD38^{-/low}$ (Morrison et al. 1995) or the SLAM family receptors (Kiel et al. 2005) and recently the CD49f (Notta et al. 2011a) has been commonly used to identify HSC populations. Other techniques to identify and isolate HSCs are based on staining patterns of fluorescent dyes, namely, Rhodamine 123 and Hoechst 3342 (Bertoncello and Williams 2004, Goodell et al. 1996, Hess et al. 2004, Kania et al. 2005, Kiel et al. 2005, Storms et al. 2000). The staining pattern is probably dependent on the existence of efflux proteins in the HSCs. A similar method uses fluorescence intensity after ultraviolet excitation to identify a rare (<0.1%) cell population termed side population (SP) that contains the vast majority of LT-HSCs (Goodell et al. 1996). Although the SP in mice is relatively homogeneous, in humans this population is phenotypically and functionally heterogeneous (Naylor et al. 2005). HSCs typically express high levels of aldehyde dehydrogenase (ALDH) enzymatic activity (Hawley et al. 2006). This characteristic has been used to identify and separate these cells by introducing a fluorescent substrate, usually densylaminoacetaldehyde (DAAA) (Jones et al. 1995, Jones et al. 1996), also termed BODIPY (Storms et al. 1999).

Quiescent HSCs occasionally divide, detach from the bone marrow, and enter the circulation (Passegue et al. 2003). For example, in a young adult mouse, about 8–10% of LT-HSCs randomly enter the circulation each day (Bradford et al. 1997, Cheshier et al. 1999). To maintain hematopoietic homeostasis in a dynamically dividing cell population, the numbers of HSCs need to be tightly regulated. Cell fate decisions, either to divide, differentiate, or undergo apoptotic death, determine the size and function of the HSC pool. Several intrinsic factors that contribute to cell fate decisions have been identified, including telomerase, which prevents telomere shortening throughout most of the HSC lifetime (Alenzi et al. 2009, Ju and Rudolph 2006), a tight DNA damage response (Alenzi et al. 2009), and expression of key transcription factors, specifically *Foxo3a* (Miyamoto et al. 2008) and *Zfx* (Galan-Caridad et al. 2007).

Extrinsic factors regulate HSC fate and function as well. The distribution of HSCs within the bone marrow is not random. HSCs reside within specialized areas of the bone marrow microenvironment, termed niches. The osteoblastic niche, localized at the inner surface of bone cavity, provides a microenvironment for the LT-HSCs where adjacent osteoblasts are believed to contribute to the quiescence state of these cells (Arai et al. 2004, Calvi et al. 2003, Perry and Li 2007, Zhang et al. 2003). The vascular niche, consisting of sinusoidal endothelial cells lining blood vessels, is believed to promote proliferation and differentiation of actively cycling ST-HSCs (Kopp et al. 2005). HSCs are capable of transitioning from one niche to another, from the bone marrow to the circulating blood, and from the blood to the bone marrow. The extent to which the microenvironment affects the HSCs or, conversely, immature or mature hematopoietic progenitor cell function, remains to be determined. It has been well established that under some conditions, those cells function in an extramedullary environment. For example, in myeloproliferative neoplasms, extramedullary hematopoiesis frequently occurs.

Leukemic Stem Cells

Based on the HSC hypothesis, a similar LSC population was assumed to exist, which consisted of rare, dormant cells with the capacity for self-renewal and proliferation and the ability to give rise to a clone of leukemia cells. For example, in acute myeloid leukemia (AML), characterized by the accumulation of blasts, most of the leukemia blasts have only a limited capacity for self-renewal, suggesting that only a small subpopulation of leukemia cells are responsible for maintaining the leukemic clone (Griffin and Lowenberg 1986, McCulloch 1983). The most convincing evidence supporting the LSC concept arose from xenotransplantation mouse models. In these experiments, immunodeficient mice—either mice with severe combined immunodeficiency disease (SCID) (Lapidot et al. 1994) or nonobese diabetic (NOD)/SCID mice (Bonnet and Dick 1997)—received transplanted leukemic cells from patients with AML. The leukemic cells engrafted, producing leukemia in the mouse with phenotypic characteristics resembling the original leukemia in the patients. Limiting dilution assays and serial injections of human AML cells demonstrating engraftment of leukemia cells in immunodeficient mice to a second and third generation revealed that only 0.2 to $100/10^6$ AML cells carry self-renewal capacity.

For the past 2 decades, xenotransplantation mouse models became the standard models of human normal hematopoietic and leukemia stem cells; however, interpretations of new data are challenging the validity of those models. For example, Kelly et al. demonstrated that as few as 100 or even 10 unfractionated mouse AML cells can form tumors (Kelly et al. 2007), suggesting that LSCs with self-renewal capacity are not as rare as previously believed. Thus, the low frequency of human AML cells engrafting in NOD/SCID mice might reflect the difficulty of human leukemia cells to engraft and proliferate in a foreign (mouse) milieu. Under conditions resembling those of human bone marrow, more leukemia cells are likely to function as LSCs than previously deduced from the original xenotransplantation mouse model studies. Whether a LSC-specific phenotypic or function, distinguish LSCs from the bulk of leukemic blasts, will be addressed in this chapter.

Several competing models have dealt with the LSC origin in AML. The first model postulates that many cells, including committed progenitors, are susceptible to leukemic transformation. Several acquired leukemogenic mutations alter the normal developmental program, and the mutated cells are blocked at a particular stage of development. This model explains the heterogeneity among the various AML patient samples: specifically, in each case, the target transformed cell was in a different stage of development (Fialkow et al. 1987, Griffin and Lowenberg 1986).

The second model assumes, in agreement with the xenotransplantation models, that the leukemogenic transformation occurs in the HSCs and that the transformation event affects the fate and the developmental program of the transformed cells. According to this model, heterogeneity in AML could be explained by a late transformation that followed the primary leukemogenic event in the HSCs. This model predicts a relatively high degree of phenotypic similarity of the leukemic blasts among the various patients (Fig. 1A) (Bonnet and Dick 1997, Lapidot et al. 1994, McCulloch 1983).

A third "intermediate" model explains the development of LSCs in two steps. The first step occurs in the stem cell compartment and is responsible for a "preleukemic" stem cell that can differentiate into mature progenitors. The second step takes place in mature progenitors with addition of transformation events that finally mature into a full functioning leukemic blast (Jordan and Guzman 2004).

While most investigators suggest that the leukemogenic event in AML occurs in the HSCs, sometimes, and most remarkably in non-AML leukemias, LSCs can arise from a committed progenitor or even a mature cell that reacquires the stem cell characteristics such as self-renewal capacity (Passegue et al. 2003). For example, chronic myelogenous leukemia (CML) granulocyte-macrophage progenitors from patients with CML in blastic transformation were shown to form self-renewing myeloid colonies, suggesting that these are the leukemia stem cells (Jamieson et al. 2004).

Similarity, it has been shown that the fusion gene *PML/RARα*, the characteristic translocation of acute promyelocytic leukemia (APL), is present in the $CD34^-CD38^+$ cells, but not in the $CD34^+CD38^-$ cells believed to be the typical LSC population. Moreover, even in AML, the leukemogenic event might occur in a committed progenitor cell. This was clearly demonstrated in studies of enforced expression of human genes often activated as a result of a chromosomal translocation in leukemic cells. In one of these studies, the MLL gene, whose activation is believed to be highly leukemogenic, was transfected into committed murine progenitor cells and, as a result, these cells became AML cells. The gene expression profile of these cells was similar to that of the progenitor cell from which they were derived, but these cells also expressed genes typically expressed in HSCs. Thus, overexpression of the MLL gene activated the self-renewal machinery that is usually found in HSCs (Cozzio et al. 2003, DiMartino et al. 2002, Krivtsov et al. 2006). Therefore, the LSC can be perceived as a hematopoietic cell that is capable of regaining self-renewal capacity and plasticity, enabling the cell to function as a stem cell (Fig. 1B).

AML LSCs

Great effort has been invested in exploring the phenotypic characteristics that distinguish LSCs from the entire population of AML blasts and from normal HSCs. The first evidence that these LSCs express a unique surface phenotype different from the phenotype of the remainder AML cells emerged from experiments in which fractions of AML cells were purified on the basis of variable expression of CD34 and CD38 surface antigens and were

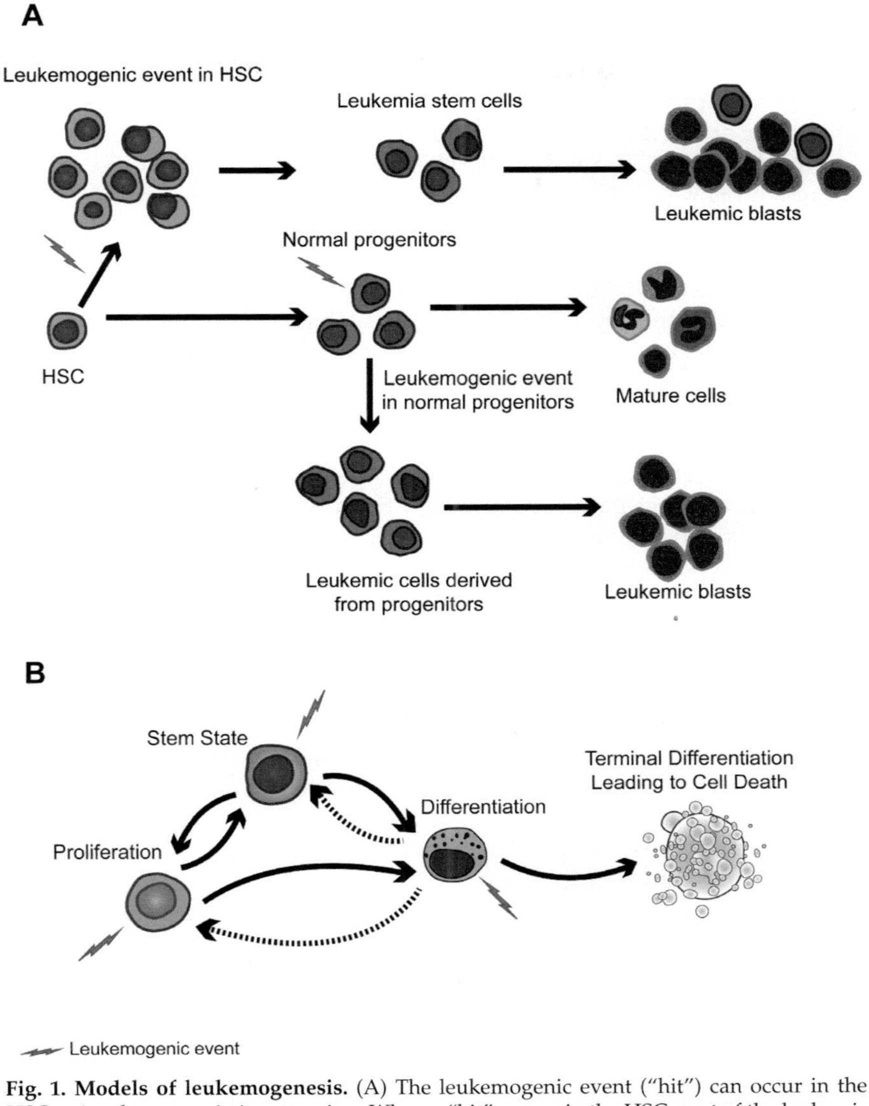

Fig. 1. Models of leukemogenesis. (A) The leukemogenic event ("hit") can occur in the HSC or in a hematopoietic progenitor. When a "hit" occurs in the HSC, part of the leukemic progeny carry stem cell characteristics. The transformed cells give rise to the leukemic blast population, some of which harbor LSC capacity (upper path). If the leukemogenic event occurs in a progenitor cell, none of the leukemic blasts carry LSC capacity (lower path) (Bonnet 2005, Dick 2008). (B) The link between stem cells and proliferating cells is not linear but rather bidirectional. The leukemogenic event can occur in a quiescent stem cell or in a proliferating cell. Any cell, unless destined to die, is capable of regaining stem cell capacity (Zipori 2004).

Color image of this figure appears in the color plate section at the end of the book.

then transplanted into NOD/SCID mice. Similar to normal HSC phenotypic expression, SCID leukemia-initiating cells (SL-ICs) were consistently found within the CD34+CD38- population, regardless of AML subtype (Bonnet and Dick 1997, Lapidot et al. 1994). There was no engraftment of CD34+CD38+ cells, even at cell doses of up to 5×10^5 (Lapidot et al. 1994).

In addition to CD34 and CD38, other surface molecules that differentially express on HSCs and LSCs were reported. These molecules have a direct clinical application and can serve as therapeutic agents that selectively target the LSCs.

One such candidate, Thy-1 (CD90), was reported to be present in HSCs but absent in LSCs by some investigators (Blair et al. 1997) but not by others (Brendel et al. 1999). Another surface marker, c-kit (CD117), is found more consistently only in HSCs but can be found in some LSCs (Blair and Sutherland 2000). In contrast, the L3-receptor α (CD123), usually detected in dendritic cells, is homogeneously expressed in LSCs but not in normal HSCs (Jordan et al. 2000).

Several investigators referred to SL-ICs as LSCs. However, this obvious data interpretation as evidence that SL-ICs reside only within the CD34+CD38 population was recently questioned. Taussig et al. reported that mouse anti-human CD38 antigen is responsible for the removal of CD38+ cells from the blood of the mouse model and that some CD34+CD38+ cells, in fact, possess self-renewal capacity. Furthermore, CD34+CD38+ cells engrafted successfully after the recipient mice were treated with immunosuppressive drugs (Taussig et al. 2008). Thus, although important insights have been gained from the xenotransplantation mouse models, the study by Taussig et al. emphasized the limitation of this model in which the leukemia cells have to overcome the cross-species barrier during the engraftment process, thereby tremendously influencing the final results and limiting their interpretation.

Two separate studies demonstrated that CD34-negative cells can also engraft in NOD/SCID mice (Terpstra et al. 1996, Blair and Sutherland 2000). Because this finding contradicted the current dogma, it was suggested that the CD34+ LSCs had lost the expression of CD34 sometime during the evolution of the disease or, conversely, that there are true stem cells or LSCs within the CD34-negative cell population. In support of this latter assumption, Taussig and colleagues demonstrated that a subpopulation of AML LSCs carrying a mutated nucleophosmin (NPM) do not express CD34 antigen (Taussig et al. 2010). In yet another study, they showed that both HSCs and LSCs express markers of differentiated "mature" myeloid cells such as the monocyte (CD13) and dendritic cell (CD123) antigens (Taussig et al. 2005). Remarkably, some investigators detected the mature myeloid cell marker CD33 on both LSCs and normal HSCs (Pearce et al. 2006).

Taken together, these studies suggest that LSCs cannot be reliably identified by surface marker expression since the phenotype of LSCs is more heterogeneous than previously realized. Furthermore, similar to the normal HSCs, there is also functional heterogeneity within the LSC compartment with different self-renewal and engraftment capacity. Using a lentivirus-transduced SL-IC that enabled tracking of these cells at a single-cell level, Hope and colleagues revealed distinct classes of LSCs, including SL-ICs that can repopulate for only the short-term, whereas other long-term SL-ICs arose later or persisted throughout the entire length of the graft's life, similar to long-term reconstituting HSCs (Hope et al. 2004). Similar to normal HSCs, a side population (SP) with aberrant markers that distinguish LSCs from normal HSC could be identified in AML bone marrow cells. This sp was enriched with LSCs. However, like normal marrow sp cells, the leukemia SP is not a pure population of LSCs (Moshaver et al. 2008b). Thus, human LSCs cannot be identified or tagged reliably by using specific phenotypic characteristics. Rather, they should be defined on the basis of cellular pathways that dictate their functional state (Table 1). Hence, a model defining the LSC as a "state" or "function" rather than a "portrait" appears to be more appropriate (Fig. 1B) (Zipori 2004). The LSC "state" is likely influenced by intracellular and extracellular cues such as its environment.

The hematopoietic stem cell (HSC) niche was coined by Schofield and suggests that without the bone marrow structure and function, lifelong hematopoiesis would not persist, despite the inherent potential of the HSC to self-renew, proliferate, and differentiate (Schofield 1978). Since the introduction of the HSC niche concept, a growing body of evidence supports both the anatomical locations and the physiological role of the HSC niche (Bianco 2011). Analogous to the normal HSC, the bone marrow microenvironment is believed to affect both the survival and function of the LSC; hence the term LSC niche was introduced (Lane et al. 2009). Similar to normal hematopoiesis, in which the endosteal region is considered the anatomical niche, human CD34+ CD38- AML cells were recently shown to home to the endosteal niche of NOD/SCID/IL2R-null mice (Ishikawa et al. 2007).

The importance of the microenvironment in determining the disease phenotype has been demonstrated in a series of experiments in which cord blood CD34+ cells were transduced with the fusion gene MLL-AF9 and transplanted into immunodeficient mice. In NOD/SCID/β2 microglobulin-null mice, these cells induced acute lymphoblastic or biphenotypic leukemia, whereas in NOD/SCID mice transgenic for SCF, GM-CSF, and IL-3, these cells produced AML (Wei et al. 2008). Hence, the environment into which the same transduced cells were transplanted determined the leukemia phenotype. In this regard, it is important to compare the HSC-

Table 1. Molecular pathways regulating LSC function.

References	Function in LSC	Description	Molecular pathway
(Lecuyer and Hoang 2004)	Expressed in HSCs. Abnormal up-regulation in progenitor cells initiates malignant transformation. Frequent in chromosomal rearrangement in pediatric population	Transcription factor	SCL
(de Guzman et al. 2002)	Induces self-renewal	A result of the t(8:21) translocation in AML	AML1-ETO
(Smith et al. 2011)	Expressed in HSCs. Some family members (e.g., HOX9) are dysregulated in LSCs	Transcription factor	Homebox (Hox) genes
(Wang et al. 2010)	Activation of Wntβ/catenin is required for the development of LSCs in AML	Signaling pathway	Wntβ/catenin
(Peng et al. 2010, Yilmaz et al. 2006, Zhang et al. 2006)	Pten is down-regulated in AML and CML LSCs. Down-regulation of Pten in mice induces AML.	Tumor suppressor gene	PTEN
(Chumsri et al. 2007, Lessard and Sauvageau 2003)	Essential for the function of HSC and LSC. BMI1 -/- knockout mice were unable to engraft.	Part of PRC1-like complex. Necessary for repression of transcription of many genes	BMI1
(Guzman et al. 2001)	Constitutively activated. Antiapoptotic	Transcription factor	NFkB
(McCormack et al. 2010)	Oncogene that induces self-renewal program in committed T cell	Highly preserved protein, role in hematopoietic development	LMO2
(Horn et al. 2008, Martelli et al. 2010)	Regulates maintenance and lineage commitment in normal HSCs. Dysregulation of this pathway proved to be leukemogenic in mice.	Signaling pathway	P13/AkT/mTOR
(Heuser et al. 2009)	STAT signaling is critical for AML LSC self-renewal in MN1- and HOXA9-expressing leukemias	Transcription factors	STAT

microenvironment interaction with the LSC-microenvironment interaction. Namely, under what conditions do the LSCs have survival advantage over normal HSCs? For example, in a rat model, there is marked expansion of the hypoxic area in the bone marrow (Mortensen et al. 1998), where LSCs preferentially survive (Becker and Jordan 2011). Homing and adhesion of HSCs to the bone marrow niche are well characterized. Whether those pathways provide the LSCs with proliferative advantage was investigated by various groups. For example, the interactions between stoma derived factor (SDF)-1 and CXCR4, the chemokine receptor expressed by HSCs, up-regulates the expression of vascular cell adhesion molecule-1 (VCAM-1) and very late antigen-4 (VLA-4). This facilitates the localization and retention of HSCs and hematopoietic progenitor cells in the bone marrow (Lapidot and Kollet 2002). This pathway is sought to be used by and provide survival advantage to the LSCs. Indeed, CXCR4 levels are significantly elevated in AML blasts (Mohle et al. 2000), particularly in patients with AML harboring internal tandem duplications in the FLT3 gene (Rombouts et al. 2004). Adhesion is facilitated by other factors such as the interaction between VLA-4/α4β1 integrin and its stromal ligand fibronectin. Remarkably, after chemotherapy, high levels of minimal residual disease (MRD) were detected only in patients with AML who had high levels of VLA-4 (Matsunaga et al. 2003). Hence, up-regulation of VLA-4 is yet another mechanism that may provide survival advantage to the LSCs (Lane et al. 2009). Other mechanisms involved in the interaction of HSCs with the bone marrow microenvironment are mediated by the adhesion molecule CD44, known to be expressed on leukemia cells. Anti-CD44 antibodies markedly reduced engraftment of AML cells into NOD/SCID mice (Jin et al. 2006).

In contrast, several observations challenged the validity of the HSC niche model. Extramedullary hematopoiesis, usually found in myeloproliferative neoplasms and metastatic neoplastic diseases, is part of the normal developmental program and occurs in the embryonic liver and spleen. Leukemia cells are frequently detected outside the blood or bone marrow compartments. Indeed, leukemic infiltrates of the skin (leukemia cutis), liver, thorax, kidneys, and lymph nodes have been readily described (Byrd et al. 1995). Thus, the endosteal niche is not the only niche for HSCs or LSCs. Whether HSCs and LSCs can survive without cues from the bone marrow environment or whether other niches can also support LSC function is not known. HSCs are capable of inducing the production of chemokines needed for their survival by the microenvironment as reported by Gillette et al. These investigators showed that HSC endosomes attach to the osteoblast surface through a specialized membrane domain, thereby inducing down-regulation of Smad proteins and increasing the production of SDF-1 necessary for homing and survival of HSCs (Gillette et al. 2009). Hence, while the endosteal niche is believed to function as the physiological HSC niche, the HSCs might have active role in shaping their microenvironment in other tissues.

Acute Promyelocytic Leukemia LSCs

The hallmark of acute promyelocytic leukemia (APL), also known as acute progranulocytic leukemia, is a fusion between the PML gene, encoding the transcription factor PML, and the all-trans retinoic acid receptor-α (ATRA) gene. This fusion (PML-RARα) is almost always the result of t(15;17)(q24;q21) chromosomal translocation. PML-RARα has been shown to induce differentiation arrest and accumulation of promyelocytes typically found in the bone marrow and blood of patients with this disease (Brown et al. 1997).

Xenotransplantation experiments using the NOD/SCID mouse model have suggested that the human AML LSC is a CD34+ CD38- cell in all types of AML except APL (Bonnet and Dick 1997, Lapidot et al. 1994). The interpretation of these observations is controversial. It is possible that APL arises from a mature progenitor cell and that the LSC in this disease is therefore a mature progenitor, or that there is no LSC activity in this type of leukemia (Wojiski et al. 2009). The clinical observation is that although most patients attain remission after treatment with ATRA, chemotherapy and ATRA, or arsenic trioxide and ATRA, as many as 30% experience relapse; this observation favors the hypothesis that a cell population with self-renewal capacity, necessary for LSC activity, exists in APL. By using a PML/RARα transgenic mouse model, Guibal et al. defined the APL LSC as $CD34^+$, c-kit^+, $FC\gamma RIII/II^+$, $Gr1^{in}$ -expressing (Guibal et al. 2009) and demonstrated that these cells encompassed self-renewal capacity (Wojiski et al. 2009). Whether this model mimics human APL is not clear since all cells in this model, including the CD34+ cells, harbor the PML/RARA mutation, found only in CD33+ but not in CD34+ human APL promyelocytes. Nevertheless, this model provides conceptual proof that a committed progenitor can transfer the disease and engraft in a second-generation mouse. Furthermore, in contrast to the classic LSC model, dilution analysis revealed that cells with Leukemic Initiating Capacity are not rare, and small numbers of these cells are sufficient for engraftment.

Acute Lymphoblastic Leukemia LSC

Whether the leukemogenic event(s) in acute lymphoblastic leukemia (ALL) takes place in the HSC or in a lymphoid progenitor is unclear and is a topic of ongoing debate. During lymphoid cell development, the immunoglobulin heavy chain gene (IgH) is rearranged in B cells, and the T-cell receptor (TCR) gene is rearranged in T lymphocytes. These events provide a practical tool for assessing clonality of the lymphoid lineage. Studies of IgH and TCR gene rearrangement suggested that the leukemic transformation in childhood ALL occur in lymphoid progenitor cells (Davi et al. 1996, Green et al. 1998, Langlands et al. 1993, Tycko et al. 1992). In recent years, the interpretation of these observations has been questioned, for example, as follows: (a) Clonal diversity of Ig gene rearrangement occurs throughout the course of the disease within subclones of the leukemia cells, so Ig rearrangement could be found also in primitive cells (Green et al. 1998) and (b) specific V2D3 rearrangement was detected in $CD34^+/CD38^-$ cells from patients with Philadelphia chromosome Ph+ ALL during remission, and those Ph-positive immature cells engrafted in immunodeficient mice (Green et al. 1998, McConville et al. 1998), suggesting that ALL-initiating cells arise from HSCs. Additional evidence supporting the assumption that ALL arises from primitive multipotent progenitors has been provided by xenotransplantation studies showing that the SL-ICs from all Ph+ ALL, regardless of maturation heterogeneity, were exclusively CD34+/CD38- cells (Cobaleda et al. 2000). Similarly, in B-cell ALL, the bulk of blasts expressing lineage-specific antigens (CD10 in common B-ALL and CD19 in pre-B-ALL) failed to engraft in NOD/SCID mice, whereas cells with an immature CD34+/CD10–/CD19– phenotype readily engrafted in primary and secondary recipients. Moreover, after engraftment, the immunophenotype and the karyotype were the same as those observed at diagnosis, suggesting that these cells differentiated *in vivo* (Cox et al. 2004).

In contrast, other investigators found that only CD19+ ALL bone marrow cells from patients with t(12:21) translocation possessed leukemia-initiating activity as assessed in the NOD/SCID mouse model. The same investigators identified a population of CD34+/CD38-/CD19+ cells. Hence, the CD34+/CD38- compartment encompasses a mixed population of ALL (Castor et al. 2005). Arguably, the cell surface phenotype of LSCs in ALL is complex and less reliable than in AML. In addition, it is clear that the origin of the ALL cell is not identical in all cases. In some ALLs, the initial transformation occurred at the HSC level, whereas in other ALLs, the first event occurred in a committed progenitor cell.

The perception that all blasts originate from the same clone was recently challenged. Notta et al. demonstrated that in Ph+ ALL, distinct subclones, each with its own capacity for repopulation of primary and secondary recipients, exist even at the time of diagnosis. Analyzing Single Nucleotide Polymorphism (SNP) data, they could trace the segregation of individual subclones in xenografts and reconstruct the functional genetic ancestry of the subclones. The authors described two growth characteristics: in the first group with the aggressive growth pattern, the dominant clone continued also in the xenograft mice, whereas in others, usually from the second group, the subclones outcompeted the dominant clone. Limiting dilution analysis showed the leukemia-initiating cell frequency was 80 times higher in the first group than in the second group, and the second group engrafted only in models with increasing immunodeficiency (NOD/SCID+antiCD122/NS122, NOD/SCID+ deletion of the common gamma chain /NSG) (Notta et al. 2011b).

In summary, during the past decade, several investigators used the LSC model to address fundamental questions related to the pathogenesis of ALL. Apparently, ALL is a diverse disease. In some patients, the LSCs share HSC characteristics, whereas in others, the LSCs express lymphoid-committed progenitor cell features. The work of Notta et al. (Notta et al. 2011b) has profound implications, not only for the pathogenesis of ALL but also for the clonal characteristics of all leukemic processes, thereby providing evidence for both clonal evolution and for the existence of leukemia-initiating cells in many individual subclones.

Chronic Myelogenous Leukemia LSCs

The hallmark of chronic myelogenous leukemia (CML) is genetic translocation between chromosomes 9 and 22 t(9;22), resulting in the formation of a truncated chromosome 22 known as the Philadelphia chromosome (Ph) (Moshaver et al. 2008a). As a consequence, fusion between the genes *bcr* and *abl* (bcr-abl) induces aberrant constitutive activation of the Abl tyrosine kinase in CML cells (Feinstein et al. 1992).

The first "hint" that Ph+ cells with repopulating capacity exist came from early clinical experience in neutropenic patients with leukemia who received transfusions of irradiated white blood cells from a patient with high cell count CML. Because repopulation of the hematopoietic system by Ph-positive cells was observed, it was thought that primitive Ph+ cells with self-renewal and differentiation capacity are able to repopulate the hematopoietic system. Nevertheless, the repopulation was transient, and the Ph-positive cells disappeared from the blood a few months thereafter (Graw et al. 1970, Levin et al. 1963). *In vitro* experiments showing that Ph+/Bcr-Abl–positive cells generated a progeny of CFUs for at least 6 weeks supported the hypothesis that a primitive multipotent cell, possibly a transformed HSC, is the stem/progenitor cell in CML (Petzer et al. 1996, Udomsakdi et al. 1992).

Serial transplantation of sublethally irradiated SCID or NOD/SCID mice with cells derived from patients with leukemia became the gold standard proof for the existence of LSCs (Dick 2008). These experiments mostly failed to demonstrate a permanent repopulating capacity of Ph+/Bcr-Abl+ cells derived from patients with chronic-phase CML. Most studies reported repopulation of the mice with cells derived from normal HSCs (Lewis et al. 1998, Sirard et al. 1996, Verstegen et al. 1998, Wang et al. 1998).

Arguably, the NOD/SCID mouse model is not the ideal model with which to demonstrate the repopulation capacity of the Ph+ CML LSC. Indeed, using a mouse model with a more profound immunodeficiency (the NSG mouse model), Sloma and colleagues recently reported engraftment of a durable Ph+ leukemia cell (Sloma et al. 2010), however, the experimental evidence for that is not yet published, and proof-of-principle experiments aimed at investigating the engraftment capacity of counterpart normal bone marrow mature progenitors are warranted. Remarkably, a detailed study showed that the granulocyte-macrophage progenitor (CFU-GM) is the LSC in blast-crisis CML (Jamieson et al. 2004).

A different approach included the use of transfected/manipulated cells in various animal models. Early experiments in a murine model, demonstrating that bone marrow cells transfected with the Bcr/Abl fusion gene engrafted in irradiated syngeneic recipients develop a monoclonal myeloproliferative disorder that resembles CML, suggested that a CML LSC exists (Daley et al. 1990, Pear et al. 1998). The generation of a CML-like disease appears to be dependent on the existence of primitive HSC. Hence, neither common myeloid progenitors nor granulomonocytic progenitors transduced with BCR/ABL could generate CML-like disease, and the presence of an upstream primitive population enriched in HSCs is probably present (Huntly et al. 2004).

In yet another mouse model lacking the JunB transcription factor, a myeloproliferative disease that resembles CML developed. Engraftment in serially transplanted mice occurred only when the LT-HSC were lacking the JunB and not when more mature progenitors were targeted. However, although the phenotype in these mice resembled human CML, it raised the question of whether human CML in a mouse model might be obtained by using cells that do not express the Bcr/Abl fusion gene product (Passegue et al. 2001, Passegue et al. 2004). Other studies showed that overexpression of Bcr/Abl and the NUP98/HOSA9 translocation in normal HSC transplanted into mice induced a CML-like disease (Neering et al. 2007). Taken together, those *in vivo* models provide indirect evidence that the leukemic transformation in CML is on the level of a primitive cell population, possibly the HSC itself; however, a direct link between the HSC and the CML LSC is still lacking.

The current dogma is that protein kinase inhibitors such as imatinib mesylate do not eliminate CML stem cells because discontinuation of imatinib usually results in molecular disease recurrence and clinical relapse (Savona and Talpaz 2008). Recent studies have suggested that although most CML cells are oncogene "addicted" and require the Bcr/Abl oncoprotein for survival and proliferation, CML LSCs are not; these LSCs survive and proliferate despite long-term oncogene inhibition and thereby induce clinical relapse after Tyrosine Kinase inhibitor (TKI) discontinuation

(Corbin et al. 2011). Moreover, this suggests that because quiescent CML stem cells are insensitive to TKI, other approaches designed to eradicate residual disease, which do not necessarily target BCR/ABL TK activity, are necessary to eradicate the disease (Corbin et al. 2011).

However, in the Stop Imatinib (STIM) study, imatinib was discontinued in patients who were in complete remission for more than 2 years. Of those patients, disease relapsed in 61%, and 39% remained in molecular remission for a median of 2 years (Mahon et al. 2010). One explanation for the durable remission of the group of patients whose disease did not relapse is that imatinib caused complete eradication of the CML stem cell, an assumption that is not supported by most investigators (Goldman and Gordon 2006, Redner 2010, Rice and Jamieson 2010). An alternative explanation is that eradication of the CML LSCs is not necessary to sustain prolonged clinical remission. Supporting this explanation are studies showing that a subgroup of patients with CML who attained cytogenetic and molecular remission after treatment with interferon-α and were taken off therapy did not experience relapse after years of follow-up but still harbored Ph+ colony-forming cells in their bone marrow (Talpaz et al. 1994).

The final phase of disease evolution in CML is blastic transformation that can be either myeloid or lymphoid. Clinically, this stage resembles acute leukemia. In addition to the Bcr-Abl oncogene, progression to blast crisis is believed to require additional molecular and/or cytogenetic alterations that contribute to blastic transformation.

Two proposed models explain the stage at which the second "hit" occurs. One model postulates that the second "hit" occurs in the HSCs; the other model, supported by detailed experiments, places the second mutagenic event downstream of the HSCs in the lineage-restricted progenitor compartment. In lymphoid blast crisis, the strongest evidence has been provided by experiments showing IgH gene rearrangement in leukemic blasts, suggesting that in most cases, the blast crisis occurs in B-cell precursors (Bakhshi et al. 1983). Laboratory data suggest that this is the case for myeloid blast crisis as well. As discussed above, in a series of experiments, Jamieson et al. demonstrated that the granulocyte-macrophage colony-stimulating factor (GM-CSF) progenitor cells of CML patients in myeloid blast crisis exploit the β-catenin pathway for self-renewal, thus gaining stem cell properties, whereas in a normal mouse, this pathway is used exclusively by HSCs (Jamieson et al. 2004).

Clinical Significance of the LSC Model

Currently used antineoplastic agents do not cure most patients with leukemia. It was suggested the new strategies to that specifically target LSCs would help in cases of refractory disease and to prevent relapse. Here we review the current and future implications of the LSC on prognosis and current and future therapy.

Effect on prognosis

Based on the hypothesis that LSCs are resistant to standard chemotherapy, Rhenen et al. found correlation between MRD and poor survival (Rhenen 2005). Similarly, Gentle et al. identified 52 genes whose expression distinguished LSC-enriched cells (characterized by CD34+CD38– surface marker expression) from other sorted AML cells. A high LSC signature was associated with worse event-free and relapse-free survival rates (Gentles 2011). However, because leukemia-sustaining cells are heterogeneous and patient-to-patient variation exists, these cells should be identified by their function; the measurement of "true" residual LSCs as an index for MRD might be problematic and impractical (Estrov 2010).

Targeting LSCs

The proposed hierarchical LSC differentiation model is similar to that of the HSC differentiation model (Bonnet and Dick 1997, Hope et al. 2004, Lapidot et al. 1994). Accordingly, LSCs are believed to exhibit several features characteristic of normal HSCs such as quiescence, efflux pumps, and detoxifying enzymes, thereby rendering them resistant to standard chemotherapy (Jones 2009). Although standard antineoplastic therapy eliminates the bulk of leukemic cells, LSCs are believed to remain largely intact during this treatment, thereby inducing relapse, especially in patients with AML. Given the complexity of leukemia biology and the heterogeneity of LSCs, additional studies are needed to investigate this hypothesis (Estrov 2010). Thus, improving the results of leukemia therapy requires a thorough understanding of the biology of leukemia and of LSCs and, perhaps, the development of targeted therapies against LSCs (Fig. 2).

According to the current concept, the leukemic clone is continuously replenished by rare, functionally distinct, quiescent, self-renewing LSCs, characterized by clonogenic capacity and the capability to differentiate into a hierarchy of immature and mature progenitors. The consequence of this process is the accumulation of a large number of rapidly dividing blast cells, typically found in acute leukemia (Bonnet and Dick 1997, Hope et al. 2004, Lapidot et al. 1994).

Contradicting this hypothesis are clinical data showing that a significant number of patients with leukemia have been cured with chemotherapy and several studies showing that conventional chemotherapy eradicates LSCs. For example, in a mouse model of human AML, treatment with granulocyte colony-stimulation factor (G-CSF) resulted in LSCs entering the S phase of

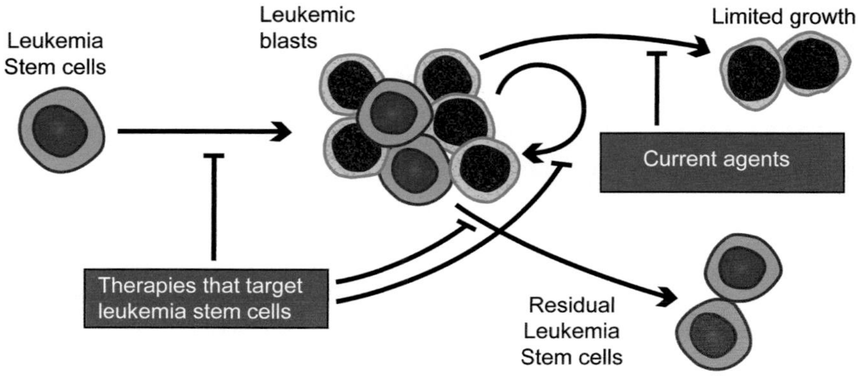

Fig. 2. Potential advantage of therapies that target the LSC. Whereas current therapies are usually effective against the bulk of the leukemic blasts, the LSC is not targeted, and residual LSCs can replenish the leukemic process. With future therapies that eradicate LSCs, this vicious cycle will be specifically targeted.

the cell cycle, thereby making them amenable to the cytotoxic effects of the cell cycle–specific drug cytarabine. Furthermore, this group showed that quiescence is not an inherent characteristic of LSCs, but is largely dependent on the location of the LSCs within the bone marrow. Whereas LSCs located in the bone marrow cavity were efficiently eliminated by cytarabine, LSCs residing adjacent to the endosteum remained intact. Treatment with G-CSF induced proliferation of endosteal LSCs and as a result, these cells became sensitive to cytarabine (Saito et al. 2010). However, clinical leukemia cell "priming" trials with G-CSF or GM-CSF yielded inconclusive results (Estey et al. 1999, Lowenberg et al. 2003, Milligan et al. 2006). Only one trial showed improvement in disease-free survival in patients with standard-risk AML (Lowenberg et al. 2003). Recently, it was shown that IFN-α affects the quiescent status of HSCs. Treatment with IFN-α induced transient proliferation of mouse HSC progenitors, which became quiescent a few hours thereafter (Essers et al. 2009). Whether IFN-α exhibits a similar effect on LSCs remains to be determined.

Targeting LSCs is especially intriguing in the setting of post-remission therapy, both as a consolidation/maintenance therapy and in the setting of MRD, where the LSCs might not have been completely eradicated by conventional chemotherapy.

If eradication of LSCs becomes the mainstay of therapy, then LSC detection methods, aimed at quantifying LSC burden before and after therapy, should become the standard of care. However, by definition, LSCs are only a minute fraction of the tumor mass. Therefore, currently available LSC detection methods are not sensitive and are unreliable overall. Phenotype heterogeneity among patients with the same disease as well as changes in leukemia cell phenotype during the course of the disease, especially after chemotherapy, makes it difficult to establish a standard method and thus assessment on an individual basis is required. Nevertheless, some investigators suggested that measuring LSC burden is plausible, for example, by monitoring aberrant cell surface marker expression on the CD34+ CD38– cells and following it at diagnosis and during remission (van Rhenen et al. 2007a).

An ideal agent would specifically target the LSCs but not the normal HSCs (Table 2). One approach has been to target surface proteins that are expressed on LSCs but not on normal HSCs. One such target is CLL-1. CLL-1 was reported to be expressed preferentially on most AML LSCs but not on normal HSCs (van Rhenen et al. 2007b). Other potential targets include CD123 (IL-3Rα) for which antibodies conjugated to ozogamicin have been shown to selectively target AML LSCs *in vitro* (Hogge et al. 2004); CD44, a transmembrane adhesion glycoprotein, is overexpressed in AML LSCs. This protein was successfully targeted by anti-CD44 monoclonal antibodies in NOD/SCID mice that received human AML cells, which markedly reduced leukemia cell repopulation. Serial transplantations demonstrated that AML LSCs have been directly targeted (Jin et al. 2006).

Targeting signaling pathways that regulate self-renewal is yet another plausible mechanism to eliminate LSCs. The Wnt signaling pathway has been targeted in AML and CML (Jamieson et al. 2004, Wang et al. 2010); the NOTCH1 signaling pathway, constitutively activated due to activating mutations in almost half of patients with T cell Acute Lymphoblastic Leukemia (T-ALL), has been targeted in ALL (Mumm et al. 2000, Weng et al. 2003). Remarkably, these pathways also regulate self-renewal in normal HSCs; therefore, such therapies might induce hematopoietic toxicity (Krause and Van Etten 2007).

The transcription factor NF-κB, constitutively activated in most AML LSCs, but not in normal unstimulated hematopoietic progenitors (Guzman, Neering et al. 2001), was suggested as a suitable target. Indeed, Guzman et al. have demonstrated that the proteasome-inhibitor MG-132, which inhibits NF-κB through stabilization of its cellular inhibitor IκB, induced apoptosis in CD34+CD38– AML cells while sparing normal primitive hematopoietic progenitors (Guzman et al. 2002).

Another potential therapeutic target is the PI3K/PTEN/Akt/mTOR pathway. This pathway has been shown to be activated, in AML cells, providing a survival advantage (Martelli et al. 2010). Indeed, inhibition of Akt or mTOR, in combination with chemotherapy, induced apoptosis of AML cells and decreased the abundance of LSCs (Xu et al. 2003, Xu et al. 2005).

Finally, the ATP-binding cassette transporters, known to be highly expressed in AML LSCs, are believed to be another specific target. ATP-

Table 2. Agents targeting LSCs.

Agent	**description**	**Primary LSC target**	**Proposed mechanisms**	**References**	**Comments**
Parthenolide (feverfew)	Plant derivative	AML, B-ALL	Induces apoptosis	(Guzman and Jordan 2005, van Rhenen et al. 2007b)	In clinical trials
DMAPT (dimethylaminoparthenolide)	Small molecule (parthenolide derivative)	AML	Targets specifically CD34+ AML cells	(Jordan 2008)	In clinical trials
MG-132	Proteasome inhibitor	AML, ALL	Inhibits the NF-κB pathway	(Colado et al. 2008, Guzman et al. 2001, Guzman et al. 2002)	
Bortezomib	Proteasome inhibitor	AML	Inhibits the NF-κB pathway and induces apoptosis	(Colado et al. 2008)	In clinical trials
ABT-737	Small molecule	AML	Inhibits bcl-2	(Kohl et al. 2007, Konopleva et al. 2006)	
TDZD-8 (4-benzyl,2-methyl,1,2,4-thiadiazolidine,3,5 dione)	Small molecule	AML	Compromises membrane integrity	(Guzman et al. 2007)	
CD44 antibody	Monoclonal antibody	AML, CML	Interrupts the interaction between stem cells and their microenvironmental niche	(Jin et al. 2006)	
CD123 antibody	Monoclonal antibody	AML	Targets IL-3Rα thought to be preferentially expressed on LSCs and to a lesser extent on HSCs	(Jin et al. 2009)	In clinical trials
Cll-1 antibody	Monoclonal antibody targeting the C-type lectin-like molecule-1	AML	Targets Cll-1 expressed on AML LSCs but not on HSCs	(Zhao et al. 2010)	

Indomethacin	Nonsteroidal anti-inflammatory drug (NSAID)	AML	Blocks the Wnt/β-catenin pathway	(Wang et al. 2010)	
Lovastatin	Statins	AML	Proapoptotic.	(Misaghian et al. 2009, Wu et al. 2004)	
CEP-701 (lestaurtinib)	Inhibits Flt-3	AML	Inhibits Flt-3 signaling and the Wnt signaling pathway	(Fathi and Chabner 2011)	In clinical trials
Rapamycin	Antibiotics	AML	Inhibits mTOR	(Recher et al. 2005)	
DT388-IL3 fusion protein	A truncated diptheria toxin fused to IL-3	AML	Targets CD123 expressed on LSCs but not HSCs	(Black et al. 2003)	In clinical trials
DAPT	γ Secretase inhibitor	T-ALL, AML	γ secretase inhibitor of the NOTCH signaling pathway	(Gal et al. 2006)	
Myelotarg	Monoclonal antibody targeting CD33 fused with calicheamicin	APL, CD33+ AML	Targets CD33-positive LSCs	(Sperr et al. 2005, Tsimberidou et al. 2006)	In 2010, was withdrawn from market at FDA request
Arsenic trioxide	Inorganic compound	APL	Induce apoptosis	(Zheng et al. 2007)	In routine clinical use
INF-α	Cytokine	CML	Believed to activate dormant HSCs by affecting the stem cell niche	(Essers et al. 2009, Simonsson et al. 2011)	In clinical trials for CML

binding cassette transporter receptors are responsible for the efflux of cholesterol and in a substantial number of AML cells, marked elevation of intracellular cholesterol was found after exposure to cytotoxic drugs. This response, not detected in normal CD34+ cells, contributed to AML cell survival. When this response was blocked, AML cells showed increased chemosensitivity (Li et al. 2003). These results formed the basis for clinical trials incorporating statins in AML therapy. Statins, known to block cholesterol formation, are well tolerated and possess additional antioxidative and anti-inflammatory effects (Li et al. 2003, Stirewalt et al. 2003).

Closing Remarks

The LSC model dissects the morphologically homogeneous leukemic blasts into two functional populations of "mature" blasts and cells with LSC capacity.

Since its introduction in the mid-1990s, the LSC model has become widely accepted. For some it has gained almost mythical status and became an inescapable fact. Yet over time, almost every aspect of the original model has been questioned. Like HSCs, LSCs are characterized by the ability to self-renew, differentiate, and repopulate the hematopoietic system with leukemic blasts (Dick 2008). Ignoring the fact that patients with acute leukemia are cured with current regimens, the LSC model provides an elegant hypothesis for relapse according to which the bulk of the blast population is eliminated by chemotherapy and the dormant LSCs are left intact, thereby inducing disease relapse.

Therefore, it has been suggested that LSCs should be directly targeted, preferentially during remission as part of a consolidation or maintenance therapy strategy. However, one should keep in mind that elimination of the LSC population should not be a target of its own. The success of chemotherapy in childhood ALL suggests that either the chemotherapy eradicates LSCs or that, as in CML, complete eradication of leukemia cells and/or LSCs is not needed either because the innate immune system is capable of controlling a minute amount of MRD or the remaining leukemia cells/LSCs are not capable of bringing back the disease. Monitoring may be important, but equally important is to know when to intervene and when not to. Criteria for that have not been established.

Xenotransplantation models became the gold standard of the LSC model. These experiments typically showed that human AML (excluding the APL type) bone marrow cells engraft in sublethally irradiated immunodeficient mice and induce leukemia in second- and third-generation mice. Similar xenotransplantation models were difficult to reproduce in other types of leukemia such as ALL and CML (Bonnet and Dick 1997,

Lapidot et al. 1994). As a result, doubts regarding the phenotype of SL-IC cells and whether these cells are indeed "true" LSCs have been raised. To successfully engraft in an immunodeficient mouse, a human leukemia cell has to overcome species barrier, a demanding challenge that also depends on the immunodeficient mouse model used. Furthermore, by showing that human CD38+ cells are eliminated by antibodies in NOD-SCID mice, Taussig et al. provided an explanation of why CD38+ human AML cells did not engraft, thereby questioning the validity of the classic LSC phenotype as being CD34+CD38 (Taussig et al. 2008). Remarkably, in similar experiments with mouse leukemic cells, the number of cells needed for engraftment was much smaller (as low as 10 leukemic cells) than the number of human leukemia cells needed for engraftment, suggesting that the rarity of human SC-IC cells is a result of the difficulty of human leukemia cells to engraft and survive in a foreign milieu. Furthermore, the fact that only 10 acute leukemia cells were needed for engraftment suggests that it is unlikely that LSCs are rare cells with unique features but rather a commonly detected blast that is capable of regaining stem cell capacity.

Thus, at a functional level, the analogy between the LSCs and the HSCs is limited. A leukemia cell can function as a LSC if it gains, regains, or maintains the capacity for self-renewal. Furthermore, this function might be affected by environmental pressure and stochastic events at the molecular and cellular levels. A more realistic definition of LSCs should be based on function rather than on specific cell surface marker phenotype or level of maturation.

The existence of different leukemia clones at diagnosis, clearly demonstrated in recent work on Ph+ ALL, adds another level of complexity. Here, several leukemic clones were identified at diagnosis. This clones engrafted in immunodeficient mice. Each clone had a different growth pattern and its own SL-ICs (Notta et al. 2011b). The expected LSC hierarchy, sought to be identified according to the LSC differentiation model (Bonnet and Dick 1997), was not found. These data should be taken into consideration and further investigated because the LSC theory has profound implications on the potential success of current and future targeted therapies.

References

Alenzi, F.Q., B.Q. Alenazi, S.Y. Ahmad, M.L. Salem, A.A. Al-Jabri, and R.K. Wyse. 2009. The haemopoietic stem cell: between apoptosis and self renewal. Yale J Biol Med 82: 7–18.

Arai, F., A. Hirao, M. Ohmura, H. Sato, S. Matsuoka, K. Takubo, K. Ito, G.Y. Koh, and T. Suda. 2004. Tie2/angiopoietin-1 signaling regulates hematopoietic stem cell quiescence in the bone marrow niche. Cell 118: 149–161.

Bakhshi, A., J. Minowada, A. Arnold, J. Cossman, J.P. Jensen, J. Whang-Peng, T.A. Waldmann, and S.J. Korsmeyer. 1983. Lymphoid blast crises of chronic myelogenous leukemia represent stages in the development of B-cell precursors. N Engl J Med 309: 826–831.

Becker, M.W., and C.T. Jordan. 2011. Leukemia stem cells in 2010: current understanding and future directions. Blood Rev 25: 75–81.

Bertoncello, I., and B. Williams. 2004. Hematopoietic stem cell characterization by Hoechst 33342 and rhodamine 123 staining. Methods Mol Biol 263: 181–200.

Bianco, P. 2011. Bone and the hematopoietic niche: a tale of two stem cells. Blood 117: 5281–5288.

Black, J.H., J.A. McCubrey, M.C. Willingham, J. Ramage, D.E. Hogge, and A.E. Frankel. 2003. Diphtheria toxin-interleukin-3 fusion protein (DT(388)IL3) prolongs disease-free survival of leukemic immunocompromised mice. Leukemia 17: 155–159.

Blair, A., D.E. Hogge, L.E. Ailles, P.M. Lansdorp, and H.J. Sutherland. 1997. Lack of expression of Thy-1 (CD90) on acute myeloid leukemia cells with long-term proliferative ability *in vitro* and *in vivo*. Blood 89: 3104–3112.

Blair, A., and H.J. Sutherland. 2000. Primitive acute myeloid leukemia cells with long-term proliferative ability *in vitro* and *in vivo* lack surface expression of c-kit (CD117). Exp Hematol 28: 660–671.

Bonnet, D. 2005. Normal and leukaemic stem cells. Br J Haematol 130: 469–479.

Bonnet, D., and J.E. Dick. 1997. Human acute myeloid leukemia is organized as a hierarchy that originates from a primitive hematopoietic cell. Nat Med 3: 730–737.

Bradford, G.B., B. Williams, R. Rossi, and I. Bertoncello. 1997. Quiescence, cycling, and turnover in the primitive hematopoietic stem cell compartment. Exp Hematol 25: 445–453.

Brendel, C., B. Mohr, C. Schimmelpfennig, J. Muller, M. Bornhauser, M. Schmidt, M. Ritter, G. Ehninger, and A. Neubauer. 1999. Detection of cytogenetic aberrations both in CD90 (Thy-1)-positive and (Thy-1)-negative stem cell (CD34) subfractions of patients with acute and chronic myeloid leukemias. Leukemia 13: 1770–1775.

Brown, D., S. Kogan, E. Lagasse, I. Weissman, M. Alcalay, P.G. Pelicci, S. Atwater, and J.M. Bishop. 1997. A PMLRARalpha transgene initiates murine acute promyelocytic leukemia. Proc Natl Acad Sci USA 94: 2551–2556.

Byrd, J.C., W.J. Edenfield, D.J. Shields, and N.A. Dawson. 1995. Extramedullary myeloid cell tumors in acute nonlymphocytic leukemia: a clinical review. J Clin Oncol 13: 1800–1816.

Calvi, L.M., G.B. Adams, K.W. Weibrecht, J.M. Weber, D.P. Olson, M.C. Knight, R.P. Martin, E. Schipani, P. Divieti, F.R. Bringhurst, L.A. Milner, H.M. Kronenberg, and D.T. Scadden. 2003. Osteoblastic cells regulate the haematopoietic stem cell niche. Nature 425: 841–846.

Castor, A., L. Nilsson, I. Astrand-Grundstrom, M. Buitenhuis, C. Ramirez, K. Anderson, B. Strombeck, S. Garwicz, A.N. Bekassy, K. Schmiegelow, B. Lausen, P. Hokland, S. Lehmann, G. Juliusson, B. Johansson, and S.E. Jacobsen. 2005. Distinct patterns of hematopoietic stem cell involvement in acute lymphoblastic leukemia. Nat Med 11: 630–637.

Cheshier, S.H., S.J. Morrison, X. Liao, and I.L. Weissman. 1999. *In vivo* proliferation and cell cycle kinetics of long-term self-renewing hematopoietic stem cells. Proc Natl Acad Sci USA 96: 3120–3125.

Chumsri, S., P. Phatak, M.J. Edelman, N. Khakpour, A.W. Hamburger, and A.M. Burger. 2007. Cancer stem cells and individualized therapy. Cancer Genomics Proteomics 4: 165–174.

Cobaleda, C., N. Gutierrez-Cianca, J. Perez-Losada, T. Flores, R. Garcia-Sanz, M. Gonzalez, and I. Sanchez-Garcia. 2000. A primitive hematopoietic cell is the target for the leukemic transformation in human philadelphia-positive acute lymphoblastic leukemia. Blood 95: 1007–1013.

Colado, E., S. Alvarez-Fernandez, P. Maiso, J. Martin-Sanchez, M.B. Vidriales, M. Garayoa, E.M. Ocio, J.C. Montero, A. Pandiella, and J.F. San Miguel. 2008. The effect of the proteasome inhibitor bortezomib on acute myeloid leukemia cells and drug resistance associated with the CD34+ immature phenotype. Haematologica 93: 57–66.

Corbin, A.S., A. Agarwal, M. Loriaux, J. Cortes, M.W. Deininger, and B.J. Druker. 2011. Human chronic myeloid leukemia stem cells are insensitive to imatinib despite inhibition of BCR-ABL activity. J Clin Invest 121: 396–409.

Cox, C.V., R.S. Evely, A. Oakhill, D.H. Pamphilon, N.J. Goulden, and A. Blair. 2004. Characterization of acute lymphoblastic leukemia progenitor cells. Blood 104: 2919–2925.

Cozzio, A., E. Passegue, P.M. Ayton, H. Karsunky, M.L. Cleary, and I.L. Weissman. 2003. Similar MLL-associated leukemias arising from self-renewing stem cells and short-lived myeloid progenitors. Genes Dev 17: 3029–3035.

Daley, G.Q., R.A. Van Etten, and D. Baltimore. 1990. Induction of chronic myelogenous leukemia in mice by the P210bcr/abl gene of the Philadelphia chromosome. Science 247: 824–830.

Davi, F., C. Gocke, S. Smith, and J. Sklar. 1996. Lymphocytic progenitor cell origin and clonal evolution of human B-lineage acute lymphoblastic leukemia. Blood 88: 609–621.

de Guzman, C.G., A.J. Warren, Z. Zhang, L. Gartland, P. Erickson, H. Drabkin, S.W. Hiebert, and C.A. Klug. 2002. Hematopoietic stem cell expansion and distinct myeloid developmental abnormalities in a murine model of the AML1-ETO translocation. Mol Cell Biol 22: 5506–5517.

Dick, J.E. 2008. Stem cell concepts renew cancer research. Blood 112: 4793–4807.

DiMartino, J.F., P.M. Ayton, E.H. Chen, C.C. Naftzger, B.D. Young, and M.L. Cleary. 2002. The AF10 leucine zipper is required for leukemic transformation of myeloid progenitors by MLL-AF10. Blood 99: 3780–3785.

Essers, M.A., S. Offner, W.E. Blanco-Bose, Z. Waibler, U. Kalinke, M.A. Duchosal, and A. Trumpp. 2009. IFNalpha activates dormant haematopoietic stem cells *in vivo*. Nature 458: 904–908.

Estey, E.H., P.F. Thall, S. Pierce, J. Cortes, M. Beran, H. Kantarjian, M.J. Keating, M. Andreeff, and E. Freireich. 1999. Randomized phase II study of fludarabine + cytosine arabinoside + idarubicin +/- all-trans retinoic acid +/- granulocyte colony-stimulating factor in poor prognosis newly diagnosed acute myeloid leukemia and myelodysplastic syndrome. Blood 93: 2478–2484.

Fathi, A.T., and B.A. Chabner. 2011. FLT3 inhibition as therapy in acute myeloid leukemia: a record of trials and tribulations. Oncologist 16: 1162–1174.

Feinstein, E., G. Cimino, R.P. Gale, and E. Canaani. 1992. Initiation and progression of chronic myelogenous leukemia. Leukemia 6 Suppl 1: 37–43.

Fialkow, P.J., J.W. Singer, W.H. Raskind, J.W. Adamson, R.J. Jacobson, I.D. Bernstein, L.W. Dow, V. Najfeld, and R. Veith. 1987. Clonal development, stem-cell differentiation, and clinical remissions in acute nonlymphocytic leukemia. N Engl J Med 317: 468–473.

Gal, H., N. Amariglio, L. Trakhtenbrot, J. Jacob-Hirsh, O. Margalit, A. Avigdor, A. Nagler, S. Tavor, L. Ein-Dor, T. Lapidot, E. Domany, G. Rechavi, and D. Givol. 2006. Gene expression profiles of AML derived stem cells; similarity to hematopoietic stem cells. Leukemia 20: 2147–2154.

Galan-Caridad, J.M., S. Harel, T.L. Arenzana, Z.E. Hou, F.K. Doetsch, L.A. Mirny, and B. Reizis. 2007. Zfx controls the self-renewal of embryonic and hematopoietic stem cells. Cell 129: 345–357.

Gillette, J.M., A. Larochelle, C.E. Dunbar, and J. Lippincott-Schwartz. 2009. Intercellular transfer to signalling endosomes regulates an *ex vivo* bone marrow niche. Nat Cell Biol 11: 303–311.

Goldman, J., and M. Gordon. 2006. Why do chronic myelogenous leukemia stem cells survive allogeneic stem cell transplantation or imatinib: does it really matter? Leuk Lymphoma 47: 1–7.

Goodell, M.A., K. Brose, G. Paradis, A.S. Conner, and R.C. Mulligan. 1996. Isolation and functional properties of murine hematopoietic stem cells that are replicating *in vivo*. J Exp Med 183: 1797–1806.

Graw, R.G., Jr., C.D. Buckner, J. Whang-Peng, B.G. Leventhal, G. Kruger, C. Berard, and E.S. Henderson. 1970. Complication of bone-marrow transplantation. Graft-versus-host disease resulting from chronic-myelogenous-leukaemia leucocyte transfusions. Lancet 2: 338–341.

Green, E., C.M. McConville, J.E. Powell, J.R. Mann, P.J. Darbyshire, A.M. Taylor, and T. Stankovic. 1998. Clonal diversity of Ig and T-cell-receptor gene rearrangements identifies a subset of childhood B-precursor acute lymphoblastic leukemia with increased risk of relapse. Blood 92: 952–958.

Griffin, J.D., and B. Lowenberg. 1986. Clonogenic cells in acute myeloblastic leukemia. Blood 68: 1185–1195.

Guibal, F.C., M. Alberich-Jorda, H. Hirai, A. Ebralidze, E. Levantini, A. Di Ruscio, P. Zhang, B.A. Santana-Lemos, D. Neuberg, A.J. Wagers, E.M. Rego, and D.G. Tenen. 2009. Identification of a myeloid committed progenitor as the cancer-initiating cell in acute promyelocytic leukemia. Blood 114: 5415–5425.

Guzman, M.L., and C.T. Jordan. 2005. Feverfew: weeding out the root of leukaemia. Expert Opin Biol Ther 5: 1147–1152.

Guzman, M.L., X. Li, C.A. Corbett, R.M. Rossi, T. Bushnell, J.L. Liesveld, J. Hebert, F. Young, and C.T. Jordan. 2007. Rapid and selective death of leukemia stem and progenitor cells induced by the compound 4-benzyl, 2-methyl, 1,2,4-thiadiazolidine, 3,5 dione (TDZD-8). Blood 110: 4436–4444.

Guzman, M.L., S.J. Neering, D. Upchurch, B. Grimes, D.S. Howard, D.A. Rizzieri, S.M. Luger, and C.T. Jordan. 2001. Nuclear factor-kappaB is constitutively activated in primitive human acute myelogenous leukemia cells. Blood 98: 2301–2307.

Guzman, M.L., C.F. Swiderski, D.S. Howard, B.A. Grimes, R.M. Rossi, S.J. Szilvassy, and C.T. Jordan. 2002. Preferential induction of apoptosis for primary human leukemic stem cells. Proc Natl Acad Sci USA 99: 16220–16225.

Hawley, R.G., A. Ramezani, and T.S. Hawley. 2006. Hematopoietic stem cells. Methods Enzymol 419: 149–179.

Hendrix, M.J., E.A. Seftor, R.E. Seftor, J. Kasemeier-Kulesa, P.M. Kulesa, and L.M. Postovit. 2007. Reprogramming metastatic tumour cells with embryonic microenvironments. Nat Rev Cancer 7: 246–255.

Hess, D.A., T.E. Meyerrose, L. Wirthlin, T.P. Craft, P.E. Herrbrich, M.H. Creer, and J.A. Nolta. 2004. Functional characterization of highly purified human hematopoietic repopulating cells isolated according to aldehyde dehydrogenase activity. Blood 104: 1648–1655.

Heuser, M., L.M. Sly, B. Argiropoulos, F. Kuchenbauer, C. Lai, A. Weng, M. Leung, G. Lin, C. Brookes, S. Fung, P.J. Valk, R. Delwel, B. Lowenberg, G. Krystal, and R.K. Humphries. 2009. Modeling the functional heterogeneity of leukemia stem cells: role of STAT5 in leukemia stem cell self-renewal. Blood 114: 3983–3993.

Hogge, D.E., M. Feuring-Buske, B. Gerhard, and A.E. Frankel. 2004. The efficacy of diphtheria-growth factor fusion proteins is enhanced by co-administration of cytosine arabinoside in an immunodeficient mouse model of human acute myeloid leukemia. Leuk Res 28: 1221–1226.

Hope, K.J., L. Jin, and J.E. Dick. 2004. Acute myeloid leukemia originates from a hierarchy of leukemic stem cell classes that differ in self-renewal capacity. Nat Immunol 5: 738–743.

Horn, S., U. Bergholz, M. Jucker, J.A. McCubrey, L. Trumper, C. Stocking, and J. Basecke. 2008. Mutations in the catalytic subunit of class IA PI3K confer leukemogenic potential to hematopoietic cells. Oncogene 27: 4096–4106.

Huntly, B.J., H. Shigematsu, K. Deguchi, B.H. Lee, S. Mizuno, N. Duclos, R. Rowan, S. Amaral, D. Curley, I.R. Williams, K. Akashi, and D.G. Gilliland. 2004. MOZ-TIF2, but not BCR-ABL, confers properties of leukemic stem cells to committed murine hematopoietic progenitors. Cancer Cell 6: 587–596.

Ishikawa, F., S. Yoshida, Y. Saito, A. Hijikata, H. Kitamura, S. Tanaka, R. Nakamura, T. Tanaka, H. Tomiyama, N. Saito, M. Fukata, T. Miyamoto, B. Lyons, K. Ohshima, N. Uchida, S. Taniguchi, O. Ohara, K. Akashi, M. Harada, and L.D. Shultz. 2007. Chemotherapy-resistant human AML stem cells home to and engraft within the bone-marrow endosteal region. Nat Biotechnol 25: 1315–1321.

Jamieson, C.H., L.E. Ailles, S.J. Dylla, M. Muijtjens, C. Jones, J.L. Zehnder, J. Gotlib, K. Li, M.G. Manz, A. Keating, C.L. Sawyers, and I.L. Weissman. 2004. Granulocyte-macrophage progenitors as candidate leukemic stem cells in blast-crisis CML. N Engl J Med 351: 657–667.

Jin, L., K.J. Hope, Q. Zhai, F. Smadja-Joffe, and J.E. Dick. 2006. Targeting of CD44 eradicates human acute myeloid leukemic stem cells. Nat Med 12: 1167–1174.

Jin, L., E.M. Lee, H.S. Ramshaw, S.J. Busfield, A.G. Peoppl, L. Wilkinson, M.A. Guthridge, D. Thomas, E.F. Barry, A. Boyd, D.P. Gearing, G. Vairo, A.F. Lopez, J.E. Dick, and R.B. Lock. 2009. Monoclonal antibody-mediated targeting of CD123, IL-3 receptor alpha chain, eliminates human acute myeloid leukemic stem cells. Cell Stem Cell 5: 31–42.

Jones, R.J., J.P. Barber, M.S. Vala, M.I. Collector, S.H. Kaufmann, S.M. Ludeman, O.M. Colvin, and J. Hilton. 1995. Assessment of aldehyde dehydrogenase in viable cells. Blood 85: 2742–2746.

Jones, R.J., M.I. Collector, J.P. Barber, M.S. Vala, M.J. Fackler, W.S. May, C.A. Griffin, A.L. Hawkins, B.A. Zehnbauer, J. Hilton, O.M. Colvin, and S.J. Sharkis. 1996. Characterization of mouse lymphohematopoietic stem cells lacking spleen colony-forming activity. Blood 88: 487–491.

Jordan, C.T. 2008. Can we finally target the leukemic stem cells? Best Pract Res Clin Haematol 21: 615–620.

Jordan, C.T., and M.L. Guzman. 2004. Mechanisms controlling pathogenesis and survival of leukemic stem cells. Oncogene 23: 7178–7187.

Jordan, C.T., D. Upchurch, S.J. Szilvassy, M.L. Guzman, D.S. Howard, A.L. Pettigrew, T. Meyerrose, R. Rossi, B. Grimes, D.A. Rizzieri, S.M. Luger, and G.L. Phillips. 2000. The interleukin-3 receptor alpha chain is a unique marker for human acute myelogenous leukemia stem cells. Leukemia 14: 1777–1784.

Ju, Z., and K.L. Rudolph. 2006. Telomeres and telomerase in stem cells during aging and disease. Genome Dyn 1: 84–103.

Kania, G., D. Corbeil, J. Fuchs, K.V. Tarasov, P. Blyszczuk, W.B. Huttner, K.R. Boheler, and A.M. Wobus. 2005. Somatic stem cell marker prominin-1/CD133 is expressed in embryonic stem cell-derived progenitors. Stem Cells 23: 791–804.

Kelly, P.N., A. Dakic, J.M. Adams, S.L. Nutt, and A. Strasser. 2007. Tumor growth need not be driven by rare cancer stem cells. Science 317: 337.

Kiel, M.J., O.H. Yilmaz, T. Iwashita, C. Terhorst, and S.J. Morrison. 2005. SLAM family receptors distinguish hematopoietic stem and progenitor cells and reveal endothelial niches for stem cells. Cell 121: 1109–1121.

Kohl, T.M., C. Hellinger, F. Ahmed, C. Buske, W. Hiddemann, S.K. Bohlander, and K. Spiekermann. 2007. BH3 mimetic ABT-737 neutralizes resistance to FLT3 inhibitor treatment mediated by FLT3-independent expression of BCL2 in primary AML blasts. Leukemia 21: 1763–1772.

Konopleva, M., R. Contractor, T. Tsao, I. Samudio, P.P. Ruvolo, S. Kitada, X. Deng, D. Zhai, Y.X. Shi, T. Sneed, M. Verhaegen, M. Soengas, V.R. Ruvolo, T. McQueen, W.D. Schober, J.C. Watt, T. Jiffar, X. Ling, F.C. Marini, D. Harris, M. Dietrich, Z. Estrov, J. McCubrey, W.S. May, J.C. Reed, and M. Andreeff. 2006. Mechanisms of apoptosis sensitivity and resistance to the BH3 mimetic ABT-737 in acute myeloid leukemia. Cancer Cell 10: 375–388.

Kopp, H.G., S.T. Avecilla, A.T. Hooper, and S. Rafii. 2005. The bone marrow vascular niche: home of HSC differentiation and mobilization. Physiology (Bethesda) 20: 349–356.

Krause, D.S., and R.A. Van Etten. 2007. Right on target: eradicating leukemic stem cells. Trends Mol Med 13: 470–481.

Krivtsov, A.V., D. Twomey, Z. Feng, M.C. Stubbs, Y. Wang, J. Faber, J.E. Levine, J. Wang, W.C. Hahn, D.G. Gilliland, T.R. Golub, and S.A. Armstrong. 2006. Transformation

from committed progenitor to leukaemia stem cell initiated by MLL-AF9. Nature 442: 818–822.
Lane, S.W., D.T. Scadden, and D.G. Gilliland. 2009. The leukemic stem cell niche: current concepts and therapeutic opportunities. Blood 114: 1150–1157.
Langlands, K., J.I. Craig, R.S. Anthony, and A.C. Parker. 1993. Clonal selection in acute lymphoblastic leukaemia demonstrated by polymerase chain reaction analysis of immunoglobulin heavy chain and T-cell receptor delta chain rearrangements. Leukemia 7: 1066–1070.
Lapidot, T., and O. Kollet. 2002. The essential roles of the chemokine SDF-1 and its receptor CXCR4 in human stem cell homing and repopulation of transplanted immune-deficient NOD/SCID and NOD/SCID/B2m(null) mice. Leukemia 16: 1992–2003.
Lapidot, T., C. Sirard, J. Vormoor, B. Murdoch, T. Hoang, J. Caceres-Cortes, M. Minden, B. Paterson, M.A. Caligiuri, and J.E. Dick. 1994. A cell initiating human acute myeloid leukaemia after transplantation into SCID mice. Nature 367: 645–648.
Lecuyer, E., and T. Hoang. 2004. SCL: from the origin of hematopoiesis to stem cells and leukemia. Exp Hematol 32: 11–24.
Lessard, J., and G. Sauvageau. 2003. Bmi-1 determines the proliferative capacity of normal and leukaemic stem cells. Nature 423: 255–260.
Levin, R.H., J. Whang, J.H. Tjio, P.P. Carbone, E. Frei, 3rd, and E.J. Freireich. 1963. Persistent Mitosis of Transfused Homologous Leukocytes in Children Receiving Antileukemic Therapy. Science 142: 1305–1311.
Lewis, I.D., L.A. McDiarmid, L.M. Samels, L.B. To, and T.P. Hughes. 1998. Establishment of a reproducible model of chronic-phase chronic myeloid leukemia in NOD/SCID mice using blood-derived mononuclear or CD34+ cells. Blood 91: 630–640.
Li, H.Y., F.R. Appelbaum, C.L. Willman, R.A. Zager, and D.E. Banker. 2003. Cholesterol-modulating agents kill acute myeloid leukemia cells and sensitize them to therapeutics by blocking adaptive cholesterol responses. Blood 101: 3628–3634.
Lowenberg, B., W. van Putten, M. Theobald, J. Gmur, L. Verdonck, P. Sonneveld, M. Fey, H. Schouten, G. de Greef, A. Ferrant, T. Kovacsovics, A. Gratwohl, S. Daenen, P. Huijgens, and M. Boogaerts. 2003. Effect of priming with granulocyte colony-stimulating factor on the outcome of chemotherapy for acute myeloid leukemia. N Engl J Med 349: 743–752.
Mahon, F.X., D. Rea, J. Guilhot, F. Guilhot, F. Huguet, F. Nicolini, L. Legros, A. Charbonnier, A. Guerci, B. Varet, G. Etienne, J. Reiffers, and P. Rousselot. 2010. Discontinuation of imatinib in patients with chronic myeloid leukaemia who have maintained complete molecular remission for at least 2 years: the prospective, multicentre Stop Imatinib (STIM) trial. Lancet Oncol 11: 1029–1035.
Martelli, A.M., C. Evangelisti, F. Chiarini, C. Grimaldi, A. Cappellini, A. Ognibene, and J.A. McCubrey. 2010. The emerging role of the phosphatidylinositol 3-kinase/Akt/mammalian target of rapamycin signaling network in normal myelopoiesis and leukemogenesis. Biochim Biophys Acta 1803: 991–1002.
Matsunaga, T., N. Takemoto, T. Sato, R. Takimoto, I. Tanaka, A. Fujimi, T. Akiyama, H. Kuroda, Y. Kawano, M. Kobune, J. Kato, Y. Hirayama, S. Sakamaki, K. Kohda, K. Miyake, and Y. Niitsu. 2003. Interaction between leukemic-cell VLA-4 and stromal fibronectin is a decisive factor for minimal residual disease of acute myelogenous leukemia. Nat Med 9: 1158–1165.
Maximow, A. 1909. Der lymphozyte als gemeinsame stmmzelle der vershiedenen blutelemente in der embryonalen entwicklung nd im poetfetalen leben der saugetiere. Demonstrationsvortag, gehalten in der asserodentlichen sizung der Berliner Hamatologoschen Gesellshaftam.
McCormack, M.P., L.F. Young, S. Vasudevan, C.A. de Graaf, R. Codrington, T.H. Rabbitts, S.M. Jane, and D.J. Curtis. 2010. The Lmo2 oncogene initiates leukemia in mice by inducing thymocyte self-renewal. Science 327: 879–883.
McCulloch, E.A. 1983. Stem cells in normal and leukemic hemopoiesis (Henry Stratton Lecture, 1982). Blood 62: 1–13.

Milligan, D.W., K. Wheatley, T. Littlewood, J.I. Craig, and A.K. Burnett. 2006. Fludarabine and cytosine are less effective than standard ADE chemotherapy in high-risk acute myeloid leukemia, and addition of G-CSF and ATRA are not beneficial: results of the MRC AML-HR randomized trial. Blood 107: 4614–4622.

Misaghian, N., G. Ligresti, L.S. Steelman, F.E. Bertrand, J. Basecke, M. Libra, F. Nicoletti, F. Stivala, M. Milella, A. Tafuri, M. Cervello, A.M. Martelli, and J.A. McCubrey. 2009. Targeting the leukemic stem cell: the Holy Grail of leukemia therapy. Leukemia 23: 25–42.

Miyamoto, K., T. Miyamoto, R. Kato, A. Yoshimura, N. Motoyama, and T. Suda. 2008. FoxO3a regulates hematopoietic homeostasis through a negative feedback pathway in conditions of stress or aging. Blood 112: 4485–4493.

Mohle, R., M. Schittenhelm, C. Failenschmid, F. Bautz, K. Kratz-Albers, H. Serve, W. Brugger, and L. Kanz. 2000. Functional response of leukaemic blasts to stromal cell-derived factor-1 correlates with preferential expression of the chemokine receptor CXCR4 in acute myelomonocytic and lymphoblastic leukaemia. Br J Haematol 110: 563–572.

Morrison, S.J., N. Uchida, and I.L. Weissman. 1995. The biology of hematopoietic stem cells. Annu Rev Cell Dev Biol 11: 35–71.

Mortensen, B.T., P.O. Jensen, N. Helledie, P.O. Iversen, E. Ralfkiaer, J.K. Larsen, and M.T. Madsen. 1998. Changing bone marrow micro-environment during development of acute myeloid leukaemia in rats. Br J Haematol 102: 458–464.

Moshaver, B., A. van Rhenen, A. Kelder, M. van der Pol, M. Terwijn, C. Bachas, A.H. Westra, G.J. Ossenkoppele, S. Zweegman, and G.J. Schuurhuis. 2008a. Identification of a small subpopulation of candidate leukemia-initiating cells in the side population of patients with acute myeloid leukemia. Stem Cells 26: 3059–3067.

Moshaver, B., A. van Rhenen, A. Kelder, M. van der Pol, M. Terwijn, C. Bachas, A.H. Westra, G.J. Ossenkoppele, S. Zweegman, and G.J. Schuurhuis. 2008b. Identification of a Small Subpopulation of Candidate Leukemia Initiating Cells in the Side Population (SP) of Patients with Acute Myeloid Leukemia. Stem Cells.

Mumm, J.S., E.H. Schroeter, M.T. Saxena, A. Griesemer, X. Tian, D.J. Pan, W.J. Ray, and R. Kopan. 2000. A ligand-induced extracellular cleavage regulates gamma-secretase-like proteolytic activation of Notch1. Mol Cell 5: 197–206.

Naylor, C.S., E. Jaworska, K. Branson, M.J. Embleton, and R. Chopra. 2005. Side population/ ABCG2-positive cells represent a heterogeneous group of haemopoietic cells: implications for the use of adult stem cells in transplantation and plasticity protocols. Bone Marrow Transplant 35: 353–360.

Neering, S.J., T. Bushnell, S. Sozer, J. Ashton, R.M. Rossi, P.Y. Wang, D.R. Bell, D. Heinrich, A. Bottaro, and C.T. Jordan. 2007. Leukemia stem cells in a genetically defined murine model of blast-crisis CML. Blood 110: 2578–2585.

Notta, F., S. Doulatov, E. Laurenti, A. Poeppl, I. Jurisica, and J.E. Dick. 2011a. Isolation of single human hematopoietic stem cells capable of long-term multilineage engraftment. Science 333: 218–221.

Notta, F., C.G. Mullighan, J.C. Wang, A. Poeppl, S. Doulatov, L.A. Phillips, J. Ma, M.D. Minden, J.R. Downing, and J.E. Dick. 2011b. Evolution of human BCR-ABL1 lymphoblastic leukaemia-initiating cells. Nature 469: 362–367.

Ogawa, M., P.N. Porter, and T. Nakahata. 1983. Renewal and commitment to differentiation of hemopoietic stem cells (an interpretive review). Blood 61: 823–829.

Pappenheim, A. 1917. Prinzipen der neuren morphologichschen haematozytologie mach zytogenetischer grundlage. Folia Haematol: 91–101.

Passegue, E., C.H. Jamieson, L.E. Ailles, and I.L. Weissman. 2003. Normal and leukemic hematopoiesis: are leukemias a stem cell disorder or a reacquisition of stem cell characteristics? Proc Natl Acad Sci USA 100 Suppl 1: 11842–11849.

Passegue, E., W. Jochum, M. Schorpp-Kistner, U. Mohle-Steinlein, and E.F. Wagner. 2001. Chronic myeloid leukemia with increased granulocyte progenitors in mice lacking junB expression in the myeloid lineage. Cell 104: 21–32.

Passegue, E., E.F. Wagner, and I.L. Weissman. 2004. JunB deficiency leads to a myeloproliferative disorder arising from hematopoietic stem cells. Cell 119: 431–443.

Pear, W.S., J.P. Miller, L. Xu, J.C. Pui, B. Soffer, R.C. Quackenbush, A.M. Pendergast, R. Bronson, J.C. Aster, M.L. Scott, and D. Baltimore. 1998. Efficient and rapid induction of a chronic myelogenous leukemia-like myeloproliferative disease in mice receiving P210 bcr/abl-transduced bone marrow. Blood 92: 3780–3792.

Pearce, D.J., D.C. Taussig, and D. Bonnet. 2006. Implications of the expression of myeloid markers on normal and leukemic stem cells. Cell Cycle 5: 271–273.

Peng, C., Y. Chen, D. Li, and S. Li. 2010. Role of Pten in leukemia stem cells. Oncotarget 1: 156–160.

Perry, J.M., and L. Li. 2007. Disrupting the stem cell niche: good seeds in bad soil. Cell 129: 1045–1047.

Petzer, A.L., C.J. Eaves, P.M. Lansdorp, L. Ponchio, M.J. Barnett, and A.C. Eaves. 1996. Characterization of primitive subpopulations of normal and leukemic cells present in the blood of patients with newly diagnosed as well as established chronic myeloid leukemia. Blood 88: 2162–2171.

Recher, C., O. Beyne-Rauzy, C. Demur, G. Chicanne, C. Dos Santos, V.M. Mas, D. Benzaquen, G. Laurent, F. Huguet, and B. Payrastre. 2005. Antileukemic activity of rapamycin in acute myeloid leukemia. Blood 105: 2527–2534.

Redner, R.L. 2010. Why doesn't imatinib cure chronic myeloid leukemia? Oncologist 15: 182–186.

Rice, K.N., and C.H. Jamieson. 2010. Molecular pathways to CML stem cells. Int J Hematol 91: 748–752.

Rombouts, E.J., B. Pavic, B. Lowenberg, and R.E. Ploemacher. 2004. Relation between CXCR-4 expression, Flt3 mutations, and unfavorable prognosis of adult acute myeloid leukemia. Blood 104: 550–557.

Saito, Y., N. Uchida, S. Tanaka, N. Suzuki, M. Tomizawa-Murasawa, A. Sone, Y. Najima, S. Takagi, Y. Aoki, A. Wake, S. Taniguchi, L.D. Shultz, and F. Ishikawa. 2010. Induction of cell cycle entry eliminates human leukemia stem cells in a mouse model of AML. Nat Biotechnol 28: 275–280.

Savona, M., and M. Talpaz. 2008. Getting to the stem of chronic myeloid leukaemia. Nat Rev Cancer 8: 341–350.

Schofield, R. 1978. The relationship between the spleen colony-forming cell and the haemopoietic stem cell. Blood Cells 4: 7–25.

Simonsson, B., H. Hjorth-Hansen, O.W. Bjerrum, and K. Porkka. 2011. Interferon alpha for treatment of chronic myeloid leukemia. Curr Drug Targets 12: 420–428.

Sirard, C., T. Lapidot, J. Vormoor, J.D. Cashman, M. Doedens, B. Murdoch, N. Jamal, H. Messner, L. Addey, M. Minden, P. Laraya, A. Keating, A. Eaves, P.M. Lansdorp, C.J. Eaves, and J.E. Dick. 1996. Normal and leukemic SCID-repopulating cells (SRC) coexist in the bone marrow and peripheral blood from CML patients in chronic phase, whereas leukemic SRC are detected in blast crisis. Blood 87: 1539–1548.

Sloma, I., X. Jiang, A.C. Eaves, and C.J. Eaves. 2010. Insights into the stem cells of chronic myeloid leukemia. Leukemia 24: 1823–1833.

Smith, L.L., J. Yeung, B.B. Zeisig, N. Popov, I. Huijbers, J. Barnes, A.J. Wilson, E. Taskesen, R. Delwel, J. Gil, M. Van Lohuizen, and C.W. So. 2011. Functional crosstalk between Bmi1 and MLL/Hoxa9 axis in establishment of normal hematopoietic and leukemic stem cells. Cell Stem Cell 8: 649–662.

Sperr, W.R., S. Florian, A.W. Hauswirth, and P. Valent. 2005. CD 33 as a target of therapy in acute myeloid leukemia: current status and future perspectives. Leuk Lymphoma 46: 1115–1120.

Stirewalt, D.L., F.R. Appelbaum, C.L. Willman, R.A. Zager, and D.E. Banker. 2003. Mevastatin can increase toxicity in primary AMLs exposed to standard therapeutic agents, but statin efficacy is not simply associated with ras hotspot mutations or overexpression. Leuk Res 27: 133–145.

Storms, R.W., M.A. Goodell, A. Fisher, R.C. Mulligan, and C. Smith 2000. Hoechst dye efflux reveals a novel CD7(+)CD34(-) lymphoid progenitor in human umbilical cord blood. Blood 96: 2125–2133.

Storms, R.W., A.P. Trujillo, J.B. Springer, L. Shah, O.M. Colvin, S.M. Ludeman, and C. Smith. 1999. Isolation of primitive human hematopoietic progenitors on the basis of aldehyde dehydrogenase activity. Proc Natl Acad Sci USA 96: 9118–9123.

Talpaz, M., Z. Estrov, H. Kantarjian, S. Ku, A. Foteh, and R. Kurzrock. 1994. Persistence of dormant leukemic progenitors during interferon-induced remission in chronic myelogenous leukemia. Analysis by polymerase chain reaction of individual colonies. J Clin Invest. 94: 1383–1389.

Taussig, D.C., F. Miraki-Moud, F. Anjos-Afonso, D.J. Pearce, K. Allen, C. Ridler, D. Lillington, H. Oakervee, J. Cavenagh, S.G. Agrawal, T.A. Lister, J.G. Gribben, and D. Bonnet. 2008. Anti-CD38 antibody-mediated clearance of human repopulating cells masks the heterogeneity of leukemia-initiating cells. Blood 112: 568–575.

Taussig, D.C., D.J. Pearce, C. Simpson, A.Z. Rohatiner, T.A. Lister, G. Kelly, J.L. Luongo, G.A. Danet-Desnoyers, and D. Bonnet. 2005. Hematopoietic stem cells express multiple myeloid markers: implications for the origin and targeted therapy of acute myeloid leukemia. Blood 106: 4086–4092.

Taussig, D.C., J. Vargaftig, F. Miraki-Moud, E. Griessinger, K. Sharrock, T. Luke, D. Lillington, H. Oakervee, J. Cavenagh, S.G. Agrawal, T.A. Lister, J.G. Gribben, and D. Bonnet. 2010. Leukemia-initiating cells from some acute myeloid leukemia patients with mutated nucleophosmin reside in the CD34(-) fraction. Blood 115: 1976–1984.

Terpstra, W., A. Prins, R.E. Ploemacher, B.W. Wognum, G. Wagemaker, B. Lowenberg, and J.J. Wielenga. 1996. Long-term leukemia-initiating capacity of a CD34-subpopulation of acute myeloid leukemia. Blood 87: 2187–2194.

Till, J.E., and C.E. Mc. 1961. A direct measurement of the radiation sensitivity of normal mouse bone marrow cells. Radiat Res 14: 213–222.

Tsimberidou, A.M., F.J. Giles, E. Estey, S. O'Brien, M.J. Keating, and H.M. Kantarjian. 2006. The role of gemtuzumab ozogamicin in acute leukaemia therapy. Br J Haematol 132: 398–409.

Tycko, B., J. Ritz, S. Sallan, and J. Sklar. 1992. Changing antigen receptor gene rearrangements in a case of early pre-B cell leukemia: evidence for a tumor progenitor cell with stem cell features and implications for monitoring residual disease. Blood 79: 481–488.

Udomsakdi, C., C.J. Eaves, P.M. Lansdorp, and A.C. Eaves. 1992. Phenotypic heterogeneity of primitive leukemic hematopoietic cells in patients with chronic myeloid leukemia. Blood 80: 2522–2530.

van Rhenen, A., B. Moshaver, A. Kelder, N. Feller, A.W. Nieuwint, S. Zweegman, G.J. Ossenkoppele, and G.J. Schuurhuis. 2007a. Aberrant marker expression patterns on the CD34+CD38- stem cell compartment in acute myeloid leukemia allows to distinguish the malignant from the normal stem cell compartment both at diagnosis and in remission. Leukemia 21: 1700–1707.

van Rhenen, A., G.A. van Dongen, A. Kelder, E.J. Rombouts, N. Feller, B. Moshaver, M. Stigter-van Walsum, S. Zweegman, G.J. Ossenkoppele, and G. Jan Schuurhuis. 2007b. The novel AML stem cell associated antigen CLL-1 aids in discrimination between normal and leukemic stem cells. Blood 110: 2659–2666.

Verstegen, M.M., P.B. van Hennik, W. Terpstra, C. van den Bos, J.J. Wielenga, N. van Rooijen, R.E. Ploemacher, G. Wagemaker, and A.W. Wognum. 1998. Transplantation of human umbilical cord blood cells in macrophage-depleted SCID mice: evidence for accessory cell involvement in expansion of immature CD34+CD38- cells. Blood 91: 1966–1976.

Virchow. 1858. Cellular PathologyBerlin.

Wang, J.C., T. Lapidot, J.D. Cashman, M. Doedens, L. Addy, D.R. Sutherland, R. Nayar, P. Laraya, M. Minden, A. Keating, A.C. Eaves, C.J. Eaves, and J.E. Dick. 1998. High level engraftment of NOD/SCID mice by primitive normal and leukemic hematopoietic cells from patients with chronic myeloid leukemia in chronic phase. Blood 91: 2406–2414.

Wang, Y., A.V. Krivtsov, A.U. Sinha, T.E. North, W. Goessling, Z. Feng, L.I. Zon, and S.A. Armstrong. 2010. The Wnt/beta-catenin pathway is required for the development of leukemia stem cells in AML. Science 327: 1650–1653.

Wei, J., M. Wunderlich, C. Fox, S. Alvarez, J.C. Cigudosa, J.S. Wilhelm, Y. Zheng, J.A. Cancelas, Y. Gu, M. Jansen, J.F. Dimartino, and J.C. Mulloy. 2008. Microenvironment determines lineage fate in a human model of MLL-AF9 leukemia. Cancer Cell 13: 483–495.

Weng, A.P., Y. Nam, M.S. Wolfe, W.S. Pear, J.D. Griffin, S.C. Blacklow, and J.C. Aster. 2003. Growth suppression of pre-T acute lymphoblastic leukemia cells by inhibition of notch signaling. Mol Cell Biol 23: 655–664.

Wojiski, S., F.C. Guibal, T. Kindler, B.H. Lee, J.L. Jesneck, A. Fabian, D.G. Tenen, and D.G. Gilliland. 2009. PML-RARalpha initiates leukemia by conferring properties of self-renewal to committed promyelocytic progenitors. Leukemia 23: 1462–1471.

Wu, J., W.W. Wong, F. Khosravi, M.D. Minden, and L.Z. Penn. 2004. Blocking the Raf/MEK/ERK pathway sensitizes acute myelogenous leukemia cells to lovastatin-induced apoptosis. Cancer Res 64: 6461–6468.

Xu, Q., S.E. Simpson, T.J. Scialla, A. Bagg, and M. Carroll. 2003. Survival of acute myeloid leukemia cells requires PI3 kinase activation. Blood 102: 972–980.

Xu, Q., J.E. Thompson, and M. Carroll. 2005. mTOR regulates cell survival after etoposide treatment in primary AML cells. Blood 106: 4261–4268.

Yilmaz, O.H., R. Valdez, B.K. Theisen, W. Guo, D.O. Ferguson, H. Wu, and S.J. Morrison. 2006. Pten dependence distinguishes haematopoietic stem cells from leukaemia-initiating cells. Nature 441: 475–482.

Zhang, J., J.C. Grindley, T. Yin, S. Jayasinghe, X.C. He, J.T. Ross, J.S. Haug, D. Rupp, K.S. Porter-Westpfahl, L.M. Wiedemann, H. Wu, and L. Li. 2006. PTEN maintains haematopoietic stem cells and acts in lineage choice and leukaemia prevention. Nature 441: 518–522.

Zhang, J., C. Niu, L. Ye, H. Huang, X. He, W.G. Tong, J. Ross, J. Haug, T. Johnson, J.Q. Feng, S. Harris, L.M. Wiedemann, Y. Mishina, and L. Li. 2003. Identification of the haematopoietic stem cell niche and control of the niche size. Nature 425: 836–841.

Zhao, X., S. Singh, C. Pardoux, J. Zhao, E.D. Hsi, A. Abo, and W. Korver. 2010. Targeting C-type lectin-like molecule-1 for antibody-mediated immunotherapy in acute myeloid leukemia. Haematologica 95: 71–78.

Zheng, X., A. Seshire, B. Ruster, G. Bug, T. Beissert, E. Puccetti, D. Hoelzer, R. Henschler, and M. Ruthardt. 2007. Arsenic but not all-trans retinoic acid overcomes the aberrant stem cell capacity of PML/RARalpha-positive leukemic stem cells. Haematologica 92: 323–331.

Zipori, D. 2004. The nature of stem cells: state rather than entity. Nat Rev Genet 5: 873–878.

Chapter 5

CD44 Integrates Signaling in Stem Cell Microenvironments[#]

Karin Williams, Premkumar Vummidi Giridhar and Susan Kasper[a,*]

Introduction

CD44 is a cell-surface glycoprotein which binds hyaluronic acid (HA) to mediate cell–cell and cell-matrix interactions, cell migration and tumor growth and progression. CD44 also binds other molecules including osteopontin, collagens, integrins and matrix metalloproteinases (MMPs) and heterodimerizes with c-Met and growth factor receptors including fibroblast growth factor receptor (FGFR). Because all of these interactions can occur within the stem cell niche, it is plausible that CD44 would be critical for mediating stem cell behavior.

Recent studies suggest that CD44 moderates stem cell maintenance, proliferation, and differentiation. Knocking down CD44 expression in $CD44^+$ primary breast cancer stem cells down-regulates stem cell markers, e.g., Myc and LEF1 (Pham et al. 2011), loss of the characteristic long G2/M phase, an increased S phase consistent with non-stem cell proliferation (Tirino et al. 2008), and cell differentiation (Pham et al. 2011). A surprising observation is that CD44-null mice are viable, fertile and appear to lack obvious morphological defects (Olaku et al. 2011). However, ICAM-1 expression could compensate in part for embryological loss of CD44 (Olaku et al. 2011). In the K5-CD44AS knock-down model, CD44 expression is

Department of Environmental Health, University of Cincinnati, Cincinnati, OH 45267, USA.
[a]E-mail: susan.kapser@uc.edu
*Corresponding author
[#]The authors have no conflicting financial interests.

specifically down-regulated in keratinocytes by antisense CD44 transgene expression driven by the tissue-specific promoter keratin 5. The K5-CD44AS mice exhibited fragile skin, decreased keratinocyte proliferation, and accumulation of HA in the epidermis, suggesting that ICAM did not appear to compensate for loss of CD44 expression (Kaya et al. 1997). These studies demonstrate that regulation of CD44 expression occurs in a developmental and temporal manner.

Here we review the genomic and protein structure of CD44 and its variants in addition to the mechanisms by which they integrate signaling in several stem cell microenvironments, including normal stem cell, CSC and (pre)metastatic niches.

Embryonic, Adult and Cancer Stem Cells and Their Niches

Stem cells are capable of going through numerous cycles of cell division while maintaining an undifferentiated state. Embryonic stem (ES) cells isolated from the inner cell mass of the blastocyst are pluripotent; that is, they can give rise to all three germ layers (ectoderm, endoderm, and mesoderm) and differentiate into nearly all cells derived from any of the three germ layers. ES cells are typically used to generate transgenic mouse models.

Adult stem cells share the properties of self-renewal and differentiation potential with ES cells; however they are multipotent, being able to give rise to a more limited number of cell lineages. Hematopoietic stem cells give rise to several blood cell lineages and account for successful bone marrow transplants. In addition, adult stem cells are important in tissue maintenance and repair. A subpopulation of adult stem cells residing at the base of an intestinal crypt give rise to transit amplifying cells which expand through multiple rounds of cell division (Bach et al. 2000). These cells will continue to differentiate as they migrate along the inner surface of the crypt until they reach the villus as mature epithelial cells. A rapid turnover rate of approximately 5 days is required to replace the spent cells of the small intestine and colon throughout its lifetime (Bach et al. 2000).

When cancer cells acquire the properties of self-renewal and differentiation potential, they are considered cancer stem cells (CSCs) (Jordan et al. 2006). CSCs are thought to play critical roles in cancer progression, metastasis and the development of therapeutic-resistant disease (Jordan et al. 2006). The mechanism(s) by which CSCs acquire stem cell characteristics is still controversial. One theory proposes that adult stem cells acquire multiple mutations throughout their lifetime which activate oncogenes or inhibit tumor suppressor genes. The resulting CSCs promote tumor growth and potentiate metastatic spread despite therapeutic intervention (Jordan et al. 2006). A second hypothesis states that cancer cells dedifferentiate and thereby reactivate part of the stem cell program (Chaffer et al. 2011, Daley 2008). These cells acquire the stem cell property of self-renewal and the potential to differentiate into the various tumor cell populations of a given cancer.

Embryonic, adult and cancer stem cells reside in highly specialized microenvironments or "niches" located in specific anatomic locations (Sneddon and Werb 2007). The niche provides a microenvironment of physiological signals which direct stem cell fate and where stem cells are guarded from excessive "noise" or signals which could disrupt fate decisions. Factors within the niche mediate stem cell viability and self-renewal processes as well as provide signals for differentiation in a developmental and temporal manner. They arise from: a) interactions between stem cells and extracellular matrix (ECM) components mediated by adhesion molecules present on the stem cell (Ohlstein and Spradling 2006, Scadden 2006), b) interactions between stem cells and other intrinsic niche factors including growth factors, cytokines, pH, oxygen tension and ionic strength (e.g., Ca^{2+} concentration) (Jang and Sharkis 2007, Nakamura et al. 2010), c) cell–cell interactions between neighboring stem cells (Parekkadan et al. 2008), and d) interactions between stem cells and adjacent differentiated cells (Sato et al. 2011, Voog et al. 2008).

Identification and characterization of stem cell niches has been hindered by lack of niche-specific markers. The most basic niche described to date contains stem cells and ECM (Scadden 2006). *Drosophila* posterior hind gut, stem cells are separated from surrounding muscle cells by a basement membrane, suggesting that the basement membrane itself constitutes part of the niche (Ohlstein and Spradling 2006). In the *Drosophila* male gonad, the stem cell niche is comprised of germ line and somatic stem cells surrounding the apical hub which is a cluster of somatic cells required for stem cell self-renewal and maintenance. The somatic stem cells give rise to both apical hub and somatic cyst cell lineages (Voog et al. 2008). Similarly, Lgr5 (leucine-rich repeat containing G-protein-coupled receptor 5) stem cells in mammalian small intestine give rise to Paneth cells which become niche cells for Lgr5 stem cells (Sato et al. 2011). Therefore, stem cells are capable of giving rise to their own niche cells which in turn, continue to support stem cell self-renewal and maintenance.

Isolating stem cells remains a major challenge. Many stem cell markers including the classical Yamanaka factors are primarily transcription factors and therefore not readily available for fluorescence-activated cell sorting (FACS) or magnetic cell sorting (MACS) of stem and progenitor cells (Takahashi and Yamanaka 2006). The rodent-specific cell surface marker Sca1 is typically used to isolate rodent stem cells (Spangrude et al. 1988). However, most cell surface markers only enrich for stem cells. Therefore, they have been used in combination to isolate enriched populations of hematopoietic, solid tissue and cancer stem cells (Hock 2010). The cell surface protein CD44 is expressed on a number of stem cell types. However, CD44 is also expressed by a range of cells [reviewed by Underhill (Underhill 1992)], including squamous and glandular epithelium (Heider et al. 1993) and fibroblasts (Green et al. 1988). However, CD44 has become a marker of choice for FACS and MACS methods to enrich/isolate normal stem cells and CSCs from blood and tissues including colorectal (Dalerba et al. 2007, Du et al. 2008), breast (Al-Hajj et al. 2003, Ponti et al. 2005), pancreas (Li et al. 2007, Li et al. 2009), head and neck (Prince et al. 2007) and prostate (Collins et al. 2005, Patrawala et al. 2006, Rybak et al. 2011).

Summary of the Cellular Functions of CD44

CD44 exhibits a broad range of functions that can be subdivided into several categories.

1) CD44 is an ECM-interacting protein that binds HA, collagen, laminin, and fibronectin (Ishii et al. 1993, Ponta et al. 2003) via its extracellular domain. CD44 also interacts with growth factors, binding them via the heparin sulphate and sugar moieties present on CD44.
2) CD44 is a membrane-bound signaling protein that heterodimerizes with numerous membrane bound receptors including integrins and receptor tyrosine kinases (RTKs) (Ponta et al. 2003). In addition, ankyrin and ezrin family members [ezrin/radixin/moesin (ERM)] crosslink CD44 via its short intracellular tail with the actin-based cytoskeleton, providing a mechanism for anchoring signaling complexes into discreet clusters which promote dimerization and signaling.
3) CD44 is a transcription factor. Gamma-secretase induced cleavage of CD44 releases the CD44 intracellular domain (CD44-ICD). The CD44-ICD subsequently translocates into the nucleus were it binds STAT3 to enhance transcription from the c-myc and Twist1 promoters (Su et al. 2011) Transcriptional regulation of c-myc expression by CD44-ICD reprograms cancer cells to a more stem cell-like phenotype (Su et al. 2011).

Genomic Structure and Function Domains of CD44 and CD44 Variants

In humans, the highly conserved CD44 gene spans approximately 60 kilobases (kb) of genomic DNA on chromosome 11. The full-length CD44 gene is composed of 20 exons and 19 introns and undergoes extensive alternative splicing. Figure 1 summarizes the genomic structure of CD44.

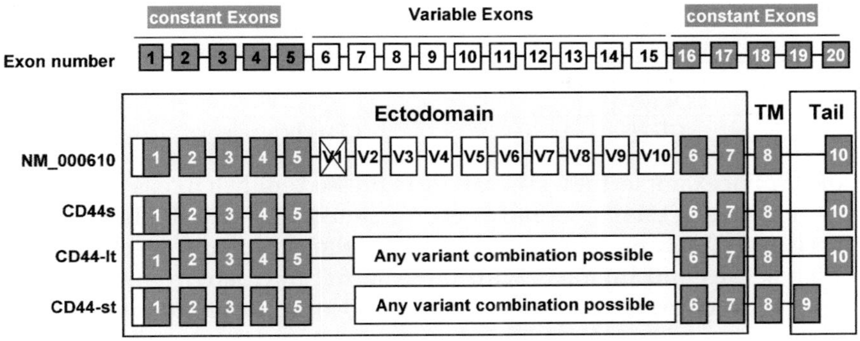

Fig. 1. Genomic structure of the human CD44 gene and its splice variants.

Top panel: The CD44 gene contains 20 exons and 19 introns. CD44s contains the constant exons 1–5 and 16–20 (blue boxes). This region is also included in all CD44 mRNAs and proteins. The variable exons 6 to 15 encode for v1 through v10 (white boxes). They are alternatively spliced to generate CD44 variant proteins containing one or multiple variable exons in tandem. An exception is human v1 which contains an in-frame termination codon and is therefore not included in any human CD44v molecules. In other species, however, CD44v could contain v1.

Bottom panel: Examples of alternatively spliced CD44 isoforms. Note that exon 19 is included in short-tail CD44v molecules whereas exon 20 is included in long-tail CD44v molecules.

Color image of this figure appears in the color plate section at the end of the book.

The extracellular domain of CD44 is encoded by the first five constant exons. The remaining constant exons express the non-variant stem (exon 16 and 17) and the transmembrane (TM) domain (exon 18). Exons 19 and 20 undergo alternative splicing, resulting in a short-tail (exon 19) or the more common long-tail (exon 20) isoform (Table 1). CD44 standard (CD44s) is generated from the constant exons (corresponding to exons 1–5) and 6–10 (corresponding to exons 16–18+20). All CD44 molecules contain the standard exons 1–5 with exception of CD44sol which is a soluble non membrane bound version of CD44.

CD44 variants arise through splicing events that insert additional exons into the CD44s molecule. These exons introduce a variable stalk region in the extracellular region juxtaposed to the TM domain. The variable exons are typically numbered V1–V10; however, they correspond to the genomic exons 6–15 (Fig. 1, open boxes). In humans, the V1 exon is not encoded due to an in-frame termination codon, however it is expressed in non-human mammals (Screaton et al. 1993). The V1 to V10 exons are alternatively spliced and incorporated into the variable region either singly or in combination. While hundreds of combinations could potentially be generated, only 12 isoforms are commonly observed. The modern nomenclature system uses CD44vx-x, where "x" represents a variable exon number ranging from v1 to v10, to indicate the combination of variable exons expressed in a given

Table 1. Commonly occurring CD44v isoforms.

Exon Usage	Historical Nomenclature
CD44v3–10	CD44 Keratinocyte
CD44v8–10	CD44 Epithelial/CD44E
CD44s	CD44 hematopoietic/CD44H

isoform. For example, the CD44 variant expressed epithelial cells historically designated as CD44E, has now been renamed CD44v8–10 (Table 1). All CD44s and CD44v proteins contain an extracellular domain (or ectodomain) which senses and interacts with the external microenvironment; a TM domain which provides a platform for CD44/cofactor/adaptor protein interactions; and an intracellular domain (ICD) that is cleaved to generate CD44-ICD which then translocates into the nucleus to regulate transcription (Table 2).

The extracellular domain of CD44 is the most diverse region of the molecule. Alternative splicing of the variant exons enlarges the stem domain dramatically while use of alternative splice acceptor sites generates a soluble version of CD44 [CD44RC, CD44sol] (Chiu et al. 1999). In addition, the CD44 ectodomain can be cleaved from the TM domain by either membrane-type 1 matrix metalloprotease (MT1-MMP) (Kajita et al. 2001) or by MMP-9 (Chetty et al. 2012). Cleaved CD44 ectodomain is then capable of binding MMP-9 or MT1-MMP (Marrero-Diaz et al. 2009, Peng et al. 2007), with CD44:MMP-9 localizing to the invadopodia (Lagarrigue et al. 2010) and CD44:MT1-MMP localizing at the trailing edge of invading tumor cells (Marrero-Diaz et al. 2009). In addition, the extracellular domain can also be cleaved by two membrane bound members of the disintegrin and metalloprotease super family ADAM10 and ADAM17 (Nagano et al. 2004, Nakamura et al. 2004). The cleavage sites for these proteases are not well mapped; however they are located within the CD44 stem domain encoded by non-variant exons 16 and 17 at the membrane-proximal region of the ectodomain. This cleavage is required for regulating CD44:ECM interactions during cell migration (Okamoto et al. 1999). In addition, cleaved CD44 ectodomain interacts with membrane bound CD44 and inhibits attachment to ECM resulting in decreased tumor growth (Sy et al. 1992, Yu et al. 1997). Taken together, these observations indicate that the cleaved CD44 ectodomain is required to regulate specific CD44 activities.

In contrast, membrane-bound CD44s/CD44v transduces ligand signaling through coregulator interactions within the intracellular domain and recruitment of the actin cytoskeleton signaling cascade (Bourguignon 2008). A percentage of membrane-bound CD44 is present within detergent resistant, lipid rich domains or 'lipid rafts' in the plasma membrane (Neame et al. 1995, Perschl et al. 1995). These lipid rafts compartmentalize signaling molecules into signal transduction hubs (Janes et al. 2000, Korade and

Table 2. Summary of the functions of the extracellular, transmembrane and cytoplasmic domains of CD44.

		CD44 Domain Function
Extracellular Domain	**Globular domain**	The extracellular domain of CD44 has a globular conformation which is formed by inter-chain disulphide bonds between six highly conserved cysteine residues. (Banerji et al. 1998) The hyaluronan binding domain [(HARD) aa20-169] contains a consensus HA binding motif [**B**(X_7)**B**] in the amino terminus of CD44 ^{38}**K**NGRYSIS**R**TEAADLC**K**^{54} (Teriete et al. 2004, Yang et al. 1994) Within the HARD domain is the LINK domain [aa 32–120] that contains HA and other glycosaminoglycan binding sites and shares similarities with other HA binding proteins (Banerji et al. 1998, Day & Prestwich 2002) The basic motif extends beyond the LINK domain from aa 150–158. It contains basic amino acids that are thought to contribute to HA binding through as yet unknown mechanisms. This has been called into question by the crystal structure (Ponta et al. 2003, Screaton et al. 1992)
	stem	The stem structure of CD44s consists of 46 aa encoded by the invariant exons 5, 6 and 7 O-glycosylation of the stem forms a stalk-like structure that is accessible for interactions with proteases (Ponta et al. 2003) The variable region encoded by exons v1–v10 resides in the stem domain Insertion of the variant exons can add up to 381 additional amino acids
TM		The hydrophobic 23aa single pass transmembrane domain is highly conserved between species. The cysteine residue located within the TM participates in CD44 oligomerization (Liu & Sy 1997) A γ-secretase cleavage site at is located at Ala^{280}/Ile^{287}. Cleavage results in release of the CD44-ICD (Brown et al. 2000, Thorne et al. 2004)
Cytoplasmic tail		Alternative splicing of exon 19 and 20 generates the CD44 "short tail' and 'long-tail' isoforms 'short-tail' [CD44-st (exon18/19)] isoform contains only three intracellular amino acids the ^{292}RRR^{295} encoded by exon 18. Exon 19 contributes an in-frame termination signal (Brown et al. 2000, Thorne et al. 2004) 'long-tail' [CD44-It (exon 18/20)] isoform contains a 72 aa intracellular tail (Lee et al. 2009, Matteucci et al. 2009) A nuclear localization signal (NLS) ^{292}RRRCGQKK^{300} is located immediately after the TM (Legg & Isacke 1998) ERM binding sites are located in the long tail at positions ^{292}RRR^{295}/^{289}KKK^{300} (Ponta et al. 2003) A second site of actin crosslinking with CD44 occurs via the 14 $\alpha\alpha$ ankyrin binding domain ^{304}NSGNGAVEDRKPSGL^{318} situated immediately adjacent to the ERM binding site (Ponta et al. 2003) The terminal end of CD44 contains a consensus PZD-domain binding motif at ^{358}KIGV^{362} leukemia-associated RhoGEF binds the PZD domain in response to HA binding (Bourguignon et al. 2006, Thorne et al. 2004)

Kenworthy 2008, Pike 2009, Schmitz and Grandl 2008). These hubs serve as organizing centers for the assembly of receptor and signaling complexes (Simons and Toomre 2000). Within the signaling hubs, CD44 heterodimerizes with many diverse signaling molecules located to mediate cell adhesion, cell function, transcription and migration (Wakahara et al. 2005).

CD44 also exists in soluble, long-tail and short-tail isoforms. The soluble CD44RC isoform enhances binding of membrane-bound CD44 to HA, thereby increasing cell adhesion as well as CD44 clustering and signaling (Chiu et al. 1999). The long-tail CD44/CD44v isoform terminates in exon 20. Gamma-secretase cleavage at position Ala^{280}/Ile^{287} liberates CD44-ICD from the TM domain (Brown et al. 2000, Thorne et al. 2004). CD44-ICD translocates into the nucleus (Nagano and Saya 2004) to mediate transcription through TPA-responsive elements (Okamoto et al. 2001). CD44-ICD target genes include the CD44 gene itself (Lee et al. 2009, Nagano and Saya 2004, Okamoto et al. 2001) and cyclin D1(Okamoto et al. 2001, Takahashi et al. 2011). Thus, CD44 promotes cell growth via cyclin D1 and increased CD44 signaling through a positive feedback mechanism.

The short-tail CD44 (CD44-st) isoform terminates in exon 19 and therefore does not contain the protein domains required for cytoskeletal interactions and signaling (Mori et al. 2008, Thorne et al. 2004). CD44-st isoforms are generally present in very low abundance compared with the long-tail CD44 isoform (Goldstein and Butcher 1990). In chondrocytes, CD44-st acts as a dominant negative receptor to block CD44:CD44 interactions (Jiang et al. 2001) while inhibition of CD44-st enhances CD44-mediated HA internalization (Jiang et al. 2001). Furthermore, inhibiting CD44-st reduces pericellular matrix formation, suggesting that CD44-st expression levels regulate chondrocyte microenvironment.

CD44 Interactions with ECM Components

CD44 not only regulates matrix formation but also interacts with components in the ECM. ECM is composed of polysaccharides and fibrous proteins and functions as a compressible scaffold to regulate rigidity of tissues and provide a matrix for cell adherence (Parekh and Weaver 2009). A common ECM component is heparan sulfate, a linear polysaccharide which associates with proteins to form heparan sulfate proteoglycan. Heparan sulfate acts as a catayst to crosslink and enmesh heparan sulfate proteoglycan into the ECM (Bernfield et al. 1999, Taipale and Keski-Oja 1997). Through its heparan sulfate moiety, CD44v3 is capable of binding HGF (van der Voort et al. 1999), FGF2 (Grimme et al. 1999) and HB-EGF (Bennett et al. 1995). Upon binding, the growth factors are presented to their cognate receptors which then heterodimerize with CD44 to induce signaling (Table 3).

Table 3. Interaction of extracellular growth factors with the CD44 extracellular domain to promote stem cell and niche cell survival and proliferation.

Ligand	CD44:Growth Factor Receptor Activity
FGF-2	Required for ES cell culture (Levenstein et al. 2006)
TGF-β	Latent TGFβ is cleaved to active TGFβ by a CD44/MMP-9 cell surface complex and promote tumor survival (Yu & Stamenkovic 2004) TGFβ are key morphogens of gut endodermal stem cells
PDGF	Induces stromal niche cells to release morphogens responsible for CSC growth (Mishra et al. 2005)
HGF	Activates cMet via CD44v6 (Okuda et al. 2011) cMet is a marker of pancreatic stem cells (Li et al. 2011) c-Met expression/function correlates stem cell phenotype in glioblastoma neurospheres (Li et al. 2011)
HB-EGF	Expands mesenchymal stem cells pool in bone marrow (Krampera et al. 2005)
VEGF	controls survival of hematopoietic stem cells, colony formation and *in vivo* repopulation (Gerber et al. 2002)
IGF	pluripotent and self-renewal properties of human ES cells are depend on IGF-2 produces from ES derived niche support cells (Bendall et al. 2007)

Growth factor precursors in ECM are processed by CD44 activity and released by protease activity for rapid cell signaling without *de novo* synthesis. CD44 heparan sulfate proteoglycan recruits MMP-7 and heparin-binding epidermal growth factor precursor (pro-HB-EGF) to form a CD44/MMP-7/pro-HB-EGF/ErbB4 complex on the surface of cells including tumor cells, uterine and mammary gland epithelium, and uterine smooth muscle (Yu et al. 2002). The HB-EGF precursor is processed by MMP-7 proteolytic activity into the mature HB-EGF molecule which then activates the ErbB4 receptor to promote cell survival (Yu et al. 2002). Interestingly, co-immunoprecipitation (co-IP) of MMP-7 and pro-HB-EGF only occurred in cells expressing CD44v3–10 and mature HB-EGF was not observed in cells expressing CD44v7–10, implying that v3 is required for pro-HB-EGF processing and ErbB4 activation (Yu et al. 2002). ECM also contains other large structural proteins that interact with each other either through protein:protein and/or polysaccharide interactions. Of these, CD44 binds laminin, fibronectin, heparan sulfate, and collagen (Ishii et al. 1993) (Table 4). Collectively, these observations indicate that interaction of CD44 and CD44v with ECM components play key roles in growth factor activation.

CD44:HA Interactions in Preparing a Stem Cell/CSC Niche

The principal ligand for CD44 is hyaluronic acid (HA, hyaluronan), a polysaccharide composed of alternative residues of D-glucuronic acid and N-acetylglucosamine (Preston and Sherman 2011). It is synthesized on the inner surface of the plasma membrane by a family of membrane

Table 4. Extracellular matrix components that bind and promote CD44 activity.

Ligand	CD44:ligand-mediated Activity	Niche
HA	HA synthesized by the ES cells is present in regions of low cell density and surrounding clusters of epithelial cells (Ishii et al. 1993)	embryoid body
	HA maintains ES cells in an undifferentiated state during ES culture (Gerecht et al. 2007, Kim et al. 2010, Pasquinelli et al. 2008)	HA-ES cell culture
	chondrogenic differentiation of encapsulated hMSCs (Chung & Burdick 2009)	HA-Hydrogel culture
		HA-Hydrogel culture 'cartilage niche'
OPN	demonstrated to contribute to hematopoietic stem cell regulation (Stier et al. 2005)	hematopoietic stem cell niche
α4β1 integrin/ VLA-4	α4β1/CD44 mediates adhesion of hematological progenitor cells (Verfaillie et al. 1994) Triggers G-protein signaling in mesenchymal stem cells increasing cell adhesion & transmigration through the vascular endothelium (Thankamony & Sackstein)	Fibronectin and bone marrow environment
Galectin9	stimulates CD44/BMPcomplexes inducing osteoblast differentiation (Tanikawa et al. 2010)	Bone

bound hyaluronan synthases (HAS1, HAS2, and HAS3) which repeatedly add sugar residues to the growing HA chain. HA chain size is synthase dependent (Spicer and McDonald 1998) and ranges from 10^4 to 10^7 Daltons (Lapcik et al. 1998). HA exits the cell via an ABC-transporter which threads HA through a pore into the pericellular space (Schulz et al. 2007) where it forms a coat on the surface of the cell. HA either remains tethered to HAS or it is released into the pericellular space were it is captured by HA-binding receptors including CD44, CD44v, and RHAMM.

HA is found at high levels in the stem cell/CSC niche (Solis Mairim Alexandra et al. 2011) where it is produced by oocytes, stem cells and stem cell niche support cells (Grskovic et al. 2006, Schraufstatter et al. 2010, Shukla et al. 2010). Therefore, HA is frequently used to supplement cell culture medium and HA-hydrogels to maintain stem cells and human embryonic stem cells in a dedifferentiated or totipotent state respectively, while promoting cell growth (Gerecht et al. 2007, Kim et al. 2010, Pasquinelli et al. 2008). In contrast, adding growth factors to HA-hydrogels will induce differentiation of embryonic stem cells (Kim et al. 2010). CD44 transcript levels remain constant during differentiation of embryonic stem to embryoid bodies (Nairn et al. 2007), suggesting that its expression is required for both processes. Collectively, these observations imply that CD44:HA interactions maintain a stem cell phenotype while CD44:growth factor receptor heterodimerization, e.g., CD44:cMet, promote differentiation (Matzke et al. 2007). The role of the CD44v isoforms in maintaining stem cells in a dedifferentiated state or in promoting differentiation has yet to been analyzed.

CSCs also synthesize HA which have the ability to recruit tumor-associated macrophages (TAMs) into the CSC niche (Okuda et al. 2011). Upon interacting with CSCs, TAMs secrete platelet derived growth factor isoform BB (PDGF-BB) (Okuda et al. 2011). PDGF-BB binds to its receptor located on adjacent stromal cells and recruits these cells into the niche. The stromal cells then contribute to the niche by secreting FGFs (Okuda et al. 2011). Thus CSCs create an integrated HA-rich microenvironment which promotes CSC self-renewal and expansion as well as increased tumor growth.

Mechanisms of CD44-mediated Signaling in Stem Cells and CSCs

The mechanisms by which CD44 regulates intracellular signaling in stem cells or CSCs remain unclear. The following mechanisms have been identified in mature cells and may pertain to stem cells/CSCs.

CD44 regulates a range of cellular processes in addition to growth factor receptor signaling. The high levels of HA dictate that CD44:HA interactions be strictly regulated. Regulation of HA binding is therefore controlled by the differential glycosylation of CD44 which either activate or inhibit ligand recognition (Katoh et al. 1995). Inflammation processes have also been reported to enhance HA:CD44 interactions (Lesley et al. 2004, Milner and Day 2003, Zhuo et al. 2006, de La Motte et al. 1999, de la Motte et al. 2003). Once a threshold level of CD44 binding is achieved, HA binding increases with increasing CD44 expression (English et al. 1998).

The mechanism of 'inside-out' signaling occurs when a TM or cytoplasmic domain induces conformational or distributional changes in a receptor via actin cytoskeleton interactions. These interactions propagate signals to the external domain of the receptor and alter ligand recognition or binding affinity (Lesley et al. 1992, Lesley and Hyman 1998). An example of this mechanism is CD44 recognition of HA induced by CD44 clustering (Lesley et al. 1995). CD44 clustering caused by CD44/cytoskeleton interactions may enhance avidity of CD44 for HA (Liu et al. 1998) or induce conformational changes in CD44 that increase its affinity for HA (Banerji et al. 2007). The CD44 cytoplasmic tail interacts with the actin cytoskeleton to increase HA binding (Bourguignon et al. 1993, Lesley et al. 1992) through phosphorylation of ERM proteins (Tsukita et al. 1994), while disruption of this interaction reduces HA binding and decreases CD44 signaling (Brown et al. 2005, Liu et al. 1996). Upon HA binding, CD44 homodimerization occurs through formation of disulphide bonds between cysteine residues in the TM domain of adjacent receptors (Liu and Sy 1996). In summary, the CD44 extracellular, TM and cytoplasmic domains are all required to initiate signaling (Lokeshwar et al. 1994, Pure et al. 1995).

CD44:Steroid Hormone Receptor Interactions

CD44 is both a co-repressor and co-activator of steroid hormone receptor activity. In primary dermal fibroblasts, CD44 binds the nuclear receptor co-repressor SMRT (Skandalis et al. 2010) to suppress estrogen receptor (ER)-mediated transcription (Varlakhanova et al. 2010). In SK-OV-3.ipl human ovarian tumor cells however, CD44 signaling activates ERα/Elk-1-mediated transcription (Bourguignon et al. 2005).

Steroid hormones function as differentiation signals and therefore it is not surprising that stem cells do not express these receptors. Prostate stem cells and prostate CSC-like cells exhibit an androgen receptor (AR)-negative phenotype (Gu et al. 2007, Kasper 2009). Induction of AR expression in a developmental and temporal manner is required for prostate organogenesis and glandular development (Cunha et al. 2004). Similarly, mammary stem cells and mammary CSCs do not express ERα or the progesterone receptor (PR) (Asselin-Labat et al. 2006, Russo and Russo 2011, Sleeman et al. 2007). The context-specific expression of steroid hormone receptors may sustain CSC self-renewal and proliferation, promote tumor growth and contribute to the development of therapeutic-resistant disease.

CD44:Integrin Interactions

CD44 and integrins occupy the same stem cell niches, yet little is known on CD44:integrin interactions. Integrins are cell adhesion receptors that link extracellular adhesion molecules with the intracellular actin cytoskeleton (Delon and Brown 2007). There are 24 known integrin heterodimers consisting of one of 18 α subunits and one of 8 β subunits and their expression in stem cell populations ranges from human ES cells to hematopoietic, neural and mesenchymal adult stem cells (Prowse et al. 2011). In these cell types, integrin signaling induced by the local ECM extracellular matrix controls a diverse range of functions including cell–cell adhesion, growth factor receptor signaling as well as cell lineage specification, differentiation, survival, proliferation and migration. The role of integrins in stem cells has been reviewed in detail (Mafi et al. 2011, Prowse et al. 2011).

In human mesenchymal stem cells, CD44 and α4β1 heterodimerize to activate G-protein coupled signaling and increase cell adhesion to, and transmigration through, vascular endothelium (Thankamony and Sackstein 2011). In colorectal cancer initiating cells (CIC), association of CD44v6 with $\alpha 6\beta 4$ promotes binding to, and activation of, Src, ankyrin and ERM proteins and down-stream signaling pathways (Zöller and Jung 2011). These interactions promote CIC maintenance, cell migration, and apoptosis resistance (Zöller and Jung 2011).

Collectively, CD44 and integrins regulate cell-cell adhesion, growth factor receptor signaling, stem cell survival, proliferation and migration. Therefore, it is plausible that they could coordinately anchor stem cells/ CSCs in their niche as well as regulate stem cell/CSC fate and/or migration out of the niche.

Differential CD44:Growth Factor Receptor Interactions in Regulating Stem Cell and CSC Characteristics

CD44 interacts with a range of membrane bound receptors, including tyrosine kinases (RTK) (Table 5). In addition, CD44 acts as an anchored reservoir for growth factor receptor ligands, e.g., FGF2 and HGF, which activate their cognate receptors upon CD44:RTK heterodimerization (Couchman 2010). Many of these growth factors moderate stem cell function including self-renewal and stem cell fate. However their interactions with CD44 in these processes have not been investigated.

CD44v in CSC Cell Migration, Preparation of the (pre) Metastatic Niche, and Metastasis

The function of CD44v in stem cells and CSCs has remained understudied. This is, in part, due to pan CD44-specific antibodies which only recognize the extracellular non-variant domain of CD44 present in all CD44s and CD44v molecules. The importance of CD44v in tumor progression is well documented (Naor et al. 2008). Since CD44v and HA are components of the stem cell niche, they may play a fundamental roles in promoting tumor growth, metastatic disease, and resistance to therapy.

The inclusion of exon v6 imparts a novel function to CD44v6 in that the v6 region permits binding of HGF to CD44v6 (Jung et al. 2009, Zoller 2011). CD44v6 presents HGF to c-Met which activates c-Met signaling to promote metastasis (Orian-Rousseau et al. 2002, Recio and Merlino 2003). CD44v6 is required for secretion and/or assembly of an adhesive matrix containing CD44v6, c-Met, HGF, MMP13, uPA, urokinase-type plasminogen activator receptor and HAS3 (Jung et al. 2009). Knock down of CD44v6 [CD44v(kd)] results in production of a matrix which does not support metastatic growth of the highly metastatic rat adenocarcinoma cell line AMSL (Klingbeil et al. 2009). In these studies only 12.5% (3/24) rats injected with [CD44v(kd)] ASML cells exhibited visible lung metastasis, and at greatly reduced rates

Table 5. Receptor Tyrosine Kinase (RTK):CD44 heterodimerization mediates stem cell and CSC activities including transcription, cell differentiation and cell migration.

RTK	CD44:RTK Activity	Protein:Protein Interaction Assay
EGFR/ErbB1/HER1	associated with fibroblast to myofibroblast activation, migration and growth during normal wound healing (Meran et al. 2011) Upon TGF-beta stimulation co-localizes with CD44 and results in ERK1/2 phosphorylation and activation (Meran et al. 2011)	co-IP (Tsatas et al. 2002) Confocal microscopy (Casalino-Matsuda et al. 2009)
FGFR	chondrosarcoma HCS-2/8 cells containing CD44s and CD44v9 co-IP with FGFR. The interaction is inhibited by Bikunin (Wakahara et al. 2005).	co-IP (Wakahara et al. 2005)
VEGFR2	Activated by CD44v6 acting as co-receptor were they promote migration, spheroid sprouting, and tubule formation (Tremmel et al. 2009)	co-IP (Tremmel et al. 2009)
TGB-βRI	Co-localized with CD44 in MDA-MB-231 (Bourguignon et al. 2002)	co-IP (Bourguignon et al. 2002)
BMP-R2	Forms a complex with CD44 and leads to downstream Smad phosphorylation (Tanikawa et al. 2010)	co-IP (Tanikawa et al. 2010)
ErbB2/Her2/Neu	Forms complex with CD44, resulting in HA-induced kinase activity Activated by CD44/HA, ErbB2 phosphorylates beta-catenin which thereupon translocates into the nucleus, stimulating LEF-1/TCF-4 transcriptional activity (Bourguignon et al. 2007, Bourguignon et al. 1997)	co-IP (Tsatas et al. 2002) Confocal microscopy (Slomiany et al. 2009, Tsatas et al. 2002)
ErbB4/HER4	Phosphorylated by CD44v3/HB-EGF complex (Yu et al. 2002)	co-IP (Yu et al. 2002)
PBGFR	CD44 suppresses PDGFRβ signaling (Yu et al. 2002)	co-IP (Li et al. 2006)
cMet	CD44v6 is required for c-Met activation by HGF in carcinoma cells, and primary keratinocytes (Damm et al. 2010)	co-IP (Orian-Rousseau et al. 2002)

CD44, CD24 and ALDH1 expression are used to identify CSCc in breast cancer populations. Analysis of the CD44v isoforms CD44S, CD44v2–v10, CD44v3–v10 and CD44v8–v10 in several breast cancer cell lines determined that CD44v2–v10, CD44v3–v10, CD44v8–v10 expression increased in cells cultured as mammospheres. High CD44v8–v10 expression correlated with $EGFR^{+}/HER2^{-/low}$ status and a basal-like cell subtype, suggesting that EGFR signaling preferentially cooperated with CD44v8–v10 (Olsson et al. 2011). In another a subgroup of basal-like tumors, high CD44s expression correlated with high HER2 expression in addition to expression of the CSC marker ALDH1, suggesting that HER2 signaling preferentially cooperated with CD44s (Olsson et al. 2011). Stratification of patients by CD44 isoform gene expression data suggested that the group which expressed CD44s and CD44v8–v10 (Cluster D) demonstrated consistently lower overall survival over a 10 year period (Olsson et al. 2011). These observations suggest that CD44 isoform and growth factor receptor expression may be CSC subtype specific and predict the overall survival of cancer patients.

compared to wild type controls (Klingbeil et al. 2009). Furthermore, the [CD44v(kd)] matrix could not support homing to lymph nodes *in vivo* (Jung et al. 2009).

CD44v6 has been implicated in establishment of the cancer stem niche. Exosomes released from tumor cells diffuse to distant sites and subsequently interact with local fibroblasts to create a (pre)metastatic niche to which tumor cells can home (Jung et al. 2009). While exosomes themselves are CD44v6-independent, they require CD44v6 to activate c-Met and urokinase-type plasminogen activator receptor and assemble a soluble matrix. Then exosomes in cooperation with the soluble matrix promote leukocyte, stroma, and endothelial cell activation in the (pre)metastatic organ (Jung et al. 2009).

In CSCs, CD44v6 binds HGF and c-Met to activate MEK and ERK signaling and promote CSC migration and invasion (Orian-Rousseau et al. 2002, Zoller 2011). Expression of c-Met is required by stem and progenitor cells in embryogenesis, branching morphogenesis and wound healing (reviewed in (Boccaccio and Comoglio 2006). In mesenchymal cells, c-Met activation promotes chemotactic migration but not proliferation (Neuss et al. 2004). Similarly, CSCs use these mechanisms to promote metastasis. While specific antibodies to v3 and v6 antibodies are available, antibodies to identify and characterize other exon products in more complex CD44v, e.g., CD44v2–10 are not available. Even protein size is not conclusive since CD44 is extensively glycosylated. Therefore the characterization of most CD44v has remained limited (Table 6).

Table 6. Functional activity and biological significance of the CD44 variants (CD44v).

Variant	Biological Activity
CD44v1	Mouse and rat CD44v1 contains a 42αα variant sequence that encodes a WIWSWF domain which has weak homology with several unrelated proteins (Screaton et al. 1992) Human homologue of v1 contains an in-frame termination codon after αα17 (Screaton et al. 1993)
CD44v3	Contains the consensus addition site for Heparin Sulfate (HS) Binds PDGF, bFGF, TGF-β, VEGF and HGF (Greenfield et al. 1999, Houck et al. 1992, Mangrulkar et al. 1995, Rider 2006, Sakata et al. 1997) Binds αMβ2 integrin via its heparan sulfate moiety (Zen et al. 2009)
CD44v5	is involved in metastasis (Cheng & Sharp 2006) Exon v5 splicing is induced by Ras–Raf–MEK–ERK signaling (Cheng & Sharp 2006, Matter et al. 2002)
CD44v6	Transiently expressed during lymphocyte activation (Arch et al. 1992) Induces metastasis in a rat non tumorigenic pancreatic line (Gunthert et al. 1991) Forms a complex with cMet and HGF (Orian-Rousseau et al. 2002) Binds VEGF and VEGFR2 (Tremmel et al. 2009)
CD44v10	interacts with phospho-cMet and CD44s in lipid rafts following HGF treatment (Singleton et al. 2007)

Conclusions and Perspective

The stem cell niche is a unique microenvironment in which self-renewal and differentiation of stem cells and CSCs are moderated. CD44 regulates processes that impact the composition of the niche. CD44 is required for production, secretion and assembly of the cell matrix and loss of CD44 expression alters the composition of the matrix and changes CSC behavior. CD44 is expressed by stem cells and CSC where it interacts with a diverse range of RTKs, ECM components, HA, integrins, growth factors and other ligands sequestered in the ECM to regulate processes including cell–cell adhesion, growth factor receptor signaling, cell survival, proliferation and migration. Therefore, CD44 appears to be a master regulator of stem cell and CSC function.

Many fundamental questions still need to be explored. For example, why is CD44 so commonly expressed on stem cell populations? Which CD44-mediated mechanisms maintain stem cells in an undifferentiated state? Do CD44:RTK interactions direct cell lineage specification? Which mechanisms control CD44 isoform expression and are these mechanisms stem cell or CSC subtype specific? What is the function of variant exons 2 through 10 and could these, alone or in combination, serve as targets for cancer therapy?

The investigation of CD44v as targets for cancer therapy is currently underway. In a phase I clinical trial using bivatuzumab mertansine (CD44v6 antibodies linked to the cytotoxic drug mertansine), 10% of patients with head and neck squamous cell carcinoma (HNSCC) experienced regression of tumors (Orian-Rousseau 2010). However, the trial was terminated because of skin-related toxicities and one patient death (Orian-Rousseau 2010). Since CD44v frequently exhibit multiple tumor-promoting activities, there remains continued interest in this class of molecules. CD44v6 binds both HGF and VEGF, activating c-Met and VEGFR signaling and promoting metastatic spread and tumor angiogenesis respectively. If both of these signaling pathways were targeted in combination, it could result in blocking tumor blood supply and inhibiting metastasis. Mapping CD44 and CD44v structures and elucidating their mechanisms of action will need to be explored and optimized before CD44-based therapies could be used to eliminate metastatic disease.

Acknowledgements

Funding for this work was provided by the National Institute of Diabetes & Digestive & Kidney Diseases (R01 DK60957 to S.K.) and the Frances Williams Preston Laboratories of the T.J. Martell Foundation (to S.K.).

References

Al-Hajj, M., M.S. Wicha, and A. Benito-Hernandez. 2003. Prospective identification of tumorigenic breast cancer cells. Proc Natl Acad Sci USA 100: 3983–3988.

Asselin-Labat, M.L., M. Shackleton, and J. Stingl. 2006. Steroid hormone receptor status of mouse mammary stem cells. J Natl Cancer Inst 98: 1011–1014.

Bach, S.P., A.G. Renehan, and C.S. Potten. 2000. Stem cells: the intestinal stem cell as a paradigm. Carcinogenesis 21: 469–476.

Banerji, S., A.J. Wright, and M. Noble. 2007. Structures of the Cd44-hyaluronan complex provide insight into a fundamental carbohydrate-protein interaction. Nat Struct Mol Biol 14: 234–239.

Bennett, K.L., D.G. Jackson, and J.C. Simon. 1995. CD44 isoforms containing exon V3 are responsible for the presentation of heparin-binding growth factor. J Cell Biol 128: 687–698.

Bernfield, M., M. Gotte, and P.W. Park. 1999. Functions of cell surface heparan sulfate proteoglycans. Annu Rev Biochem 68: 729–777.

Boccaccio, C., and P.M. Comoglio. 2006. Invasive growth: a MET-driven genetic programme for cancer and stem cells. Nat Rev Cancer 6: 637–645.

Bourguignon, L.Y. 2008. Hyaluronan-mediated CD44 activation of RhoGTPase signaling and cytoskeleton function promotes tumor progression. Semin Cancer Biol 18: 251–259.

Bourguignon, L.Y., E. Gilad, and K. Rothman. 2005. Hyaluronan-CD44 interaction with IQGAP1 promotes Cdc42 and ERK signaling, leading to actin binding, Elk-1/estrogen receptor transcriptional activation, and ovarian cancer progression. J Biol Chem 280: 11961–11972.

Brown, K.L., D. Birkenhead, and J.C. Lai. 2005. Regulation of hyaluronan binding by F-actin and colocalization of CD44 and phosphorylated ezrin/radixin/moesin (ERM) proteins in myeloid cells. Exp Cell Res 303: 400–414.

Brown, M.S., J. Ye, and R.B. Rawson. 2000. Regulated intramembrane proteolysis: a control mechanism conserved from bacteria to humans. Cell 100: 391–398.

Chaffer, C.L., I. Brueckmann, and C. Scheel. 2011. Normal and neoplastic nonstem cells can spontaneously convert to a stem-like state. Proc Natl Acad Sci USA 108: 7950–7955.

Chetty, C., S.K. Vanamala, and C.S. Gondi. 2012. MMP-9 induces CD44 cleavage and CD44 mediated cell migration in glioblastoma xenograft cells. Cell Signal.

Chiu, R.K., C. Carpenito, and S.T. Dougherty. 1999. Identification and characterization of CD44RC, a novel alternatively spliced soluble CD44 isoform that can potentiate the hyaluronan binding activity of cell surface CD44. Neoplasia 1: 446–452.

Collins, A.T., P.A. Berry, and C. Hyde. 2005. Prospective identification of tumorigenic prostate cancer stem cells. Cancer Res 65: 10946–10951.

Couchman, J.R. 2010. Transmembrane signaling proteoglycans. Annu Rev Cell Dev Biol 26: 89–114.

Cunha, G.R., W. Ricke, and A. Thomson. 2004. Hormonal, cellular, and molecular regulation of normal and neoplastic prostatic development. J Steroid Biochem Mol Biol 92: 221–236.

Dalerba, P., S.J. Dylla, and I.K. Park. 2007. Phenotypic characterization of human colorectal cancer stem cells. Proc Natl Acad Sci USA 104: 10158–10163.

Daley, G.Q. 2008. Common themes of dedifferentiation in somatic cell reprogramming and cancer. Cold Spring Harb Symp Quant Biol 73: 171–174.

de La Motte, C.A., V.C. Hascall, and A. Calabro. 1999. Mononuclear leukocytes preferentially bind via CD44 to hyaluronan on human intestinal mucosal smooth muscle cells after virus infection or treatment with poly(I.C). J Biol Chem 274: 30747–30755.

de la Motte, C.A., V.C. Hascall, and J. Drazba. 2003. Mononuclear leukocytes bind to specific hyaluronan structures on colon mucosal smooth muscle cells treated with polyinosinic acid:polycytidylic acid: inter-alpha-trypsin inhibitor is crucial to structure and function. Am J Pathol 163: 121–133.

Delon, I., and N.H. Brown. 2007. Integrins and the actin cytoskeleton. Curr Opin Cell Biol 19: 43–50.

Du, L., H. Wang, and L. He. 2008. CD44 is of functional importance for colorectal cancer stem cells. Clin Cancer Res 14: 6751–6760.

English, N.M., J.F. Lesley, and R. Hyman. 1998. Site-specific de-N-glycosylation of CD44 can activate hyaluronan binding, and CD44 activation states show distinct threshold densities for hyaluronan binding. Cancer Res 58: 3736–3742.

Gerecht, S., J.A. Burdick, and L.S. Ferreira. 2007. Hyaluronic acid hydrogel for controlled self-renewal and differentiation of human embryonic stem cells. Proc Natl Acad Sci USA 104: 11298–11303.

Goldstein, L.A., and E.C. Butcher. 1990. Identification of mRNA that encodes an alternative form of H-CAM(CD44) in lymphoid and nonlymphoid tissues. Immunogenetics 32: 389–397.

Green, S.J., G. Tarone, and C.B. Underhill. 1988. Aggregation of macrophages and fibroblasts is inhibited by a monoclonal antibody to the hyaluronate receptor. Exp Cell Res 178: 224–232.

Grimme, H.U., C.C. Termeer, and K.L. Bennett. 1999. Colocalization of basic fibroblast growth factor and CD44 isoforms containing the variably spliced exon v3 (CD44v3) in normal skin and in epidermal skin cancers. Br J Dermatol 141: 824–832.

Grskovic, B., C. Pollaschek, and M.M. Mueller. 2006. Expression of hyaluronan synthase genes in umbilical cord blood stem/progenitor cells. Biochim Biophys Acta 1760: 890–895.

Gu, G., J. Yuan, and M. Wills. 2007. Prostate cancer cells with stem cell characteristics reconstitute the original human tumor *in vivo*. Cancer Res 67: 4807–4815.

Heider, K.H., M. Hofmann, and E. Hors. 1993. A human homologue of the rat metastasis-associated variant of CD44 is expressed in colorectal carcinomas and adenomatous polyps. J Cell Biol 120: 227–233.

Hock, H. 2010. Some hematopoietic stem cells are more equal than others. J Exp Med 207: 1127–1130.

Ishii, S., R. Ford, and P. Thomas. 1993. CD44 participates in the adhesion of human colorectal carcinoma cells to laminin and type IV collagen. Surg Oncol 2: 255–264.

Janes, P.W., S.C. Ley, A.I. Magee. 2000. The role of lipid rafts in T cell antigen receptor (TCR) signalling. Semin Immunol 12: 23–34.

Jang, Y.Y., and S.J. Sharkis. 2007. A low level of reactive oxygen species selects for primitive hematopoietic stem cells that may reside in the low-oxygenic niche. Blood 110: 3056–3063.

Jiang, H., C.B. Knudson, and W. Knudson. 2001. Antisense inhibition of CD44 tailless splice variant in human articular chondrocytes promotes hyaluronan internalization. Arthritis Rheum 44: 2599–2610.

Jordan, C.T., M.L. Guzman, and M. Noble. 2006. Cancer stem cells. N Engl J Med 355: 1253–1261.

Jung, T., D. Castellana, and P. Klingbeil. 2009. CD44v6 dependence of premetastatic niche preparation by exosomes. Neoplasia 11: 1093–1105.

Kajita, M., Y. Itoh, and T. Chiba. 2001. Membrane-type 1 matrix metalloproteinase cleaves CD44 and promotes cell migration. J Cell Biol 153: 893–904.

Kasper, S. 2009. Identification, characterization, and biological relevance of prostate cancer stem cells from clinical specimens. Urol Oncol 27: 301–303.

Katoh, S., Z. Zheng, and K. Oritani. 1995. Glycosylation of CD44 negatively regulates its recognition of hyaluronan. J Exp Med 182: 419–429.

Kaya, G., I. Rodriguez, and J.L. Jorcano. 1997. Selective suppression of CD44 in keratinocytes of mice bearing an antisense CD44 transgene driven by a tissue-specific promoter disrupts hyaluronate metabolism in the skin and impairs keratinocyte proliferation. Genes Dev 11: 996–1007.

Kim, J., I.S. Kim, and T.H. Cho. 2010. *In vivo* evaluation of MMP sensitive high-molecular weight HA-based hydrogels for bone tissue engineering. J Biomed Mater Res A 95: 673–681.

Klingbeil, P., R. Marhaba, and T. Jung. 2009. CD44 variant isoforms promote metastasis formation by a tumor cell-matrix cross-talk that supports adhesion and apoptosis resistance. Mol Cancer Res 7: 168–179.

Korade, Z., and A.K. Kenworthy. 2008. Lipid rafts, cholesterol, and the brain. Neuropharmacology 55: 1265–1273.

Lagarrigue, F., S. Dupuis-Coronas, and D. Ramel. 2010. Matrix metalloproteinase-9 is upregulated in nucleophosmin-anaplastic lymphoma kinase-positive anaplastic lymphomas and activated at the cell surface by the chaperone heat shock protein 90 to promote cell invasion. Cancer Res 70: 6978–6987.

Lapcik, L.Jr., L. Lapcik, and S. De Smedt. 1998. Hyaluronan: Preparation, Structure, Properties, and Applications. Chem Rev 98: 2663–2684.

Lee, J.L., M.J. Wang, and J.Y. Chen. 2009. Acetylation and activation of STAT3 mediated by nuclear translocation of CD44. J Cell Biol 185: 949–957.

Lesley, J., N. English, and A. Perschl. 1995. Variant cell lines selected for alterations in the function of the hyaluronan receptor CD44 show differences in glycosylation. J Exp Med 182: 431–437.

Lesley, J., I. Gal, and D.J. Mahoney. 2004. TSG-6 modulates the interaction between hyaluronan and cell surface CD44. J Biol Chem 279: 25745–25754.

Lesley, J., Q. He, and K. Miyake. 1992. Requirements for hyaluronic acid binding by CD44: a role for the cytoplasmic domain and activation by antibody. J Exp Med 175: 257–266.

Lesley, J., and R. Hyman. 1998. CD44 structure and function. Front Biosci 3: d616–630

Li, C., D.G. Heidt, and P. Dalerba. 2007. Identification of pancreatic cancer stem cells. Cancer Res 67: 1030–1037.

Li, C., C.J. Lee, and D.M. Simeone. 2009. Identification of human pancreatic cancer stem cells. Methods Mol Biol 568: 161–173.

Liu, D., and M.S. Sy. 1996. A cysteine residue located in the transmembrane domain of CD44 is important in binding of CD44 to hyaluronic acid. J Exp Med 183: 1987–1994.

Liu, D., D. Zhang, and H. Mori. 1996. Binding of CD44 to hyaluronic acid can be induced by multiple signals and requires the CD44 cytoplasmic domain. Cell Immunol 174: 73–83.

Lokeshwar, V.B., N. Fregien, and L.Y. Bourguignon. 1994. Ankyrin-binding domain of CD44(GP85) is required for the expression of hyaluronic acid-mediated adhesion function. J Cell Biol 126: 1099–1109.

Mafi, P., S. Hindocha, and R. Mafi. 2011. Adult mesenchymal stem cells and cell surface characterization—a systematic review of the literature. Open Orthop J 5: 253–260.

Marrero-Diaz, R., J.J. Bravo-Cordero, and D. Megias. 2009. Polarized MT1-MMP-CD44 interaction and CD44 cleavage during cell retraction reveal an essential role for MT1-MMP in CD44-mediated invasion. Cell Motil Cytoskeleton 66: 48–61.

Matzke, A., V. Sargsyan, and B. Holtmann. 2007. Haploinsufficiency of c-Met in cd44-/- mice identifies a collaboration of CD44 and c-Met *in vivo*. Mol Cell Biol 27: 8797–8806.

Milner, C.M., and A.J. Day. 2003. TSG-6: a multifunctional protein associated with inflammation. J Cell Sci 116: 1863–1873.

Mori, T., K. Kitano, and S. Terawaki. 2008. Structural basis for CD44 recognition by ERM proteins. J Biol Chem 283: 29602–29612.

Nagano, O., D. Murakami, and D. Hartmann. 2004. Cell-matrix interaction via CD44 is independently regulated by different metalloproteinases activated in response to extracellular Ca(2+) influx and PKC activation. J Cell Biol 165: 893–902.

Nagano, O., and H. Saya. 2004. Mechanism and biological significance of CD44 cleavage. Cancer Sci 95: 930–935.

Nairn, A.V., A. Kinoshita-Toyoda, and H. Toyoda. 2007. Glycomics of proteoglycan biosynthesis in murine embryonic stem cell differentiation. J Proteome Res 6: 4374–4387.

Nakamura, H., N. Suenaga, and K. Taniwaki. 2004. Constitutive and induced CD44 shedding by ADAM-like proteases and membrane-type 1 matrix metalloproteinase. Cancer Res 64: 876–882.

Nakamura, S., T. Matsumoto, and J. Sasaki. 2010. Effect of calcium ion concentrations on osteogenic differentiation and hematopoietic stem cell niche-related protein expression in osteoblasts. Tissue Eng Part A 16: 2467–2473.

Naor, D., S.B. Wallach-Dayan, and M.A. Zahalka. 2008. Involvement of CD44, a molecule with a thousand faces, in cancer dissemination. Semin Cancer Biol 18: 260–267.

Neame, S.J., C.R. Uff, and H. Sheikh. 1995. CD44 exhibits a cell type dependent interaction with triton X-100 insoluble, lipid rich, plasma membrane domains. J Cell Sci 108 (Pt 9): 3127–3135.

Neuss, S., E. Becher, and M. Woltje. 2004. Functional expression of HGF and HGF receptor/c-met in adult human mesenchymal stem cells suggests a role in cell mobilization, tissue repair, and wound healing. Stem Cells 22: 405–414.

Ohlstein, B., and A. Spradling. 2006. The adult Drosophila posterior midgut is maintained by pluripotent stem cells. Nature 439: 470–474.

Okamoto, I., Y. Kawano, and D. Murakami. 2001. Proteolytic release of CD44 intracellular domain and its role in the CD44 signaling pathway. J Cell Biol 155: 755–762.

Okamoto, I., Y. Kawano, and H. Tsuiki. 1999. CD44 cleavage induced by a membrane-associated metalloprotease plays a critical role in tumor cell migration. Oncogene 18: 1435–1446.

Okuda, H., A. Kobayashi, and B. Xia. 2011. Hyaluronan synthase HAS2 promotes tumor progression in bone by stimulating the interaction of breast cancer stem-like cells with macrophages and stromal cells. Cancer Res.

Olaku, V., A. Matzke, and C. Mitchell. 2011. c-Met recruits ICAM-1 as a coreceptor to compensate for the loss of CD44 in Cd44 null mice. Mol Biol Cell 22: 2777–2786.

Olsson, E., G. Honeth, and P.O. Bendahl. 2011. CD44 isoforms are heterogeneously expressed in breast cancer and correlate with tumor subtypes and cancer stem cell markers. BMC Cancer 11: 418.

Orian-Rousseau, V. 2010. CD44, a therapeutic target for metastasising tumours. Eur J Cancer 46: 1271–1277.

Orian-Rousseau, V., L. Chen, and J.P. Sleeman. 2002. CD44 is required for two consecutive steps in HGF/c-Met signaling. Genes Dev 16: 3074–3086.

Parekh, A., and A.M. Weaver. 2009. Regulation of cancer invasiveness by the physical extracellular matrix environment. Cell Adh Migr 3: 288–292.

Parekkadan, B., Y. Berdichevsky, and D. Irimia. 2008. Cell–cell interaction modulates neuroectodermal specification of embryonic stem cells. Neurosci Lett 438: 190–195.

Pasquinelli, G., C. Orrico, and L. Foroni. 2008. Mesenchymal stem cell interaction with a non-woven hyaluronan-based scaffold suitable for tissue repair. J Anat 213: 520–530.

Patrawala, L., T. Calhoun, and R. Schneider-Broussard. 2006. Highly purified CD44+ prostate cancer cells from xenograft human tumors are enriched in tumorigenic and metastatic progenitor cells. Oncogene 25: 1696–1708.

Peng, S.T., C.H. Su, and C.C. Kuo. 2007. CD44 crosslinking-mediated matrix metalloproteinase-9 relocation in breast tumor cells leads to enhanced metastasis. Int J Oncol 31: 1119–1126.

Perschl, A., J. Lesley, and N. English. 1995. Transmembrane domain of CD44 is required for its detergent insolubility in fibroblasts. J Cell Sci 108 (Pt 3): 1033–1041.

Pham, P.V., N.L. Phan, and N.T. Nguyen. 2011. Differentiation of breast cancer stem cells by knockdown of CD44: promising differentiation therapy. J Transl Med 9: 209.

Pike, L.J. 2009. The challenge of lipid rafts. J Lipid Res 50 Suppl: S323–328.

Ponta, H., L. Sherman, and P.A. Herrlich. 2003. CD44: from adhesion molecules to signalling regulators. Nat Rev Mol Cell Biol 4: 33–45.

Ponti, D., A. Costa, and N. Zaffaroni. 2005. Isolation and *in vitro* propagation of tumorigenic breast cancer cells with stem/progenitor cell properties. Cancer Res 65: 5506–5511.

Preston, M., and L.S. Sherman. 2011. Neural stem cell niches: roles for the hyaluronan-based extracellular matrix. Front Biosci (Schol Ed) 3: 1165–1179.

Prince, M.E., R. Sivanandan, and A. Kaczorowski. 2007. Identification of a subpopulation of cells with cancer stem cell properties in head and neck squamous cell carcinoma. Proc Natl Acad Sci USA 104: 973–978.

Prowse, A.B., F. Chong, and P.P. Gray. 2011. Stem cell integrins: implications for *ex vivo* culture and cellular therapies. Stem Cell Res 6: 1–12.

Pure, E., R.L. Camp, and D. Peritt. 1995. Defective phosphorylation and hyaluronate binding of CD44 with point mutations in the cytoplasmic domain. J Exp Med 181: 55–62.

Recio, J.A., and G. Merlino. 2003. Hepatocyte growth factor/scatter factor induces feedback up-regulation of CD44v6 in melanoma cells through Egr-1. Cancer Res 63: 1576–1582.

Russo, J., and I.H. Russo. 2011. The role of the Basal stem cell of the human breast in normal development and cancer. Adv Exp Med Biol 720: 121–134.

Rybak, A.P., L. He, and A. Kapoor. 2011. Characterization of sphere-propagating cells with stem-like properties from DU145 prostate cancer cells. Biochim Biophys Acta 1813: 683–694.

Sato, T., J.H. van Es, and H.J. Snippert. 2011. Paneth cells constitute the niche for Lgr5 stem cells in intestinal crypts. Nature 469: 415–418.

Scadden, D.T. 2006. The stem-cell niche as an entity of action. Nature 441: 1075–1079.

Schmitz, G., and M. Grandl. 2008. Update on lipid membrane microdomains. Curr Opin Clin Nutr Metab Care 11: 106–112.

Schraufstatter, I.U., N. Serobyan, and J. Loring. 2010. Hyaluronan is required for generation of hematopoietic cells during differentiation of human embryonic stem cells. J Stem Cells 5: 9–21.

Schulz, T., U. Schumacher, and P. Prehm. 2007. Hyaluronan export by the ABC transporter MRP5 and its modulation by intracellular cGMP. J Biol Chem 282: 20999–21004.

Screaton, G.R., M.V. Bell, and J.I. Bell. 1993. The identification of a new alternative exon with highly restricted tissue expression in transcripts encoding the mouse Pgp-1 (CD44) homing receptor. Comparison of all 10 variable exons between mouse, human, and rat. J Biol Chem 268: 12235–12238.

Shukla, S., R. Nair, and M.W. Rolle. 2010. Synthesis and organization of hyaluronan and versican by embryonic stem cells undergoing embryoid body differentiation. J Histochem Cytochem 58: 345–358.

Simons, K., and D. Toomre. 2000. Lipid rafts and signal transduction. Nat Rev Mol Cell Biol 1: 31–39.

Skandalis, S.S., I. Kozlova, and U. Engstrom. 2010. Proteomic identification of CD44 interacting proteins. IUBMB Life 62: 833–840.

Sleeman, K.E., H. Kendrick, and D. Robertson. 2007. Dissociation of estrogen receptor expression and *in vivo* stem cell activity in the mammary gland. J Cell Biol 176: 19–26.

Sneddon, J.B., and Z. Werb. 2007. Location, location, location: the cancer stem cell niche. Cell Stem Cell 1: 607–611.

Solis Mairim Alexandra, Y.-H. Chen, and T.Y. Wong. 2011. Hyaluronan Regulates Cell Behavior: A Potential Niche Matrix for Stem Cells. Biochemistry research international 2011.

Spangrude, G.J., S. Heimfeld, and I.L. Weissman. 1988. Purification and characterization of mouse hematopoietic stem cells. Science 241: 58–62.

Spicer, A.P., and J.A. McDonald. 1998. Characterization and molecular evolution of a vertebrate hyaluronan synthase gene family. J Biol Chem 273: 1923–1932.

Su, Y.J., H.M. Lai, and Y.W. Chang. 2011. Direct reprogramming of stem cell properties in colon cancer cells by CD44. Embo J 30: 3186–3199.

Sy, M.S., Y.J. Guo, and I. Stamenkovic. 1992. Inhibition of tumor growth *in vivo* with a soluble CD44-immunoglobulin fusion protein. J Exp Med 176: 623–627.

Taipale, J., and J. Keski-Oja. 1997. Growth factors in the extracellular matrix. Faseb J 11: 51–59.

Takahashi, K., and S. Yamanaka. 2006. Induction of pluripotent stem cells from mouse embryonic and adult fibroblast cultures by defined factors. Cell 126: 663–676.

Takahashi, N., C.B. Knudson, and S. Thankamony. 2011. Induction of CD44 cleavage in articular chondrocytes. Arthritis Rheum 62: 1338–1348.

Thankamony, S.P., and R. Sackstein. 2011. Enforced hematopoietic cell E- and L-selectin ligand (HCELL) expression primes transendothelial migration of human mesenchymal stem cells. Proc Natl Acad Sci USA 108: 2258–2263.

Thorne, R.F., J.W. Legg, and C.M. Isacke. 2004. The role of the CD44 transmembrane and cytoplasmic domains in co-ordinating adhesive and signalling events. J Cell Sci 117: 373–380.

Tirino, V., V. Desiderio, and R. d'Aquino. 2008. Detection and characterization of CD133+ cancer stem cells in human solid tumours. PLoS One 3: e3469.

Underhill, C. 1992. CD44: the hyaluronan receptor. J Cell Sci 103 (Pt 2): 293–298.

van der Voort, R., T.E. Taher, and V.J. Wielenga. 1999. Heparan sulfate-modified CD44 promotes hepatocyte growth factor/scatter factor-induced signal transduction through the receptor tyrosine kinase c-Met. J Biol Chem 274: 6499–6506.

Varlakhanova, N., C. Snyder, and S. Jose. 2010. Estrogen receptors recruit SMRT and N-CoR corepressors through newly recognized contacts between the corepressor N terminus and the receptor DNA binding domain. Mol Cell Biol 30: 1434–1445.

Voog, J., C. D'Alterio, and D.L. Jones. 2008. Multipotent somatic stem cells contribute to the stem cell niche in the Drosophila testis. Nature 454: 1132–1136.

Wakahara, K., H. Kobayashi, and T. Yagyu. 2005. Bikunin down-regulates heterodimerization between CD44 and growth factor receptors and subsequently suppresses agonist-mediated signaling. J Cell Biochem 94: 995–1009.

Yu, Q., B.P. Toole, and I. Stamenkovic. 1997. Induction of apoptosis of metastatic mammary carcinoma cells *in vivo* by disruption of tumor cell surface CD44 function. J Exp Med 186: 1985–1996.

Yu, W.H., J.F. Woessner, Jr., and J.D. McNeish. 2002. CD44 anchors the assembly of matrilysin/MMP-7 with heparin-binding epidermal growth factor precursor and ErbB4 and regulates female reproductive organ remodeling. Genes Dev 16: 307–323.

Zhuo, L., A. Kanamori, and R. Kannagi. 2006. SHAP potentiates the CD44-mediated leukocyte adhesion to the hyaluronan substratum. J Biol Chem 281: 20303–20314.

Zoller, M. 2011. CD44: can a cancer-initiating cell profit from an abundantly expressed molecule? Nat Rev Cancer 11: 254–267.

Zöller, M., and T. Jung. 2011. The Colorectal Cancer Initiating Cell: Markers and Their Role in Liver Metastasis. Springer.

Chapter 6

The Dark Site of Conventional Cancer Therapy—Triggering the Origin of Recurrence Cancer Stem Cells

Christa Nagler,[a] Bernd Niggemann,[b] Kurt S. Zänker[c] and Thomas Dittmar[d,*]

Introduction

The Cancer Stem Cell (CSC) hypothesis, which postulates that cancer is hierarchically organized like normal tissues, namely comprising of a small number of (cancer) stem cells and bulk of tumor/differentiated cells, threw new light on the subject: which cancer cells will be the initiators or be the cause of tumor recurrences (Clarke et al. 2006, Clarke and Fuller 2006, Dittmar et al. 2009, Li et al. 2007, Nagler et al. 2011b, Wicha et al. 2006). CSCs have been referred to as cancer cells exhibiting stem cell properties, such as the capability to self-renew and to differentiate into a diversity of various cell types, thereby maintaining (and even restoring) organ tissue homeostasis (Clarke et al. 2006, Clarke and Fuller 2006, Dittmar et al. 2009, Li et al. 2007, Nagler et al. 2011b, Wicha et al. 2006). Due to their ability to initiate tumor growth it is now commonly believed that tumors arise from a single CSC or a CSC pool (Clarke et al. 2006, Clarke and Fuller 2006, Dittmar et al. 2009, Li et al. 2007, Nagler et al. 2011b, Wicha et al. 2006). Moreover,

Center for Biomedical Education and Research, Institute of Immunology, Witten/Herdecke University, Stockumer Str. 10.
[a]E-mail: thomas.dittmar@uni-wh.de
[b]E-mail: bernd.niggemann@uni-wh.de
[c]E-mail: kurt.zänker@uni-wh.de
[d]E-mail: thomas.dittmar@uni-wh.de
*Corresponding author

since as per definition only this particular cell type/population exhibits tumor-initiation capacity, CSCs have also been suggested to be the seeds of metastases (Charafe-Jauffret et al. 2009, Croker et al. 2008, Hermann et al. 2007, Kucia et al. 2005, Li et al. 2007). In fact, specific metastatic CSCs (mCSCs (Li et al. 2007)) have already been identified in pancreatic cancer (Hermann et al. 2007) and breast cancer (Charafe-Jauffret et al. 2009). Thereby, the trafficking of normal stem cells and metastasis of CSCs most likely involve similar mechanisms, including the pivotal SDF-1α/CXCR4 axis (Kucia et al. 2005), which act as a navigation system for circulating tumor (stem) cells in order to metastasize in organ-specific manner (Dittmar et al. 2008). In fact, pancreatic mCSCs have been demonstrated to be CXCR4 positive, whereas primary tumor pancreatic CSC lack the expression of the SDF-1α receptor concomitant with the absence of metastatic spreading capacity (Hermann et al. 2007). An analysis of 33 breast cancer cell lines showed that nearly two-third of them contained an ALDEFLOUR positive population that displayed stem cell characteristics *in vitro* and in NOD/SCID xenografts (Charafe-Jauffret et al. 2009). Moreover, animal studies provided evidence that only the ALDEFLOUR positive cells were responsible for mediating metastasis (Charafe-Jauffret et al. 2009) indicating the crucial role of CSCs in inducing secondary tumors. However, the authors did not investigate whether metastasizing ALDEFLOUR positive cells were CXCR4 positive.

Several lines of evidence indicated that CSCs possess enhanced drug resistance as well as a highly efficient DNA repair system (Dean et al. 2005, Eyler and Rich 2008, Ma et al. 2008, Rich 2007), which is similar to normal stem cells and which is mandatory to ensure the status quo of the stem cell population within an organ. The loss of stem cells would mean that the organ homeostasis could no longer be maintained ultimately resulting in organ disintegration. The necessity of the stem cell status quo is clearly seen in patients that have received a bone marrow/hematopoietic stem cell (HSC) transplantation, but in which the stem cells do not repopulate or solely repopulate the bone marrow in a short-term manner. These patients will die because of disintegration of the hematopoietic system. However, what is of importance for normal stem cells, namely to survive in the presence of cytotoxic compounds and/or radiation, is fatal in the context of cancer. Here, the enhanced resistance of CSCs towards chemotherapeutic compounds and radiation might enable these cells to survive cancer therapy and to reinitiate tumor growth a month or even years later. As a consequence of that, the general strategies how to treat cancer has changed in the past years. Current conventional cancer therapies are generally designed to kill the rapidly proliferating tumor cells, whereby the efficacy of how fast such therapeutic approaches will lead to a reduction of the tumor mass is generally used as a read-out: the faster, the better. However, due to their stem cell characteristics CSCs are capable of surviving conventional cancer therapies. Thus, new cancer strategies have to be developed allowing the selective elimination of CSCs. Eradication of CSCs will result in the disintegration of the tumor system. This approach has been compared to "cutting off a dandelion out of the ground" (Huff et al. 2006, Jones et al. 2004): only the removal of the unseen root, and not elimination of the visible portion of the plant, will prevent regrowth of the weed (Huff et al. 2006, Jones et al. 2004).

In the present chapter we will not focus on anti-CSC specific therapies, but will have a closer look on that type of CSCs that will reinitiate tumor growth month to years later after real cancer therapy. If we agree with the CSC hypothesis and the fact that CSCs will be capable of surviving conventional cancer therapy then it can be concluded that the recurrent cancer should be phenotypically similar or even identical to the original tumor. If so, the recurrent cancer should be treatable with the same therapy protocol that has been used for the treatment of the primary tumor. This, however, is mostly not the case—a phenomenon, which has been referred to as "oncogenic resistance" (Blagosklonny 2005). The term "oncogenic resistance" (Blagosklonny 2005) describes the properties of most recurrent cancers, namely being more aggressive than the original tumor as well as being resistant towards first line therapy. To us, the only rationale explaining this phenomenon is that recurrent cancers originate from a new type of CSCs, the so-called recurrence CSCs (rCSCs) (Dittmar et al. 2009). Moreover, since rCSCs are most likely phenotypically different to primary tumor CSCs (pCSCs (Li et al. 2007)) as well as mCSCs (Li et al. 2007), we further conclude that this particular type of tumor stem cells most likely originates during first line therapy. In the present chapter we will discuss how rCSCs might originate, whereby particular attention will be drawn to the process of cell fusion.

Recurrence Cancer Stem Cells

As described above the phenomenon of "oncogenic resistance" does not fit with the CSC hypothesis (Blagosklonny 2005, Dittmar et al. 2009). Because recurrent cancers are often phenotypically different from the original tumor a novel type of CSCs is required: the recurrence CSCs (rCSCs) (Dittmar et al. 2009). This particular CSC type most likely originates during cancer therapy and will be the seed of aggressive and first line therapy resistant cancer relapses (Dittmar et al. 2009). In fact, repeated rounds of cisplatin treatment of a murine mammary carcinoma model resulted in the selection of cisplatin refractory $CD24^{med}CD29^{hi}$ mouse mammary CSCs exhibiting genetic alterations (Shafee et al. 2008). Here, the normal stem cell marker Nanog as well as Top2A were decreased in highly cisplatin resistant murine mammary CSCs, whereby in one case Top2A down-regulation was attributed to a genomic deletion (Shafee et al. 2008). However, whole genome analyses were not performed and thus it remains unknown whether the selection for cisplatin resistant CD24^{med}CD29^{hi} mouse mammary CSC was accompanied by additional genomic alterations.

Phenotypically different rCSCs were further identified in acute lymphoblastic leukemia (ALL) and BCR-ABL1 lymphoblastic leukemia-initiated cells (Mullighan et al. 2008, Notta et al. 2011). Interestingly, both studies provided evidence that leukemia-initiating cells represents a heterogeneous pool of multiple genetically distinct subclones, whereby the specific relapse clones were often already present as minor subpopulation at diagnosis (Notta et al. 2011). Even though it remains unknown why the specific rCSC clones only contribute to leukemia relapses and not to the primary disease these data indicate that CSC biology is much more complex and dynamic than expected. This knowledge will have a marked impact on future anti-CSC strategies since only the eradication of all CSCs will guarantee therapy success.

DNA copy number analysis (CAN) profiling indicated that all subclones most likely have a common origin thus favoring a branching multi-clonal evolution model of leukemogenesis (Notta et al. 2011). Mullighan et al. proposed that the individual clones originated more due to diversity mutations than to overall genomic instability (Mullighan et al. 2008). By contrast, ploidy reductions have been observed in fusion-derived hepatocytes, which gave rise to daughter cells with one-half chromosomal content (Duncan et al. 2009). Thereby, multipolar mitotic spindles frequently formed in mouse polyploid hepatocytes, which can result in one-step ploidy reversal to generate offsprings with multiple numerical chromosome imbalances as well as uniparental origins (Duncan et al. 2010). These data indicate that ploidy reductions can generate genetic diversity and because of that it can not be ruled out that also cell fusion might have contributed to the origin of the heterogenous pool of multiple genetically distinct ALL CSC subclones.

The Biological Phenomenon of Cell Fusion

Cell fusion plays a crucial role in several physiological as well as pathophysiological processes (an overview is given in (Dittmar and Zänker 2011)). Among the physiological events the process of fertilization seems to be the most descriptive one (for review see (Gadella and Evans 2011)). However, the process of cell fusion also does take place in other physiological processes such as placenta formation (for review see: (Huppertz and Gauster 2011)), development of skeletal muscle fibers (for review see: (Simionescu and Pavlath 2011)), and tissue regeneration (for review see: (Alvarez-Dolado and Martinez-Losa 2011, Sanges et al. 2011)). Pathophysiological events, in which the process of cell fusion is involved includes the fusion of enveloped viruses, such as *Herpesviridae, Flaviviridae* or *Retroviridiae*, with host cells (Backovic and Jardetzky 2011), and cancer (Duelli and Lazebnik 2003, Lu and Kang 2009, Nagler et al. 2011b). Particular in the context of cancer, cell fusion has been associated with tumor progression since *in vitro* and *in vivo* data provided evidence that hybrid cells, which originated from tumor cell × tumor cell or tumor cell × normal cell fusion events, exhibited novel characteristics including an enhanced drug resistance, an enhanced metastatic behavior, an increased proliferation and an increased resistance to apoptosis (Chakraborty et al. 2000, Dittmar et al. 2009, Dittmar et al. 2011, Dittmar et al. 2006, Duelli and Lazebnik 2003, Duelli and Lazebnik 2007, Mekler 1971, Mekler et al. 1971, Pawelek and Chakraborty 2008, Rachkovsky et al. 1998). The hypothesis that cell fusion takes place in a cancer context was already postulated by the German pathologist Otto Aichel 100 years ago (Aichel 1911). Aichel hypothesized that not only the tumor cells aneuploid karyotype, but also the tumor cells adoption of leukocyte characteristics, such as the ability to migrate and to pass through the circulatory system, might be attributed to the fusion with a leukocyte (Aichel 1911).

Even though the phenomenon of cell fusion seems to be omnipresent in life considerably less is still known about how the merging of two membranes is facilitated with the molecules (adhesion molecules, receptors and ligands) also being involved in this process. Cell fusion can be divided into five substeps: (i) priming, (ii) chemotaxis, (iii) adhesion, (iv) fusion, and (v) post-fusion (Zhou and Platt 2011). The priming phase primes cells to be ready for cell fusion, e.g., due to up-regulation of, e.g., chemokines, chemokine receptors and adhesion molecules, which are mandatory for the subsequent steps "chemotaxis" and "adhesion". During fusion the plasma membranes of the parental cells will merge, a process, which is accompanied with massive actin cytoskeleton rearrangements. Finally, during the post-fusion phase, an adaptation process takes place within the hybrid cells, determining whether the bi-nucleated cell will undergo further cell fusion events (as is the case for synciotrophoblasts in the placenta or for osteoclasts) or not (e.g., in the process of fertilization). Molecules or molecule pairs that have been associated with cell fusion have been generally identified on cell types know to undergo cell fusion during physiological processes, including trophoblasts and macrophages. A few of them, including the macrophage fusion receptor (MFR; also known as SIRPα) and its ligands CD47 (Vignery 2005), CD44 (Vignery 2005), CD200 (Cui et al. 2007) as well as CD163 (Shabo et al. 2009, Shabo et al. 2008) and syncytin-1 (Huppertz et al. 2006), may also play a role in tumor cell fusion.

Several recently published papers provided evidence that cell fusion events were increased about 10 to 100-fold in tissues, including liver and intestine, during chronic inflammatory conditions (Fang et al. 2005,

Johansson et al. 2008, Nygren et al. 2008) suggesting that inflammation/inflammatory conditions seem to be a positive trigger for cell fusion. However, at present it remains unknown whether inflammation/inflammatory conditions favor (a) single substep(s) or may even support the entire multistep process of cell fusion.

From a physiological point of view the fact that inflammation/inflammatory conditions positively triggers cell fusion makes sense. Cell fusion is one mechanism on how cells of the myelomonocytic lineage, e.g., macrophages, and bone marrow-derived stem cells (BMDCs) adopt the function and properties of a foreign organ tissue, thereby facilitating its regeneration (Camargo et al. 2004b, Terada et al. 2002, Vassilopoulos et al. 2003, Wang et al. 2003, Willenbring et al. 2004). On the other hand, tumor tissue resembles chronically inflamed tissue and because of that it can be concluded that the inflammatory conditions within the cancerous tissues do also positively trigger the fusion between tumor cells and tumor cells or tumor cells and other cells including cells of the myelomonocytic lineage and BMDCs. As mentioned above, these cells have been described to potently restore tissue function by cell fusion (Camargo et al. 2004b, Terada et al. 2002, Vassilopoulos et al. 2003, Wang et al. 2003, Willenbring et al. 2004). However, these cells are not able to distinguish between normal chronically inflamed tissue and chronically inflamed tumor tissue. As a consequence of that, they will also fuse with tumor cells in order to regenerate "degenerated tumor tissue".

That both cell types are capable of fusing with tumor cells, thereby giving rise to stable, long-lived hybrid cells has already been demonstrated in *in vivo* studies. Rizvi and colleagues demonstrated that BMDCs can fuse with neoplastic intestinal epithelial cells, thereby giving rise to stable hybrid cells (Rizvi et al. 2006). Even though such hybrid cells did not cause tumor initiation in the applied mouse model of intestinal cancer, which was most likely attributed to the Min mice short life span and which does not allow us to evaluate the impact of cell fusion on tumor progression, the data of Rizvi et al. demonstrate that BMDCs can fuse with tumor cells (Rizvi et al. 2006). Moreover, the authors observed a long-term repopulation by donor-derived cells in all principal intestinal epithelial lineages suggesting that the fusion partners of the BMDCs were long-lived intestinal progenitors or even stem cells (Rizvi et al. 2006). From a physiological point of view the finding that BMDCs (and putatively cells of myelomonocytic origin) may predominantly fuse with progenitor cells and/or stem cells makes sense to ensure that the homeostasis of the desired tissue will not be impaired. As loss of the stem cell status quo will ultimately lead to organ tissue disintegration. Thus, to ensure the status quo of the tissue stem cell pool, these cells should be the preferred fusion partners for BMDCs and cells of the myelomonocytic lineage.

The fusion between macrophages and neoplastic intestinal tissue was recently demonstrated in a landmark study by Powell and colleagues (Powell et al. 2011). For that purpose a parabiosis-based approach was chosen in which a GFP mouse and a tumor bearing $APC^{Min/+}$; ROSA26 mouse were surgically joined, thereby facilitating a shared circulation (Powell et al. 2011). Analysis of the intestinal tissue of $APC^{Min/+}$; ROSA26 mice revealed the presence of GFP positive cells and β-Gal double positive cells (ROSA26 mice constitutively express β-Gal) indicating that these cells originated by spontaneous cell fusion (Powell et al. 2011). Identical data were obtained when tumorigenic intestinal tissue was analyzed suggesting that BMDC derived cells can fuse spontaneously with intestinal tumor cells (Powell et al. 2011). Another interesting finding was that not only macrophages (the fusogenic capacity of this particular cell type has been demonstrated in several studies (Camargo et al. 2004a, Pawelek et al. 2006, Pawelek and Chakraborty 2008, Powell et al. 2011, Rachkovsky et al. 1998, Vignery 2005, Willenbring et al. 2004)), but also B- and T-lymphocytes were capable of fusing with (neoplastic) intestinal epithelial cells (Powell et al. 2011). However, macrophage fusion was significantly more robust than B- and T-cell fusion (Powell et al. 2011). Analysis of the transcriptome of such hybrid cells revealed that a number of cells were regulated in these cells known to be modulated in metastasis thus providing an exciting explanation for how tumor cells may gain macrophage-attributed properties that play a crucial role in tumor metastasis including extravasation, migration and immune evasion (Powell et al. 2011). Moreover, gene expression analysis further revealed that each hybrid cell possessed a unique transcriptome (Powell et al. 2011), which is in accordance to Cloudman S91×macrophage hybrids (Rachkovsky et al. 1998), human breast cancer×breast epithelial cell hybrids (Dittmar et al. 2011) and murine 67NR mammary carcinoma cell×mouse BMDC hybrid clones (Nagler et al. 2011a).

Why do Hybrid Cells Possess a Unique Transcriptome?

It is still a matter of controversial debates whether aneuploidy or mutations are the driving force in tumorigenesis and the discussion sometimes resembles the question what was first: the hen or the egg? Do mutations favor aneuploidy or are mutations the result of an aneuploid karyotype? And how does the concept of "aneuploidy as a driving force in carcinogenesis" relate to findings that aneuploidy acts both oncogenically and as a tumor suppressor (Weaver et al. 2007)? So what is right and what is wrong?

In fact, there is no clear answer to these questions. Aneuploidy is a common phenomenon is virtually all cancers (Dey 2004, Duesberg et al. 2006, Duesberg et al. 1998, Lu and Kang 2009, Williams et al. 2008), whereby also microsatellite and chromosomal stable tumors, e.g., colorectal cancer (Banerjea et al. 2009, Hawkins et al. 2001), have been reported. And yes: mutations favor aneuploidy and vice versa and because of that it is considered that both processes synergistically act together in concert, thereby favoring the origin of a malignant phenotype (Pihan and Doxsey 2003).

Irrespective of the unanswered question whether aneuploidy is an inducer of tumorigenesis or not it is well recognized that aneuploidy is a strong inducer of genomic instability and as a consequence of that of genomic variability among tumor cells (Duesberg et al. 1998, Sheltzer et al. 2011, Wilkens et al. 2004). One possibility how aneuploidy can be induced in (tumor) hybrid cells is cell fusion. When two cells fuse with each other a heterokaryon is formed first representing one cell with two distinct nuclei (Bjerkvig et al. 2005, Lu and Kang 2009). In a second step both nuclei can fuse thereby giving rise to a so-called synkaryon (one hybrid cell with one nucleus) (Bjerkvig et al. 2005, Lu and Kang 2009). The transition from a heterokaryon to a synkaryon is often associated with a resorting and loss of chromosomes, thus forcing genetic instability and variability in hybrid cells. Data of Jacobsen and colleagues revealed that some hybrid cells, which derived from spontaneous *in vivo* fusion events between human breast cancer cells and mouse stroma cells not only exhibited both human and murine chromosomes, but also carried mouse/human chromosomal translocations (Jacobsen et al. 2006). We have recently demonstrated that the mean chromosomal number (MCN) of hybrid cells that derived from spontaneous fusion events between human breast epithelial cells exhibiting stem cell characteristics and human breast cancer cells varied between 78±11 to 88±13, which was nearly the sum of the MCN of the parental cells (Dittmar et al. 2009). By contrast, the MCN of mBMDC/67NR-Hyg hybrid clones was nearly the half of the sum of the MCN of the parental cells (Nagler et al. 2011a). Even though it is not yet clear whether the process of ploidy reduction (Duncan et al. 2009, Duncan et al. 2010) was responsible for this phenomenon these data nicely illustrate the random and unpredictable nature of cell fusion on the ultimate outcome and phenotype of an emerging hybrid cell. Thus, the origin of hybrid cells resembles the process of autocatalytic karyotypic evolution—a process, which was originally postulated for the correlation between aneuploidy and multidrug resistance of cancer cells (Duesberg et al. 2000, Duesberg et al. 2001, Li et al. 2005, Li et al. 2000).

Cell Fusion and Drug Resistance

As mentioned above, independent research findings revealed that cell fusion can give rise to hybrid cells exhibiting an enhanced drug resistance (Chakraborty et al. 2000, Dittmar et al. 2009, Dittmar et al. 2011, Dittmar et al. 2006, Duelli and Lazebnik 2003, Duelli and Lazebnik 2007, Mekler 1971, Mekler et al. 1971, Pawelek and Chakraborty 2008, Rachkovsky et al. 1998). For instance, the fusion of methotrexate resistant tumor cells with 5-FU resistant cancer cells gave rise to methotrexate/5-FU and mephalan resistant hybrid cells (Miller et al. 1989). By contrast, the parental cells were not resistant to mephalan (Miller et al. 1989). We have shown that hybrid cells derived from spontaneous fusion events between human breast cancer cells and human breast epithelial cells exhibiting stem cell characteristics possessed a markedly enhanced resistance towards chemotherapeutic compounds (Dittmar et al. 2011). However, our data revealed that breast tumor × breast epithelial hybrid cell clones did not exhibit an overall enhanced drug resistance, but rather was likely to possess an enhanced adaptation capacity in order to withstand to the exposure of cytotoxic drugs (Dittmar et al. 2011). For instance, hybrid clone M13MDA435-4 showed a marked resistance towards doxorubicin and paclitaxel, but the hybrid cell clone expression profile of ATP binding cassette (ABC) multidrug resistance transporter did not match with the cells resistance pattern. By contrast, mBMDC/67NR-Hyg hybrid cell clones, which derived from spontaneous fusion events between mouse BMDCs and murine 67NR-Hyg mammary carcinoma cells possessed markedly increased Abcb1a/Abcb1b expression levels (Abcb1a/Abcb1b is the murine homologue to human multidrug resistance protein 1 (MDR1), also named p-glycoprotein), which was correlated to a markedly enhanced resistance towards doxorubicin, 17-DMAG, etoposide and paclitaxel (Nagler et al. 2011a).

These data show that different mechanisms can lead to an enhanced drug resistance of hybrid cells. Cell fusion can give rise to hybrid cells already expressing increased ABC multidrug resistance transporter levels, which then contribute to drug resistance. Alternatively, hybrid cells could evolve possessing an enhanced adaptation capacity allowing these cells to withstand chemotherapeutic drugs. Even though in the latter case the adaptation process is most likely associated with an increase in ABC multidrug resistance transporter levels in dependence of cytotoxic drug concentrations (Hembruff et al. 2008) it can not be ruled that hybrid cells can resist chemotherapeutic compounds by different mechanisms, e.g., due to up-regulation of anti-apoptotic pathways. As mentioned above, cell fusion can also give rise to hybrid cells exhibiting an increased resistance to apoptosis (Duelli and Lazebnik 2003).

Cancer Therapy, Inflammation and Cell Fusion

The finding that cell fusion can give rise to drug resistant tumor hybrid cells together with the fact that the enhanced resistance towards certain drug(s) of hybrid cells might be triggered by the drug(s) itself/themselves may point to a fatal severe side effect of current cancer therapies. On the one hand, such therapies will eliminate rapidly proliferating tumor cells (and hopefully CSCs). On the other hand, destruction of tumor tissue is correlated to an inflammatory reaction within the targeted tumor tissue. Inflammation is a physiological process that is initiated upon tissue injury and which is mandatory for wound healing and tissue regeneration. Immunocompetent cells, particularly macrophages, are recruited to the inflamed tissue to remove cell debris and to induce tissue regeneration (Coussens and Werb 2002, Ricardo et al. 2008). In addition to macrophages BMDCs are also recruited to degenerated tissues via inflammatory signals for subsequent regeneration processes (Davies et al. 2009, Houghton et al. 2004, Johansson et al. 2008, Kollet et al. 2003). Conjointly, cell fusion is one mechanism how these cells adapt tissue function (Camargo et al. 2004b, Terada et al. 2002, Vassilopoulos et al. 2003, Wang et al. 2003, Willenbring et al. 2004). Last, but not least, (chronic) inflammatory conditions positively trigger the process of cell fusion (Fang et al. 2005, Johansson et al. 2008, Nygren et al. 2008) (Fig. 1).

Thus it can be concluded that tumor therapy mediated degeneration of cancerous tissue is likely to be accompanied with an enhanced cell fusion rate. Moreover, as mentioned above, BMDCs and macrophages should preferentially fuse with progenitor cells and/or stem cells in order to ensure the stem cell status quo within the particular tissue. In the context of cancerous tissue this would mean that BMDCs and macrophages would preferentially fuse with CSCs and their progenitor cells, which, due to the heterogeneous CSC pool, might ultimately result in the origin of (diverse) rCSCs (Fig. 1).

Moreover, it can not be ruled out that the cancer therapy itself would act as a selection pressure on fusing cells, thus favoring the origin of chemotherapy/radiation resistant hybrid cancer stem cells (Fig. 1). This would be in agreement with the fact that recurrences exhibiting an "oncogenic resistance" phenotype are often resistant towards first line therapy.

Conclusions

In the present chapter we discussed the role of cell fusion in the origin of recurrence CSCs. We provided evidence that rCSCs are mandatory to explain the phenomenon of oncogenic resistance and that this particular type of CSCs are most likely to originate during cancer therapy. In accordance with Darwinian evolution, cell fusion events, caused by cancer therapy mediated tumor tissue destruction concomitant with the selection pressure "first line therapy", will lead to the origin of this new CSC type. In summary, these

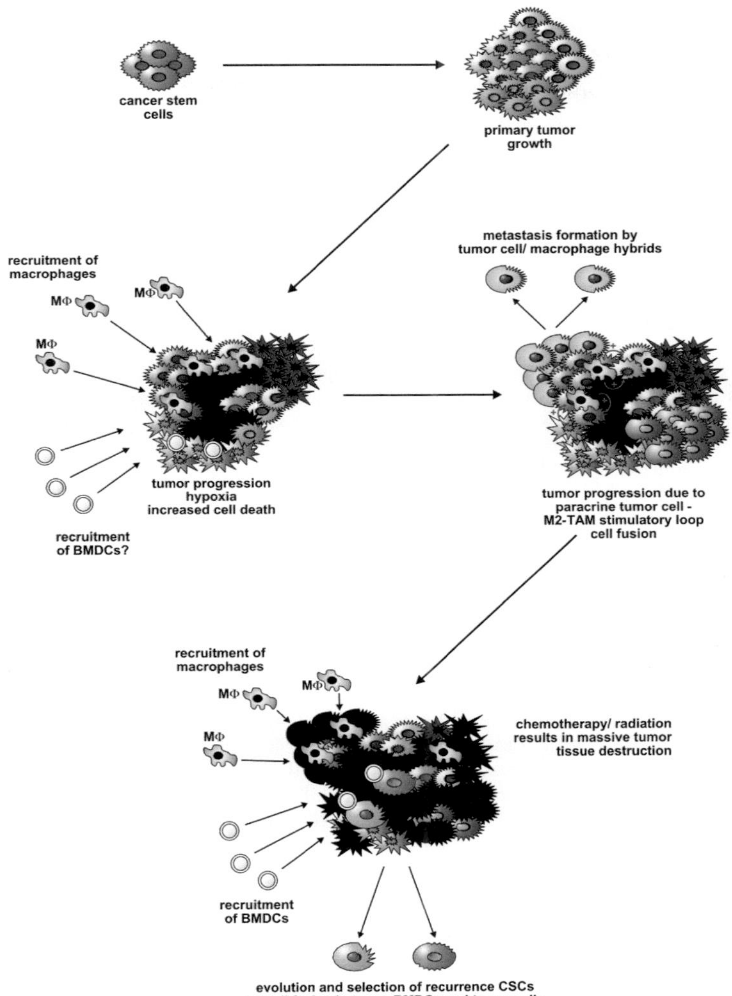

Fig. 1. Schematic overview about the cancer cycle. A primary tumor originates from CSCs. Due to increased tumor proliferation hypoxia concomitant with an enhanced loss of tumor cells occurs in the growing tumor leading to the recruitment of macrophages and BMDCs. Particularly M2-TAMs trigger tumor (cell) proliferation due to a paracrine stimulatory loop. M2-TAMs and tumor cells can fuse, thereby giving rise to hybrid cells exhibiting an enhanced metastatogenic behavior. Conventional cancer therapy, including chemotherapy and radiation, leads to a massive tumor tissue destruction (but spares CSCs) resulting in increased inflammatory conditions concomitant with an increased recruitment of BMDCs and macrophages. BMDCs might fuse with tumor cells and/or cancer stem cells, thereby giving rise to rCSCs—a process, which is possibly supported by the selection pressure conventional cancer therapy. Recurrence CSCs exhibit an "oncogenic resistance" phenotype.

Color image of this figure appears in the color plate section at the end of the book.

considerations clearly point to the dark of conventional cancer therapies, namely triggering the origin of recurrence cancer stem cells.

Acknowledgements

This work was supported by the "Verein zur Förderung der Krebsforschung e.V.", Berlin, Germany and the Fritz-Bender-Foundation, Munich, Germany.

References

Aichel, O. Über Zellverschmelzung mit quantitativ abnormer Chromosomenverteilung als Ursache der Geschwulstbildung. pp 1–115. In: Roux, W. (ed) 1911. Vorträge und Aufsätze über Entwicklungsmechanik der Organismen, Wilhelm Engelmann. Leipzig, Germany.

Alvarez-Dolado, M., and M. Martinez-Losa. 2011. Cell fusion and tissue regeneration. Adv Exp Med Biol 713: 161–175.

Backovic, M., and T.S. Jardetzky. 2011. Class III viral membrane fusion proteins. Adv Exp Med Biol 714: 91–101.

Banerjea, A., R.E. Hands, M.P. Powar, S.A. Bustin, and S. Dorudi. 2009. Microsatellite and chromosomal stable colorectal cancers demonstrate poor immunogenicity and early disease recurrence. Colorectal Dis 11: 601–608.

Bjerkvig, R., B.B. Tysnes, K.S. Aboody, J. Najbauer, and A.J. Terzis. 2005. Opinion: the origin of the cancer stem cell: current controversies and new insights. Nat Rev Cancer 5: 899–904.

Blagosklonny, M.V. 2005. Why therapeutic response may not prolong the life of a cancer patient: selection for oncogenic resistance. Cell Cycle 4: 1693–1698.

Camargo, F.D., S.M. Chambers, and M.A. Goodell. 2004a. Stem cell plasticity: from transdifferentiation to macrophage fusion. Cell Prolif 37: 55–65.

Camargo, F.D., M. Finegold, and M.A. Goodell. 2004b. Hematopoietic myelomonocytic cells are the major source of hepatocyte fusion partners. J Clin Invest 113: 1266–1270.

Chakraborty, A.K., S. Sodi, M. Rachkovsky, N. Kolesnikova, J.T. Platt, J.L. Bolognia, and J.M. Pawelek. 2000. A spontaneous murine melanoma lung metastasis comprised of host x tumor hybrids. Cancer Res 60: 2512–2519.

Charafe-Jauffret, E., C. Ginestier, F. Iovino, J. Wicinski, N. Cervera, P. Finetti, M.H. Hur, M.E. Diebel, F. Monville, J. Dutcher, M. Brown, P. Viens, L. Xerri, F. Bertucci, G. Stassi, G. Dontu, D. Birnbaum, and M.S. Wicha. 2009. Breast cancer cell lines contain functional cancer stem cells with metastatic capacity and a distinct molecular signature. Cancer Res 69: 1302–1313.

Clarke, M.F., J.E. Dick, P.B. Dirks, C.J. Eaves, C.H. Jamieson, D.L. Jones, J. Visvader, I.L. Weissman, and G.M. Wahl. 2006. Cancer stem cells--perspectives on current status and future directions: AACR Workshop on cancer stem cells. Cancer Res 66: 9339–9344.

Clarke, M.F., and M. Fuller. 2006. Stem cells and cancer: two faces of eve. Cell 124: 1111–1115.

Coussens, L.M., and Z. Werb. 2002. Inflammation and cancer. Nature 420: 860–867.

Croker, A.K., D. Goodale, J. Chu, C. Postenka, B.D. Hedley, D.A. Hess, and A.L. Allan. 2008. High aldehyde dehydrogenase and expression of cancer stem cell markers selects for breast cancer cells with enhanced malignant and metastatic ability. J Cell Mol Med.

Cui, W., E. Cuartas, J. Ke, Q. Zhang, H.B. Einarsson, J.D. Sedgwick, J. Li, and A. Vignery. 2007. CD200 and its receptor, CD200R, modulate bone mass via the differentiation of osteoclasts. Proc Natl Acad Sci USA 104: 14436–14441.

Davies, P.S., A.E. Powell, J.R. Swain, and M.H. Wong. 2009. Inflammation and proliferation act together to mediate intestinal cell fusion. PLoS ONE 4: e6530

Dean, M., T. Fojo, and S. Bates. 2005. Tumour stem cells and drug resistance. Nat Rev Cancer 5: 275–284.

Dey, P. 2004. Aneuploidy and malignancy: an unsolved equation. J Clin Pathol 57: 1245–1249.

Dittmar, T., C. Heyder, E. Gloria-Maercker, W. Hatzmann, and K.S. Zanker. 2008. Adhesion molecules and chemokines: the navigation system for circulating tumor (stem) cells to metastasize in an organ-specific manner. Clin Exp Metastasis 25: 11–32.

Dittmar, T., C. Nagler, S. Schwitalla, G. Reith, B. Niggemann, and K.S. Zanker. 2009. Recurrence cancer stem cells—made by cell fusion? Med Hypotheses 73: 542–547.

Dittmar, T., S. Schwitalla, J. Seidel, S. Haverkampf, G. Reith, S. Meyer-Staeckling, B.H. Brandt, B. Niggemann, and K.S. Zanker. 2011. Characterization of hybrid cells derived from spontaneous fusion events between breast epithelial cells exhibiting stem-like characteristics and breast cancer cells. Clin Exp Metastasis 28: 75–90.

Dittmar, T., J. Seidel, K.S. Zänker, and B. Niggemann. 2006. Carcinogenesis driven by bone marrow-derived stem cells. Contrib Microbiol 13: 156–169.

Dittmar, T., and K.S. Zänker. 2011. Cell Fusion in Health and Disease. Springer, Dordrecht, The Netherlands.

Duelli, D., and Y. Lazebnik. 2003. Cell fusion: a hidden enemy? Cancer Cell 3: 445–448.

Duelli, D., and Y. Lazebnik. 2007. Cell-to-cell fusion as a link between viruses and cancer. Nat Rev Cancer 7: 968–976.

Duesberg, P., R. Li, A. Fabarius, and R. Hehlmann. 2006. Aneuploidy and cancer: from correlation to causation. Contrib Microbiol 13: 16–44.

Duesberg, P., C. Rausch, D. Rasnick, and R. Hehlmann. 1998. Genetic instability of cancer cells is proportional to their degree of aneuploidy. Proc Natl Acad Sci USA 95: 13692–13697.

Duesberg, P., R. Stindl, and R. Hehlmann. 2000. Explaining the high mutation rates of cancer cells to drug and multidrug resistance by chromosome reassortments that are catalyzed by aneuploidy. Proc Natl Acad Sci USA 19: 14295–14300.

Duesberg, P., R. Stindl, and R. Hehlmann. 2001. Origin of multidrug resistance in cells with and without multidrug resistance genes: Chromosome reassortments catalyzed by aneuploidy. Proc Natl Acad Sci USA 98: 11283–11288.

Duncan, A.W., R.D. Hickey, N.K. Paulk, A.J. Culberson, S.B. Olson, M.J. Finegold, and M. Grompe. 2009. Ploidy reductions in murine fusion-derived hepatocytes. PLoS Genet 5: e1000385

Duncan, A.W., M.H. Taylor, R.D. Hickey, A.E. Hanlon Newell, M.L. Lenzi, S.B. Olson, M.J. Finegold, and M. Grompe. 2010. The ploidy conveyor of mature hepatocytes as a source of genetic variation. Nature 467: 707–710.

Eyler, C.E., and J.N. Rich. 2008. Survival of the fittest: cancer stem cells in therapeutic resistance and angiogenesis. J Clin Oncol 26: 2839–2845.

Fang, T.C., M.R. Alison, H.T. Cook, R. Jeffery, N.A. Wright, and R. Poulsom. 2005. Proliferation of bone marrow-derived cells contributes to regeneration after folic acid-induced acute tubular injury. J Am Soc Nephrol 16: 1723–1732.

Gadella, B.M., and J.P. Evans. 2011. Membrane fusions during Mammalian fertilization. Adv Exp Med Biol 713: 65–80.

Hawkins, N.J., I. Tomlinson, A. Meagher, and R.L. Ward. 2001. Microsatellite-stable diploid carcinoma: a biologically distinct and aggressive subset of sporadic colorectal cancer. Br J Cancer 84: 232–236.

Hembruff, S.L., M.L. Laberge, D.J. Villeneuve, B. Guo, Z. Veitch, M. Cecchetto, and A.M. Parissenti. 2008. Role of drug transporters and drug accumulation in the temporal acquisition of drug resistance. BMC Cancer 8: 318.

Hermann, P.C., S.L. Huber, T. Herrler, A. Aicher, J.W. Ellwart, M. Guba, C.J. Bruns, and C. Heeschen. 2007. Distinct populations of cancer stem cells determine tumor growth and metastatic activity in human pancreatic cancer. Cell Stem Cell 1: 313–323.

Houghton, J., C. Stoicov, S. Nomura, A.B. Rogers, J. Carlson, H. Li, X. Cai, J.G. Fox, J.R. Goldenring, and T.C. Wang. 2004. Gastric cancer originating from bone marrow-derived cells. Science 306: 1568–1571.

Huff, C.A., W. Matsui, B.D. Smith, and R.J. Jones. 2006. The paradox of response and survival in cancer therapeutics. Blood 107: 431–434.

Huppertz, B., C. Bartz, and M. Kokozidou. 2006. Trophoblast fusion: fusogenic proteins, syncytins and ADAMs, and other prerequisites for syncytial fusion. Micron 37: 509–517.

Huppertz, B., and M. Gauster. 2011. Trophoblast fusion. Adv Exp Med Biol 713: 81–95.

Jacobsen, B.M., J.C. Harrell, P. Jedlicka, V.F. Borges, M. Varella-Garcia, and K.B. Horwitz. 2006. Spontaneous fusion with, and transformation of mouse stroma by, malignant human breast cancer epithelium. Cancer Res 66: 8274–8279.

Johansson, C.B., S. Youssef, K. Koleckar, C. Holbrook, R. Doyonnas, S.Y. Corbel, L. Steinman, F.M. Rossi, and H.M. Blau. 2008. Extensive fusion of haematopoietic cells with Purkinje neurons in response to chronic inflammation. Nat Cell Biol 10: 575–583.

Jones, R.J., W.H. Matsui, and B.D. Smith. 2004. Cancer stem cells: are we missing the target? J Natl Cancer Inst 96: 583–585.

Kollet, O., S. Shivtiel, Y.Q. Chen, J. Suriawinata, S.N. Thung, M.D. Dabeva, J. Kahn, A. Spiegel, A. Dar, S. Samira, P. Goichberg, A. Kalinkovich, F. Arenzana-Seisdedos, A. Nagler, I. Hardan, M. Revel, D.A. Shafritz, and T. Lapidot. 2003. HGF, SDF-1, and MMP-9 are involved in stress-induced human CD34+ stem cell recruitment to the liver. J Clin Invest 112: 160–169.

Kucia, M., R. Reca, K. Miekus, J. Wanzeck, W. Wojakowski, A. Janowska-Wieczorek, J. Ratajczak, and M.Z. Ratajczak. 2005. Trafficking of normal stem cells and metastasis of cancer stem cells involve similar mechanisms: pivotal role of the SDF-1-CXCR4 axis. Stem Cells 23: 879–894.

Li, F., B. Tiede, J. Massague, and Y. Kang. 2007. Beyond tumorigenesis: cancer stem cells in metastasis. Cell Res 17: 3–14.

Li, R., R. Hehlman, R. Sachs, and P. Duesberg. 2005. Chromosomal alterations cause the high rates and wide ranges of drug resistance in cancer cells. Cancer Genet Cytogenet 163: 44–56.

Li, R., A. Sonik, R. Stindl, D. Rasnick, and P. Duesberg. 2000. Aneuploidy versus gene mutation hypothesis of cancer: recent study claims mutation, but is found to support aneuploidy. Proc Natl Acad Sci USA 97: 3236–3241.

Lu, X., and Y. Kang. 2009. Cell fusion as a hidden force in tumor progression. Cancer Res 69: 8536–8539.

Ma, S., T.K. Lee, B.J. Zheng, K.W. Chan, and X.Y. Guan. 2008. CD133+ HCC cancer stem cells confer chemoresistance by preferential expression of the Akt/PKB survival pathway. Oncogene 27: 1749–1758.

Mekler, L.B. 1971. [Hybridization of transformed cells with lymphocytes as 1 of the probable causes of the progression leading to the development of metastatic malignant cells]. Vestn Akad Med Nauk SSSR 26: 80–89.

Mekler, L.B., O.B. Drize, I.V. Osechinskii, and M.A. Shliankevich. 1971. [Transformation of a normal differentiated cell of an adult organism, induced by the fusion of this cell with another normal cell of the same organism but with different organ or tissue specificity]. Vestn Akad Med Nauk SSSR 26: 75–80.

Miller, F.R., A.N. Mohamed, and D. McEachern. 1989. Production of a more aggressive tumor cell variant by spontaneous fusion of two mouse tumor subpopulations. Cancer Res 49: 4316–4321.

Mullighan, C.G., L.A. Phillips, X. Su, J. Ma, C.B. Miller, S.A. Shurtleff, and J.R. Downing. 2008. Genomic analysis of the clonal origins of relapsed acute lymphoblastic leukemia. Science 322: 1377–1380.

Nagler, C., C. Hardt, K.S. Zänker, and T. Dittmar. 2011a. Co-cultivation of murine BMDCs with 67NR mouse mammary carcinoma cells give rise to highly drug resistant hybrid cells. Cancer Cell International 11: 21.

Nagler, C., K.S. Zanker, and T. Dittmar. 2011b. Cell Fusion, Drug Resistance and Recurrence CSCs. Adv Exp Med Biol 714: 173–182.

Notta, F., C.G. Mullighan, J.C. Wang, A. Poeppl, S. Doulatov, L.A. Phillips, J. Ma, M.D. Minden, J.R. Downing, and J.E. Dick. 2011. Evolution of human BCR-ABL1 lymphoblastic leukaemia-initiating cells. Nature 469: 362–367.

Nygren, J.M., K. Liuba, M. Breitbach, S. Stott, L. Thoren, W. Roell, C. Geisen, P. Sasse, D. Kirik, A. Bjorklund, C. Nerlov, B.K. Fleischmann, S. Jovinge, and S.E. Jacobsen. 2008. Myeloid and lymphoid contribution to non-haematopoietic lineages through irradiation-induced heterotypic cell fusion. Nat Cell Biol 10: 584–592.

Pawelek, J., A. Chakraborty, R. Lazova, Y. Yilmaz, D. Cooper, D. Brash, and T. Handerson. 2006. Co-opting macrophage traits in cancer progression: a consequence of tumor cell fusion? Contrib Microbiol 13: 138–155.

Pawelek, J.M., and A.K. Chakraborty. 2008. Fusion of tumour cells with bone marrow-derived cells: a unifying explanation for metastasis. Nat Rev Cancer 8: 377–386.

Pihan, G., and S.J. Doxsey. 2003. Mutations and aneuploidy: co-conspirators in cancer? Cancer Cell 4: 89–94.

Powell, A.E., E.C. Anderson, P.S. Davies, A.D. Silk, C. Pelz, S. Impey, and M.H. Wong. 2011. Fusion between Intestinal epithelial cells and macrophages in a cancer context results in nuclear reprogramming. Cancer Res 71: 1497–1505.

Rachkovsky, M., S. Sodi, A. Chakraborty, Y. Avissar, J. Bolognia, J.M. McNiff, J. Platt, D. Bermudes, and J. Pawelek. 1998. Melanoma x macrophage hybrids with enhanced metastatic potential. Clin Exp Metastasis 16: 299–312.

Ricardo, S.D., H. van Goor, and A.A. Eddy. 2008. Macrophage diversity in renal injury and repair. J Clin Invest 118: 3522–3530.

Rich, J.N. 2007. Cancer stem cells in radiation resistance. Cancer Res 67: 8980–8984.

Rizvi, A.Z., J.R. Swain, P.S. Davies, A.S. Bailey, A.D. Decker, H. Willenbring, M. Grompe, W.H. Fleming, and M.H. Wong. 2006. Bone marrow-derived cells fuse with normal and transformed intestinal stem cells. Proc Natl Acad Sci USA 103: 6321–6325.

Sanges, D., F. Lluis, and M.P. Cosma. 2011. Cell-fusion-mediated reprogramming: pluripotency or transdifferentiation? Implications for regenerative medicine. Adv Exp Med Biol 713: 137–159.

Shabo, I., H. Olsson, X.F. Sun, and J. Svanvik. 2009. Expression of the macrophage antigen CD163 in rectal cancer cells is associated with early local recurrence and reduced survival time. Int J Cancer 125: 1826–1831.

Shabo, I., O. Stal, H. Olsson, S. Dore, and J. Svanvik. 2008. Breast cancer expression of CD163, a macrophage scavenger receptor, is related to early distant recurrence and reduced patient survival. Int J Cancer 123: 780–786.

Shafee, N., C.R. Smith, S. Wei, Y. Kim, G.B. Mills, G.N. Hortobagyi, E.J. Stanbridge, and E.Y. Lee. 2008. Cancer stem cells contribute to cisplatin resistance in Brca1/p53-mediated mouse mammary tumors. Cancer Res 68: 3243–3250.

Sheltzer, J.M., H.M. Blank, S.J. Pfau, Y. Tange, B.M. George, T.J. Humpton, I.L. Brito, Y. Hiraoka, O. Niwa, and A. Amon. 2011. Aneuploidy drives genomic instability in yeast. Science 333: 1026–1030.

Simionescu, A., and G.K. Pavlath. 2011. Molecular mechanisms of myoblast fusion across species. Adv Exp Med Biol 713: 113–135.

Terada, N., T. Hamazaki, M. Oka, M. Hoki, D.M. Mastalerz, Y. Nakano, E.M. Meyer, L. Morel, B.E. Petersen, and E.W. Scott. 2002. Bone marrow cells adopt the phenotype of other cells by spontaneous cell fusion. Nature 416: 542–545.

Vassilopoulos, G., P.R. Wang, and D.W. Russell. 2003. Transplanted bone marrow regenerates liver by cell fusion. Nature 422: 901–904.
Vignery, A. 2005. Macrophage fusion: the making of osteoclasts and giant cells. J Exp Med 202: 337–340.
Wang, X., H. Willenbring, Y. Akkari, Y. Torimaru, M. Foster, M. Al-Dhalimy, E. Lagasse, M. Finegold, S. Olson, and M. Grompe. 2003. Cell fusion is the principal source of bone-marrow-derived hepatocytes. Nature 422: 897–901.
Weaver, B.A., A.D. Silk, C. Montagna, P. Verdier-Pinard, and D.W. Cleveland. 2007. Aneuploidy acts both oncogenically and as a tumor suppressor. Cancer Cell 11: 25–36.
Wicha, M.S., S. Liu, and G. Dontu. 2006. Cancer stem cells: an old idea—a paradigm shift. Cancer Res 66: 1883–1890; discussion 1895–1886.
Wilkens, L., P. Flemming, M. Gebel, J. Bleck, C. Terkamp, L. Wingen, H. Kreipe, and B. Schlegelberger. 2004. Induction of aneuploidy by increasing chromosomal instability during dedifferentiation of hepatocellular carcinoma. Proc Natl Acad Sci USA 101: 1309–1314.
Willenbring, H., A.S. Bailey, M. Foster, Y. Akkari, C. Dorrell, S. Olson, M. Finegold, W.H. Fleming, and M. Grompe. 2004. Myelomonocytic cells are sufficient for therapeutic cell fusion in liver. Nat Med 10: 744–748.
Williams, B.R., V.R. Prabhu, K.E. Hunter, C.M. Glazier, C.A. Whittaker, D.E. Housman, and A. Amon. 2008. Aneuploidy affects proliferation and spontaneous immortalization in mammalian cells. Science 322: 703–709.
Zhou, X., and J.L. Platt. 2011. Molecular and cellular mechanisms of Mammalian cell fusion. Adv Exp Med Biol 713: 33–64.

Chapter 7

Current Concepts of How to Eliminate Cancer Stem Cells

Elena Campos-Sanchez,[1] Isabel Romero-Camarero,[2] Isidro Sanchez-Garcia[2,a,*] and Cesar Cobaleda[1,b,*]

ABSTRACT

After many years of struggle against the traditional theories, cancer stem cells (CSCs) are finally coming of age as the main responsible for the maintenance and relapse of human cancer. There are two main reasons for this fact. First, the enormous amount of quickly accumulating experimental evidences proving their existence and their contribution to cancer maintenance and survival in humans. Second, the patent evidence of the very slow advance of our fight against cancer in the last 50 years. Indeed, although early detection and screening programs have increased the chances of diagnosing cancer at very early stages, the prognosis for disseminated tumors is as dismal today as it was 5 decades ago. It is therefore clear that a new conceptual framework is required in our approach to treat cancer. The acknowledgement of the existence of CSCs provides this framework and opens new avenues for exploring therapeutic approaches. In this chapter we revise the most recent discoveries in the research aimed at the targeting and elimination of CSCs.

[1]Centro de Biología Molecular Severo Ochoa, CBMSO (CSIC/UAM)
Nicolás Cabrera, n°1, Campus de Cantoblanco, 28049, Madrid, SPAIN.
[b]E-mail: ccobaleda@cbm.uam.es
[2]Experimental Therapeutics and Translational Oncology Program, Instituto de Biología Molecular y Celular del Cáncer, CSIC/ Universidad de Salamanca, Campus M. de Unamuno s/n, 37007 Salamanca, SPAIN.
[a]E-mail: isg@usal.es
*Corresponding authors

Introduction

Cancer is the second cause of death in developed countries, and its incidence is steadily on the rise. In general, since 1930, cancer death rates have increased with age for both males and females, and the age-specific occurrence of cancer was similar for both genders (Wingo et al. 2003). Furthermore, the five-year-mortality trends for the most relevant types of tumors have only slightly receded (only up to a 3.3% for prostate cancers) or even have increased (2.2% for liver cancer) in the last 40 years (Wingo et al. 2003, Marshall 2011), in spite of the enormous amounts of money invested in cancer research during this period. Most of the advances in the field of cancer treatment have been related to the early detection programs that have increased our chances of identifying cancers in very early stages (Etzioni et al. 2003). However, given the disseminated nature of the disease (Husemann et al. 2008, Sanchez-Garcia 2009), in many cases this early detection comes already too late, and the prognosis for metastatic cancers is today as dark as it was 50 years ago.

Current therapeutic approaches treat cancer as a disease of proliferation, and pay little or no attention to the internal structure of the tumor or its cellular heterogeneity. However, we now know that cancers are much more than uncontrolled cell lines, and that there is an internal hierarchical structure within the tumor, with cells with different biological properties and different sensitivities to therapy. This chapter is based on the assumption of the main postulates of the cancer stem cell (CSC) theory. A full discussion of the CSC concept and its implications is therefore out of the scope of the chapter and, for this, the reader is referred to other chapters in this book and to other recent reviews (Abollo-Jimenez et al. 2010, Castellanos et al. 2010, Greaves 2010, Hermann et al. 2010, Koch et al. 2010, Lane and Gilliland 2010, Sanchez-Garcia 2010, Shackleton 2010, Clevers 2011, Visvader 2011). We will, however, discuss, in the first part of this review, the most recent results that are finally bringing CSCs to the forefront of the study of human cancer, and that are starting to show the implications of CSCs in prognosis, malignancy and disease evolution in humans. Afterwards we will revise the most recent discoveries in the research aimed at the targeting and elimination of CSCs.

The Coming of Age of Cancer Stem Cells

At the basis of the CSC theory is the concept of cancer as a stem cell-based, hierarchically organized, aberrant tissue, in which only a subset of cells (the CSCs) have the capacity of maintaining and re-generating the tumor (Sanchez-Garcia et al. 2007, Dick 2008). In current anti-cancer therapies, most drugs are aimed at interfering, by several mechanisms, with the replication of the rapidly-dividing cells that conform the majority of the tumoral mass (Chabner and Roberts 2005). This approach, which causes many highly undesirable side-effects because of its lack of specificity, is able to eliminate the clinically visible tumor mass ("cancer cure") in many patients. However with the current detection techniques, it is impossible to detect tumoral cells when their number is below 10^9 (Frangioni 2008). Therefore, "cancer cure" refers to a state in which tumoral cells are below this detection limit... but there is a wide range between 10^9 and 0 tumoral cells, as it is clear from the high frequency of relapses in patients that had initially achieved clinical remission after chemo- or radiotherapy cycles. The high frequency of relapses indicates that the cells responsible for tumor replenishment are not being affected by the conventional therapy and, with time, they regenerate the tumor. It is true that there are examples of successful treatment of certain tumor types with current therapeutic regimes, like the high rate of cure (80%) of most types of childhood acute lymphoblastic leukemias (Pui et al. 2008). However, even in these cases, the use of highly toxic, unspecific drugs comes at a price, since most patients develop treatment-related diseases (including cancers) in their adulthood (Marder 2010).

It is therefore clear that a new conceptual framework is required in our approach to treat cancer. Cancer is a disease of genetic origin, and men and women in developed countries have a 0.5 and a 0.33 lifetime risk of developing cancer, respectively (American Cancer Society. Cancer Facts & Figures 2011. Atlanta: American Cancer Society 2011). This high incidence is possibly an indication of the fact that cancer is not just "an unfortunate accident", but rather a catastrophe waiting to happen, a latent possibility intrinsic to our cells, as the cost we have to pay for being a complex multicellular organism with many different developmental pathways controlled with "just" one genome (Huang 2011). From this point of view, cancer is an aberrant differentiation program, established as a consequence of one or a series of oncogenic alterations that open the way for a new pathologic lineage to appear and develop (Abollo-Jimenez et al. 2010, Castellanos et al. 2010). This perspective of tumorigenesis moves the focus away from proliferation and towards the alteration of the normal differentiation programs: cancer is a disease of aberrant reprogrammed differentiation, and as such must be studied and (hopefully) treated. Normal tissue development is a consequence of stem cell programming and stepwise commitment towards specific lineages. Cancer development is a consequence of the tumoral reprogramming that arises from the interaction between the driving oncogenic events and the plasticity of the cancer cell-of-origin where these events initially take place.

All these aforementioned concepts, that were explicitly or implicitly laid down by the pioneers many years ago (Fialkow et al. 1977, Hamburger and

Salmon 1977, Bonnet and Dick 1997, Dick 2008, Sanchez-Garcia 2010) have been thoroughly explored by many research groups in the 2000's decade, mainly by using xenografts of prospectively purified human tumoral cells into immunodeficient mice, or by using genetically engineered mouse models of cancer (Sanchez-Garcia et al. 2007, Cobaleda and Sanchez-Garcia 2009). Once this groundwork has being laid, in the last few years special stress has been put in demonstrating that all these biological findings are indeed of some clinical relevance to humans. Certainly one would expect that, if CSCs are essential for tumor survival, then the characteristics of these CSCs should have some impact in the tumor biology. That this is indeed the case has recently been demonstrated for several types of tumors. For example, it has been shown that an stem cell-like, self-renewing gene expression signature can predict clinical outcome in acute myeloid leukemia (AML) (Eppert et al. 2011, Greaves 2011). By analyzing the gene expression profiles of functionally proved AML CSCs, the authors show that populations enriched in either normal hematopoietic stem cells (HSCs) or CSCs shared transcriptional "self-renewal" gene expression signatures. These profiles are a reflection of their stem cell function but, more importantly, they are predictive of adverse prognosis in individuals with AML. A different study showed that high-level expression of an AML CSC-associated signature was predictive of clinical outcome in several independent patient cohorts (Gentles et al. 2010). In acute lymphoblastic leukemias (ALL), a plausible surrogate marker for stem cell activity, as it is the speed of leukemic growth in recipient mice, has been shown to be correlated with tendency to relapse and to be characterized by a specific gene expression profile (Meyer et al. 2011). It would therefore seem that a 'stemness' signature could be useful for stratifying patients and predict disease outcome. This is the conclusion of another study in which, starting from several stem cell types, ranging from normal stem cells (embryonic stem, ES) to induced pluripotent stem cells (iPSCs) and invasive tumor cells, the researchers have derived an expression signature that reflects common biological aspects of various stem-like characteristics (Shats et al. 2011). This so-called consensus stemness ranking (CSR) signature was more expressed in several cancer types in parallel with increasing tumor grade and decreased survival, and could predict clinical response of breast cancer patients to a neoadjuvant regimen that included a CSR-specific agent (Shats et al. 2011). In a mouse model of human acute myeloid leukemia (AML), CSCs are maintained by a self-renewing transcriptional subprogram similar to that of embryonic stem cells (ESCs), and this is linked to poor-prognosis human malignancies (Somervaille et al. 2009). Something similar has been found for human breast cancers (Ben-Porath et al. 2008). Furthermore, investigating changes in the CSC populations in human breast cancer, before and after primary systemic therapy, clinical evidence has been obtained showing that the putative breast cancer CSCs are chemoresistant and are associated with tumor progression (Lee et al. 2011). It has also been shown that poorly differentiated breast cancers had a higher CSC content than well-differentiated ones (Pece et al. 2010). In other very frequent type of solid tumors, colorectal cancer (CRC), it has recently been demonstrated that a specific intestinal stem cell (ISC) signature can be defined that predicts relapse in treated CRC patients (Merlos-Suarez et al. 2011). This signature is associated with both tumor-initiating and long-term self-renewal capacities.

In spite of the previously described conclusions, caution is always required when interpreting results in a research field as new and quickly evolving as this one. By way of example, it a recent study of colorectal cancer it has been shown that the expression of many canonical Wnt target genes, included in the above-mentioned intestinal stem cell (ISC) signature, was in fact inversely correlated with prognosis (de Sousa et al. 2011). In most of the reports that we have summarized above it is assumed that the association of a CSC signature with a bad prognosis is directly related to the higher number of CSCs present in the corresponding malignant tissue. However, in general, CSCs constitute a very reduced percentage of the tumor, therefore making it unlikely that they can significantly contribute in a relevant manner to a global tumoral gene expression profile. One possibility (Eppert et al. 2011) is that stem cell-specific expression profiles could persist in the proliferating non-stem tumoral cells at a population level, but that no individual differentiated cell has the full stem cell genes repertoire (since they are not CSCs). In the study from de Sousa et al. (de Sousa et al. 2011) it is apparent that the CSC-related signatures can identify poor-prognosis patients despite the fact that key CSC markers and canonical Wnt target genes inversely correlate with tumor relapse and disease stage in colorectal cancer. One proposed explanation is that CSC signatures correlate with tumors with an immature profile and a poorly differentiated histology. Accordingly, tumoral progression in this study seems to be accompanied by changing from a more ES cell-like gene expression profile (defined by genes like *SOX2*, *OCT4*, and *Nanog*) to a more intestinal-tissue-specific stem cell signature (defined by genes like *ASCL2* and *LGR5*) (de Sousa et al. 2011). A similar conclusion is drawn in a different study in which breast cancer samples are stratified using a breast-CSC signature, which also seems to correlate with basal, i.e., more immature, breast cancers (Bertucci et al. 2007).

Summarizing the relevance of these last years' findings, in spite of all the caveats, they clearly have important biological and clinical implications regarding CSCs and their role in tumor biology. First, the clear relationship between the CSCs' and other stem cells' signatures, and the fact that these signatures indeed seem to predict survival, provides evidence for the hierarchical organization of cancers according to the CSC model and confirms that CSCs are not experimental artifacts. Moreover, CSCs, defined on the basis of functional stem cell properties, are distinct from other, non-CSC tumor cells, and are clinically relevant. Therefore, it is highly likely that therapies targeting CSCs would improve patient survival, and that animal models based on the CSC theory would be useful in the preclinical evaluation of new cancer drugs. Finally, the identification of transcriptional profiles conserved between CSCs and other types of stem cells suggests that there are specific genes in charge of establishing and maintaining the stem cell state, and that they might influence clinical outcome.

Killing Cancer Stem Cells: Aiming at a Moving Target

It is clear therefore that CSCs are relevant therapeutic targets if we want to achieve definitive remissions in cancer patients. The key question is then: how to target them? There are two main aspects to this problem: i) which cells are the real CSCs? and ii) what are the specific molecular targets that we can attack to kill them or to interfere with their function without affecting normal stem cells?

Regarding this first question (which cells are the real CSCs?), until now the search for CSCs has been based on the prospective purification of cancer cell subpopulations and the determination of their cancer-transplantation capacities into immunodeficient mice. This approach, being very powerful, presents nevertheless several methodological shortcomings that have been described in detail elsewhere (Cobaleda and Sanchez-Garcia 2009). However, beyond the technical deficiencies, the biology of CSCs themselves poses many difficulties for their identification. One of the most relevant problems is the elusive nature of the CSCs. Already at the earliest stages of the tumor, the path from the cancer cell-of-origin (CCO, the cell suffering the first genetic lesion associated to cancer development) to the CSC is complex, and impossible to determine in human cancers, which are diagnosed when they have already evolved into full-blown tumors. Traditionally, the cellular origin of tumors was extrapolated to the most phenotypically similar normal cells. However, if tumors are stem cell-based tissues, clearly this association cannot be made, and the search for CSCs is providing many examples of this fact. For example, luminal cells have traditionally been assumed to be the CCOs for human prostate cancer, given the fact that these tumors are characterized by luminal cell expansion and absence of basal cells. However, it has recently been shown that human prostate basal cells can initiate prostate cancer in immunodeficient mice, therefore demonstrating that the histological appearance of cancers does not necessarily correlate with their cellular origins (Goldstein et al. 2010). Similarly, in breast cancer it has been shown that BRCA1-deficient basal-like tumors originate from luminal epithelial progenitors and not from basal stem cells, as their phenotype might have suggested (Molyneux et al. 2010). In two different mouse models of glioma it has been found that, even though the mutations are induced in neural precursor stem cells, the cells behaving as final CSCs are in fact oligodendrocyte precursors (Persson et al. 2010, Liu et al. 2011). Therefore, when the cancer cells-of-origin have the potential for phenotypic plasticity and are susceptible to oncogene-induced reprogramming, tumor phenotypes may not directly reflect normal histogenesis. Furthermore, mathematical modeling has shown that the plasticity of the CSCs is an important determinant of the prognosis of tumors and that higher levels of dedifferentiation substantially reduce the effectiveness of therapy directed at cancer stem cells by leading to higher rates of resistance (Leder et al. 2010).

Beyond the blurred origin of the CSCs, their 'mature' nature is not less complicated. In childhood B-cell acute lymphoblastic leukemias, it has been shown that CSC-activity can be identified in blasts at different maturational stages (le Viseur et al. 2008), therefore making it difficult to identify a single cellular component as the responsible for tumor maintenance. In this type of hematopoietic tumor it has been shown that the leukemia-propagating cells are genetically variegated, presenting subclonal patterns with different competitive regenerative capacities *in vivo* (Anderson et al. 2011). Genetic variegation of CSCs may represent a great block to effective treatment if the molecular targets identified are not the cancer-initiating lesions, but rather secondary mutations segregated in subclones, even when these ones finally appear as the dominant ones. In adult BCR-ABL$^+$ leukemias, the reconstruction of the subclonal genetic ancestry of the tumors has shown that they evolve following a branching multiclonal model, as it is clearly shown by analyzing samples expanded in recipient mice versus the original patient's samples (Notta et al. 2011). Also in other types of tumors, like melanoma, it seems that a large proportion of the tumoral cells has the capacity of regenerating the tumors, and that the markers defining different tumor subpopulations are reversibly expressed, suggesting a model of phenotypic and functional plasticity in which cellular heterogeneity is mainly driven by reversible changes rather than fixed by epigenetic marks (Quintana et al. 2010). Even in the cases in which potential melanoma CSC biomarkers have been identified, it still seems that their distribution in the population is not stable. Indeed, using the Histone 3 Lysine 4 (H3K4) demethylase JARID1B as a biomarker a small, melanoma stem cell-like subpopulation has been isolated that seems to be essential for continuous tumor growth (Roesch et al. 2010). However, the expression of JARID1B is dynamically regulated, in such a way that JARID1B-negative cells can become positive, therefore suggesting that tumor maintenance is a dynamic process depending on a temporarily distinct subpopulation

(Roesch et al. 2010). The potential reasons for this behavior are suggested by another study showing that the expression of the histone demethylase Jarid1A is related to the acquisition of a 'drug-tolerant' phenotype that is transiently acquired and abandoned at low frequency by individual cells within the tumoral population (Sharma et al. 2010). This would suggest that the cancer cell population might use a dynamic strategy for survival, in such a way that individual cells (with cancer-regeneration capacity) would transiently adopt a reversible drug-tolerant state to protect the population from elimination by lethal exposures (Sharma et al. 2010). This is related to the fact that mathematical modeling of tumoral progression has shown that all subpopulations of cells purified according to a given phenotype return to equilibrium proportions over time, therefore suggesting that CSC-like cells can arise de novo from non-stem-like cells (Gupta et al. 2011). Finally, and to complicate the problem even more, it is increasingly clear that there is an evolution of the nature of CSCs during the progression of the disease, including the response to treatment and final relapse, indicating that CSCs are capable of evolving ways of survival in response to the changes in the environment and other external selective pressures (Mullighan et al. 2008, Yachida et al. 2010, Anderson et al. 2011, Kuster et al. 2011, van Delft et al. 2011, Ding et al. 2012).

All these data evidence the tremendous difficulty of finding the right cellular targets at the time of treatment (diagnosis, relapse, etc.), and how much we still need to learn about the biology of cancer stem cells in order to be able to attack them. The current research in this field is tremendously active and it builds to a large degree (for both good and bad) on our previous experience in the study of cancer cells in general. Still, as we have explained before, one essential change introduced by the acceptance of the CSC theory has been to move the focus of the research in cancer therapies from proliferation to differentiation and stem cell biology. Therefore, the putative CSC-specific targets are being mainly sought, in a prospective manner, among the most relevant of the molecular signaling routes involved in the regulation of fate determination and in the specification and maintenance of the stem cell identity (NOTCH, WNT, PTEN, HEDGEHOG, POLYCOMB, HOX, etc.). These signaling routes, together with some other approaches of less relevance (due to their more restricted applicability at the moment) like differentiation therapies, constitute what we could already call the "classical" anti-CSC targets. On the other side, due to our general lack of knowledge about the biology of stem cells, research in this field is also revealing new unexpected potential anti-CSC targets related with a wide variety of cellular processes. In the next two sections we will summarize the most recent advances in these two fronts.

"Classical" Anti-Cancer Stem Cell Targets

The first rational approximation to attacking CSCs is to aim at the basic mechanisms regulating the two most important properties of their stem cell biology: self-renewal and differentiation potential (and the closely related metastatic capacity). In a recent revision (Sagrera et al. 2009) we have described in detail the molecular mechanisms of the main signaling routes involved in this process, their alterations in cancer and the rationale behind the attempts to target them. In this section, we will therefore highlight the most relevant results published in these fields in the last 3–4 years (concerning specifically targeting of CSCs, not just cancer cells in general), and we will refer the reader to general reviews for the details about the respective pathways (Warr et al. 2011).

One of the most important pathways involved in the control of stem cell self-renewal and differentiation is the Wnt signaling pathway (Barker and Clevers 2006, Barker 2008, Angers and Moon 2009). Accordingly, it is frequently deregulated in tumors of different kinds and is the focus of many drug-design activities. Regarding its role specifically in CSCs, it has been established that β-catenin, one of the main molecular effectors of the canonical Wnt pathway, mediates the drug resistance of CSCs in a mouse model of tumors driven by oncogenes related to the *mixed lineage leukemia* (*MLL*) gene (Yeung et al. 2010). Indeed, β-catenin becomes activated during the development of MLL-induced CSCs and its deletion resulted in a reversion of MLL-induced CSCs to a pre-cancerous state, abolishing their oncogenic potential (Yeung et al. 2010). Also, in mouse models of acute myelogenous leukemia (AML) the Wnt/β-catenin signaling pathway is required for self-renewal of CSCs, either if they are derived from hematopoietic stem cells (HSC) or from more differentiated granulocyte-macrophage progenitors (GMP) (Wang et al. 2010). Given the fact that this pathway is normally active in HSCs but not in GMP, these results imply that its reactivation is required for the transformation of progenitor cells to CSCs by certain oncogenes. It has not been completely clarified yet whether β-catenin is or not absolutely required for the self-renewal of adult HSCs, but if it was not, then the targeting of the Wnt/β-catenin pathway might represent a new therapeutic opportunity in AML. Furthermore, this pathway is convergent with other intracellular routes that might also be involved in tumoral progression. Wnt signaling activation is mainly mediated by β-catenin nuclear localization, a process whose underlying molecular mechanisms are not yet fully characterized. In this context, it has recently been shown that the gene *FoxM1*, which is also activated in most human tumors, is a vital constituent of Wnt signaling and actively participates in regulating β-catenin activation by directly interacting with it (Zhang et al. 2011). This interaction has been demonstrated to be required for glioma formation, and offers a new possible target for drug development. Another example of these Wnt pathway-related potential therapeutic targets is CD27, in this case in the context of chronic myelogenous leukemia (CML) (Schurch et al. 2012). CD27 is a TNF-receptor family member protein that has been found expressed in CML CSCs, where it leads to an augmented expression of Wnt target genes, that results in an increased proliferation rate (Schurch et al. 2012). The experimental interference with CD27 signaling in CSCs delayed disease progression and prolonged survival, therefore suggesting a new therapeutic approach to blocking the Wnt/ β-catenin pathway in CML.

Although, as we have seen, the canonical Wnt pathway signals *via* β-catenin, it is known that there are two other noncanonical Wnt pathways that signal independently of β-catenin; one of them is the Ca^{2+}/NFAT pathway (Veeman et al. 2003). In CML the current treatment of choice is imatinib mesylate, a drug specifically targeted against the *BCR-ABL* oncogene responsible for the CML and capable of inducing long-term remissions. However, the treatment does not achieve the elimination of the disease due to its inability of eliminating CML CSCs (Corbin et al. 2011). An RNAi-based synthetic lethal screening based on imatinib treatment was performed on CML cells (Gregory et al. 2010), and many components of the Wnt/Ca^{2+}/NFAT signaling pathway were identified. Their inhibition led to an enhanced inhibition of the activity of the *BCR-ABL* oncogene in the tumoral cells, suggesting that the addition of pharmacological agents targeting this pathway could synergize to improve the existing CML therapies and compensate for the current survival of CML CSCs.

One important note of caution in the context of the Wnt signaling pathway, which can also be extended to all other potential anti-CSC drug targets, is related to the fact that all these signaling pathways are under study precisely because they are essential for the biology of normal cells (not only stem cells). For example, it has recently been described that canonical Wnt signaling regulates different processes of the hematopoietic development in a dosage-dependent manner (Luis et al. 2011), in such a way that there are lineage-specific Wnt dosages that are different for hematopoietic stem cells (HSCs), myeloid and T lymphoid precursors during hematopoiesis. Therefore, the interference with Wnt signaling might seem to be non-toxic for some developmental stages but can be unexpectedly harmful for other ones. Therefore, a higher degree of knowledge and control are required before we can selectively target stem cell developmental signaling pathways with the aim of affecting only CSCs without interfering with the homeostasis of the normal cellular counterparts. This aspect is not new in cancer therapy, and all anti cancer drugs at use today have narrow therapeutic margins and still have serious side effects because of their lack of specificity. Still, current antiproliferative drugs leave stem cells (both normal and tumoral) largely unaffected, allowing for a regeneration of the damaged tissues (hematopoietic system, mucose epithelia, hair follicles, etc.) once the chemotherapy is interrupted. However, if normal stem cells were killed by therapy, this would lead to the death of the affected patient. The problem is that our knowledge of the normal biology of stem cells and their signaling routes is still very incomplete. Therefore, most targeting approaches will have to be, at the moment, largely empirical and will require tightly controlled clinical trials to verify their lack of toxicity to normal stem cells (see last section).

Another pathway of great importance for the biology of normal stem cells is the PI3K signaling pathway, that acts by integrating numerous different signals (growth factors, nutrients and oxygen status) to control cell proliferation, growth, and survival (Chalhoub and Baker 2009). It includes several downstream factors like Akt and mTorc1 or the FoxO family of transcription factors, and it is negatively regulated by some others like Pten, Pml or Tsc1/2. Alterations in any of these elements in this so-called PML/PTEN/Akt/mTOR/FoxO signaling network have profound consequences for the biology of stem cells, and all of them have been described as being mutated in many different types of cancers (Ito et al. 2009). Although the molecular nature of most of these signaling proteins makes them good potentially druggable targets, however, the complexity of the interrelationships of the different members of the network makes it extremely difficult to specifically interfere with CSC functions. A potentially successful example of this possibility is illustrated by the recent finding that the TGF-β–FoxO signaling maintains CSCs in chronic myeloid leukemia (Naka et al. 2010); cells with nuclear localization of Foxo3a and reduced Akt phosphorylation were found to be enriched in CSC activity. Furthermore, the capacity of Foxo3a-KO CML CSCs to cause disease is significantly decreased when compared with CSCs from a WT background. Also, it seems that TGF-β regulates Akt activation in CSCs and controls the localization of Foxo3a. Finally, in a proof-of-principle experiment, the combination of TGF-β inhibition, Foxo3a deficiency and imatinib treatment led to the efficient exhaustion of CML *in vivo*, and the treatment of human CML CSCs with a TGF-β inhibitor blocked their colony-forming capacity *in vitro* (Naka et al. 2010).

One of the most important developmental pathways affecting stem cell biology is the Hedgehog (Hh) signaling route, and it is consequently very frequently altered in cancers (Varjosalo and Taipale 2008, Ng and Curran 2011). The Hh proteins act as morphogens that control multiple developmental processes. They organize the behavior of stem cells within their target tissues, in such a way that concentration gradients of Hh proteins act in combination with other signaling pathways to control the apoptosis, proliferation and differentiation programs of stem and progenitor cells. In this manner, different populations of progenitor cells can be induced to expand or contract. This includes and affects CSCs. In small cell lung cancer (SCLC) cells the Hh pathway is activated, promoting clonogenicity of human SCLC cells *in vitro* and tumoral progression in injected mice *in vivo* (Park et al. 2011). Deletion of Hh signaling molecule smoothened (Smo) strongly interfered with tumoral spread in the mouse, therefore confirming the inhibition of the Hedgehog pathway as a potential therapeutic strategy to control the disease progression and recurrence in SCLC patients. Another potentially interesting field of application for the already-existing anti-Hh signaling pathway drugs are hematopoietic tumors, since it has recently been described in mice that Hedgehog signaling is dispensable for adult HSC function and that normal hematopoiesis can take place in the absence of functional Smo (Gao et al. 2009, Hofmann et al. 2009). Therefore, targeting of Hh pathway-dependent tumoral cells could be possible without interfering with the viability of the hematopoietic system. Since Hh signaling has been shown to be involved in the survival of BCR-ABL-dependent chronic myelogenous leukemia (Dierks et al. 2008, Zhao et al. 2009), and the Smo inhibitor cyclopamine seems to be effective in this context in experimental model systems, these findings open a promising new avenue for the treatment of this hematopoietic pathology.

Homeobox-containing genes are among the most potent regulators of cell fate, especially in the specification of the body plan in the early embryos, but also at later stages of development and in adult stem cells, and they are very frequently mutated, inactivated or deregulated in human cancers (Shah and Sukumar 2010). In a mouse model of MN1-induced acute myeloid leukemias (AML) it has recently been demonstrated that the MEIS1/AbdB-like HOX-protein complex is required for the leukemogenesis to take place (Heuser et al. 2011). Furthermore, by using a single-cell transformation assay selecting isolated hematopoietic stem and progenitor cells as cellular targets, the authors could identify a narrow window of transformation susceptibility during hematopoietic differentiation, in such a way that only common myeloid progenitors (CMP) could be transformed by MN1, but not granulocyte-macrophage progenitors (GMP). This susceptibility required the MEIS1/AbdB-like HOX-protein complex activity, and its inhibition demonstrated antileukemic activity, suggesting that propensity to transformation might be considered a therapeutic target, and it might be dependent on specific developmental regulators (Heuser et al. 2011). However, until now, no drugs have been described that can specifically and efficiently target HOX proteins, therefore making this approach unpractical for the time being.

Closely related to the homeobox-containing genes from the developmental point of view are the Polycomb family of transcriptional repressors, a group of epigenetic regulators that play essential roles in controlling cell fate determination and maintenance (Pietersen and van

Lohuizen 2008). One of the Polycomb members most frequently altered in cancer is *BMI1*, a member of the Polycomb repressive complex 1 (PRC1). Bmi1 is directly involved in the epigenetic regulation of cell proliferation and cell cycle by controlling the *Ink4a/Arf* locus encoding the $p16^{Ink4a}$ and $p19^{Arf}$ proteins and therefore, indirectly controlling the expression and function of essential tumor suppressors like Rb and p53 that regulate cell cycle, apoptosis, DNA repair and senescence. Therefore, Bmi1 is an essential regulator of the maintenance of adult HSCs, and Bmi1 KO mice develop hematopoietic failure in early adulthood (Park et al. 2003). Additionally, it is also been established that Bmi1 is required for the self-renewal of leukemic CSCs (Lessard and Sauvageau 2003). These results have recently been supported by the demonstration that, beyond controlling cell cycle regulators, Bmi1 is also essential for the efficient reprogramming of myeloid progenitors into leukemic CSCs in an MLL-AF9-dependent models, showing that leukemic fusion genes require Polycomb to establish CSC-specific transcriptional profiles (Yuan et al. 2011). However, this requirement for Bmi1 seems to be oncogene-dependent and related with the capacity of the fusion oncogenes to activate the *Hox* genes (tightly controlled by the Polycomb genes) in the reprogramming process (Smith et al. 2011). Indeed, the oncogenes AML1-ETO and PLZF-RARα, which do not activate Hox, when introduced in cells lacking Bmi1, only were capable of triggering senescence. However, the chimeric oncogene MLL-AF9, which can induce the expression of *Hoxa7* and *Hoxa9*, could efficiently transform *Bmi1*-KO cells. MLL-AF9 expression could not induce leukemia in double *Bmi1*, *Hoxa9* KO mice, but any of these two genes was enough to rescue the capacity of MLL-AF9 to establish CSCs in this background. These results reveal the functional relationship between MLL/Hox and Bmi1 in regulating cellular senescence versus CSC development, and suggest a synergistic targeting of both molecules as a potential way of eradicating leukemic CSCs (Smith et al. 2011).

Among the developmental mechanisms deregulated in cancer, the Notch signaling pathway is one of the best studied ones (Lobry et al. 2011). This, like the Hedgehog and the Wnt pathways, is dependent on extrinsic signals that interact with the extracellular portion of the Notch protein, causing a proteolytic cleavage that leads to the internalization of the active intracellular domain of Notch (NICD), that will translocate to the nucleus where it will interact with coactivators like RBPJk to control the transcription of many target genes. The Notch pathway is highly evolutionarily conserved and it is implicated in many processes during embryogenesis and in the self-renewing of tissues in the adult organism. Its functions include stem cell maintenance, cell fate specification, proliferation, and apoptosis. In the adult, it is especially important for the hematopoietic system, and it is frequently altered in T acute lymphoblastic leukemias

(T-ALL), where NOTCH1 becomes constitutively activated as a consequence of chromosomal translocations (Ellisen et al. 1991) or activating mutations (Weng et al. 2004). However, aberrant Notch signaling has been described in many other types of tumors (Ranganathan et al. 2011). Notch drives tumorigenesis mainly by stimulating cell cycle progression and inhibiting apoptosis, effects mediated by the transcriptional regulation of its target genes. However, it can also work as a tumor suppressor in some other contexts (Ranganathan et al. 2011). In transgenic T-ALL mouse models, it has been shown that the inhibition of Notch signaling by inhibitors of the γ-secretase (the protease that releases the active Notch form) significantly reduced or eliminated leukemic CSCs and prolonged the survival of the animals (Tatarek et al. 2011). Another possible way of interfering with this pathway is to target the process of Notch dimerization, since it has been demonstrated that dimeric Notch complexes are not necessary for T-cell fate specification but are required for T-cell leukemic transformation, so this process could be selectively inhibited (Liu et al. 2010). Finally, instead of trying to interfere with the Notch pathway inside the cells, one could conceivably block the extracellular signals that activate it (in cases where there is not intracellularly constitutively active). This has been achieved in early-passage colon tumor xenograft models treated with antibodies against the Delta-like 4 ligand (DLL4, one of the signals that activates Notch cleavage in mammals). The treatment inhibited the expression of Notch target genes, reduced tumoral cell proliferation and reduced CSC frequency in this model, in an effect that was not related to the inhibition of angiogenesis (Hoey et al. 2009). Therefore, and as we can see, our detailed knowledge of the Notch signaling pathway is providing us with many tools to try and target different functional processes in a CSC-specific way.

Other pathways involved in self-renewal or maintenance of stem cell identity have already been described in detail previously (Sagrera et al. 2009) and only discrete advances have been achieved regarding them in the last years. For example, TGF-β has been shown to cooperate with the leukemia inhibitory factor (LIF) in the maintenance and self-renewal of glioma CSCs (Penuelas et al. 2009), therefore identifying LIF as a potential therapeutic target against glioma CSCs and providing a molecular mechanism through which anti-TGF-β therapies might be successful against glioma. Other proteins belonging to the TGF-β superfamily are Nodal and Activin, both of them important regulators of embryonic stem (ES) cell fate. Both have been shown to regulate the self-renewal of pancreatic CSCs (Donahue and Dawson 2011, Lonardo et al. 2011), since they are highly expressed in these cells in contrast to their low levels in differentiated tumor cells. Their pharmacological inhibition abolished self-renewal and tumorigenicity of the treated cells, therefore opening the door to new therapeutic approaches to treat this devastating cancer. Regarding other molecule involved in stem cell biology, telomerase, its inhibition using imetelstat (a oligonucleotide with high affinity for the template region of the RNA component of telomerase that works as an inhibitor of the reverse transcriptase activity of telomerase) resulted in and inhibition of the clonogenic capacity of the putative multiple myeloma CSCs (Brennan et al. 2010).

Summarizing this section, we can see that the search for specific and effective anti-CSC drugs by researching into the stem cell-related molecular mechanisms has fallen short of its initial promises, at least as of today. Basically, although many interesting and potentially useful discoveries are being made, our knowledge of the signaling pathways is still not precise enough (in general) to guarantee the transfer to the clinical trials to provide anti-CSC success and specificity. Nevertheless, new compounds are constantly being discovered, and it is to be expected that one day we will understand these processes well enough, and then many of these new drugs will find their therapeutic window of opportunity.

New Emerging Anti-Cancer Stem Cell Molecular Targets

The enormous amount of active research taking place in the field of CSC biology is shedding light on previously unconsidered cellular mechanisms that could open new potential therapeutic anti-CSC approaches. In this last section we will go through several of these recent findings to provide the reader with a general impression of what might be the future aspects to pay attention to. Given the diversity of the biological functions involved, any attempt of classification would be very artificial, so we will proceed to list and briefly describe and discuss the different selected findings. Many new advances are taking place in the field of cancer therapy in general, but we will restrict our selection to therapeutic approaches focused against CSCs.

One interesting aspect that is becoming the focus of much attention is that of the particular requirements of tumor energetic metabolism. For example, the glycine decarboxylase (GLDC) enzyme has been found to be essential for CSCs in non-small cell lung cancer (NSCLC), and its ectopic overexpression promotes cellular transformation (Zhang et al. 2012). GLDC tumorigenicity is mediated by the induction of severe changes in the metabolism of glycine and serine, and in glycolysis, globally leading to an aberrant pyrimidine metabolism. These changes are also detected in the patients, where aberrant GLD activation correlates with a worse prognosis. Since enzymes are in general easily druggable targets, this recent finding might provide an interesting therapeutic approach in the near future (Zhang et al. 2012). Another aspect of metabolism that is affected in tumors is the generation of reactive oxygen species (ROS). Normal stem cells in different locations, like the central nervous system or the HSC, contain lower levels of ROS than differentiated cells, and it seems that this aspect is related to their stem cell nature (see (Diehn et al. 2009) and references therein). In the case of the mammary tissue, both normal mammary epithelial stem cells and human and mouse breast CSCs contain lower ROS levels that the corresponding differentiated (normal or tumoral) cells. Since ROS are critical mediators of cell death induced by ionizing radiation, it is tempting to speculate that this could be one reason why CSCs develop less DNA damage and are therefore more resistant to death after radiotherapy (Baumann et al. 2008, Diehn et al. 2009, Koch et al. 2010). This is demonstrated by the fact that depletion of ROS scavengers in CSCs by using specific inhibitors results in their radiosensitization and in decreased clonogenicity (Diehn et al. 2009). Also, hypoxia-inducible factor 2α (HIF2α) has been shown to be critical for the tumorigenicity of glioma stem cells (Li et al. 2009), and targeting HIFs inhibits their self-renewal, proliferation, and survival *in vitro*, and diminishes tumor initiation potential *in vivo*. Furthermore, HIF2A expression correlates with poor survival in glioma patients (Li et al. 2009). Also in hematological malignancies it has been reported that Hif1α maintains mouse lymphoma CSCs by repressing a negative feedback loop in the Notch pathway, and that this route can be targeted with the HIF-inhibitor echinomycin to selectively eliminate these cells (Wang et al. 2011). The oxygen levels are not only relevant for the survival of CSCs in the primary tumor, but they are also important for the initiation of metastases in distal organs, since this process is facilitated by the existence of the so-called premetastatic niches that generate a permissive environment for the tumoral cells. In this context, it has recently been found that the lysyl oxidase (LOX) enzyme is secreted by hypoxic tumor cells, and that it modifies the extracellular matrix and induces the recruitment of $Mac1^+$ myeloid cells, therefore finally leading to an increased growth of the metastatic tumor (Erler et al. 2009). These characteristics identify LOX as a feasible target for the treatment and prevention of tumoral metastases. Also in this context of metastatic colonization, many other signals have been shown to play important roles in the initiation of the process. Recently, a specific component of the extracellular matrix, periostin (POSTN), has been described to be expressed in the stroma of the primary tumor (Malanchi et al. 2012). Infiltrating metastatic cells need to induce POST expression in the target organ to successfully initiate colonization, since POSTN seems to be necessary for CSC maintenance by recruiting Wnt ligands and therefore increasing Wnt signaling in these cells. According to these data, preventing the initiation of the niche generation could be a new strategy to treat metastatic disease (Malanchi et al. 2012).

Another relevant aspect concerning the biology of CSCs is that of stem cell dormancy (Goss and Chambers 2010, Uhr and Pantel 2011). According to the CSC hypothesis, CSC dormancy would be the responsible for the long-term relapses of many cancer patients who spent many years in remission ("cancer cure"). Dormancy is a normal aspect of stem cell biology that forms part of their capacity of sustaining the maintenance of a tissue for the whole life of the organism. In the case of the hematopoietic system, dormant HSCs are induced to proliferate in response to injuries to reestablish tissue homeostasis. It has recently been described that one of the main signals involved in the exit from dormancy is Interferon-α (IFNα), that induces the exit of HSCs from G0 (Essers et al. 2009). Dormant HSCs are resistant to the antiproliferative drug 5-fluoro-uracil (5FU). However, when pretreated with IFNα they are induced to proliferate and become therefore sensitive to 5FU. In the case of AML, it was demonstrated that $CD34^{+}CD38^{-}$ human primary AML CSCs reside in the endosteal region of the bone marrow and are chemotherapy resistant (Ishikawa et al. 2007). Recently it has been shown that these cells are quiescent, and that they can be induced to enter the cell cycle by treatment with granulocyte colony-stimulating factor (G-CSF) (Saito et al. 2010). This G-CSF treatment, in combination with cell cycle-dependent chemotherapy, significantly increases the elimination of human primary AML stem cells i*n vivo* in a mouse transplantation model (Goff and Jamieson 2010, Saito et al. 2010). These results suggest therefore that forcing CSCs out of dormancy (at least, for hematopoietic tumors) could help to making them susceptible to conventional chemotherapeutic agents (Goss and Chambers 2010). However, it must be taken into account that the failure of both conventional and targeted therapies in use today is not necessarily due to the fact that the CSCs are hidden away from the drugs used. Actually, it seems that, in some cases, the problem is that the drug does not have the desired effect in CSCs. The most paradigmatic case is that of the anti-CML drug imatinib mesylate (Gleevec®) which targets the BCR-ABL oncogene, and has transformed CML from a life-threatening disease into a chronic condition. However, the patients still carry residual leukemic cells, and disease reappears when the treatment is interrupted. The reason for this is not that the drug cannot enter CSCs, but rather that they are insensitive to the drug. Indeed, CML CSCs are not addicted to the BCR-ABL oncogene and are therefore insensitive to the imatinib treatment (Corbin et al. 2011). And all this in spite of the fact that CML is most probably the best studied type of tumor, and research in CML has usually paved the way for the findings in many other cancers: it was in CML where the first translocation was described, and it has been the first cancer to have a specific, oncogene-targeted blockbuster drug (imatinib, that actually works very well against differentiated tumor cells… but it does not kill CSCs). Therefore, several potential new strategies have been proposed to try to overcome the CSC problem in the case of CML. For example, it has recently been found that BCR-ABL controls the expression of the cytokine Interleukin-6 (IL6), creating a paracrine feedback loop that redirects CML multipotent progenitors towards the myeloid lineage at the expense of lymphoid differentiation (Reynaud et al. 2011), therefore suggesting that therapeutically blocking IL6 might help to control CML CSCs. Also, this study has found that CML CSCs are hyperproliferative versus normal HSCs due to major alterations in cell cycle controls and changes in the Notch pathway, two aspects that could also be potentially targeted. In a different study it has been shown that the arachidonate 5-lipoxygenase (5-LO) gene (Alox5) is a vital regulator of CML CSCs. In its absence, BCR-ABL could not induce leukemia in mice, and treatment of a CML mouse model with a 5-LO inhibitor impaired CSC function and prolonged mice survival (Chen et al. 2009, Chen et al. 2009), therefore adding a new potential target to the list. We have described in the previous section how Foxo3a is required for CML CSCs maintenance (Naka et al. 2010). New results have recently shed more light into the downstream molecular mechanisms responsible for this observation, since it has been found that the BCL6 transcription factor is a critical effector of FoxO in the self-renewal signaling in CML CSCs (Hurtz et al. 2011). BCL6 represses $p19^{ARF}$ and *TP53* in CML cells and is required for leukemia initiation. Furthermore, peptide inhibition of BCL6 in human CML cells selectively eradicates $CD34^+CD38^-$ CML CSCs from patient-derived samples (Hurtz et al. 2011). Another strategy based on drugs of wider spectrum is to treat CML CSCs with histone deacetylase inhibitors (HDACi) to induce genome-wide epigenetic changes in the target cells. Treatment with HDACi in combination with imatinib was significantly more effective in inducing apoptosis in CML CSCs when compared with treatment with imatinib alone (Zhang et al. 2010). Therefore, as we can see, since today's standard treatment against CML is imatinib, most new findings are based on compound treatments including this drug. However, in the context of elimination of CML CSCs, exploring other approaches could also be potentially useful. For example, it has been described that the new drug omacetaxine mepesuccinate inhibited the proliferation of both CML and B-ALL CSCs and provided a significant survival benefit in mouse models (Chen et al. 2009). The mechanism of action of omacetaxine is still unknown, but it is different from that of tyrosine kinase inhibitors, opening new molecular routes to the treatment of CML (Chen et al. 2009).

Most of the targets that we have described up to know are intracellular molecules, or the intracellular signaling portion of developmental pathways. This is due to the fact that, in general, the studies aimed at prospectively enriching and purifying CSCs populations have taught us that most CSCs cannot be identified by the presence of a unique, specific surface marker. Indeed (and very much like any other stem cell in the organism), the combination of several present and absent markers is required to define populations that are enriched (not even 100% purified) on CSCs. This makes it difficult to use external surface markers as CSC targets. However, there are certain contexts in which this possibility could be considered. For example, a neutralizing antibody targeted against the Interleukin-3 (IL-3) receptor α chain (CD123) managed to attack AML CSCs, impairing their homing to the bone marrow (BM) in a transplantation setting into nonobese diabetic/severe-combined immunodeficient (NOD/SCID) mice (Jin et al. 2009). Also in AML, and in a context more closely related with the human patient situation, the group of Irving Weissman has identified the surface protein CD47 as a potential anti-CSC therapeutic target (Jaiswal et al. 2009, Majeti et al. 2009, Ritchie and Smyth 2009). In normal conditions, CD47 expression protects normal HSCs from phagocytosis by macrophages. The action of mobilizing cytokines induces a transient upregulation of CD47 on HSCs and progenitors, just prior to and during their migratory phase, to protect them from macrophage engulfment. However, CD47 also becomes upregulated on myeloid leukemias, and its overexpression increases pathogenicity by allowing CSCs to avoid phagocytosis. Treating mice that had been engrafted with human AML CSCs with an anti-CD47 antibody led to AML depletion by targeting the CSCs (Jaiswal et al. 2009, Majeti et al. 2009, Ritchie and Smyth 2009). Another surface molecular with potential anti-CSC therapeutic use is Integrin-$\alpha 6$ in the context of glioblastoma (Lathia et al. 2010), where it is expressed together with conventional CSC markers, and its targeting inhibits self-renewal, proliferation, and tumor formation capacity. A different surface molecule recently reported to be a promising target against AML CSCs is T cell immunoglobulin mucin-3 (TIM-3) (Kikushige et al. 2010), which has been found to be expressed in the CSCs of most AML types, but not in normal HSCs. TIM-3$^+$ AML cells successfully transplanted human AML into recipient mice, but TIM-3$^-$ cells could not do it. Furthermore, an anti-TIM-3 antibody blocked the transplantation capability of human AML cells, and not of normal human HSCs, into recipient mice. Finally, treatment of human-AML engrafted mice with this antibody reduced tumor burden and eliminated the capacity of tumor cells to engraft into secondary recipients (Kikushige et al. 2010). Another surface marker that is frequently used for the prospective purification of CSCs from solid tumors is CD44. CD44 is an adhesion molecule expressed in many other non-tumoral cellular contexts. However, it has recently been described that a specific variant of CD44 (CD44v) controls the intracellular level of reduced glutathione (GSH) by interacting with xCT, a glutamate-cystine transporter (Ishimoto et al. 2011). In human gastrointestinal cancer cells, a high level of CD44 expression is related with an increased GSH synthesis and defense against ROS. The elimination of CD44 induced loss of xCT from the cell surface and suppressed tumor growth in a transgenic mouse model of gastric cancer. These results also provide an example of how CSCs might make use of molecules abundantly expressed in many tissues for functions that can be specifically targeted in an anti-CSC therapeutic approach (Zoller 2011).

As new research clarifies the molecular details of the processes of differentiation, new regulators of development are being identified for many tissues, and, as it could not be otherwise, very often they are also found to be involved in cancer development. In melanoma, recent results show that the inhibition of the microphthalmia associated transcription factor (Mitf), the master regulator of melanocyte differentiation, increases the tumorigenic potential of melanoma cells and upregulates the expression of stem cell-specific transcription factors like Oct4 and Nanog (Cheli et al. 2011). In melanoma cells, a slow-growing population can be identified that presents low Mitf levels and high tumorigenic potential, and the elimination of this population decreases tumor formation. It seems that Mitf, together with the cell-cycle regulator p27, acts as a molecular switch controlling the transition between melanoma CSCs and their tumoral differentiated progeny, and that elimination of low-Mitf cells might be a potentially applicable strategy to cure melanoma (Cheli et al. 2011). In glioblastomas, the tyrosine kinase c-Met is activated in neurospheres enriched for glioblastoma CSCs, and c-Met expression correlates with stem cell marker expression. Apparently, c-Met induces the expression of reprogramming factors and counteracts differentiation-promoting factors in neurospheres, therefore enhancing the population of glioblastoma CSCs (Li et al. 2011).

Finally, to close down this heterogeneous series of new emerging (potential) anti-CSC target molecules, we have the microRNAs. These molecules have essential functions as regulators of most processes in cellular biology, and their roles in cancer and (potentially) in cancer therapy are being confirmed constantly by new published results (Lee and Dutta 2009, Kasinski and Slack 2011). However, their involvement in CSC biology, although for sure it will be proven unquestionably important in a few years, is still not fully characterized. There are nonetheless some examples in the literature already: microRNA29a is highly expressed in HSCs and becomes down-regulated upon differentiation (Han et al. 2010). When it is ectopically expressed in mouse HSCs leads to an AML-like myeloproliferative disorder by inducing the *de novo* acquisition of self-renewal capacity by myeloid progenitors. Furthermore, this miRNA has been detected in human AML primary cells, suggesting that it could play a similar role in human pathology, by converting myeloid progenitors into self-renewing AML CSCs (Han et al. 2010). In the context of pancreatic cancer, the microRNA-34a (miR-34a), a transcriptional target of p53, has been shown to be down-regulated. The restoration of miR-34a expression levels in human pancreatic CSCs by epigenetic means (using the demethylating agent 5-Aza-2'-deoxycytidine (5-Aza-dC) and the histone deacetylase inhibitor Vorinostat (SAHA)) strongly inhibited cell proliferation, cell cycle progression, self-renewal, epithelial to mesenchymal transition (EMT) and invasion, therefore suggesting that miR-34a regulates pancreatic cancer by modulating its CSC characteristics (Nalls et al. 2011). Something similar has been shown for miR-200 in breast cancer. Indeed, miR-200c was down-regulated in human breast CSCs and in normal human and mouse mammary stem/progenitor cells (Shimono et al. 2009). BMI1, that, as we have already discussed, is a known regulator of stem cell self-renewal, was modulated by miR-200c, and miR-200c expression inhibited the clonal expansion of breast cancer cells and suppressed tumor formation driven by human breast CSCs *in vivo* (Shimono et al. 2009). It seems that the role of miR-200 in CSC survival is largely mediated by regulation of epigenetic modifications since, in addition to controlling BMI1, it has also been shown to directly target Suz12, a subunit of the Polycomb repressor complex 2 (PRC2). Loss of miR-200 during the formation of CSCs increases Suz12 expression and, through Histone 3 Lysine 37 (H3K37) trimethylation, leads to repression of E-cadherin, therefore increasing the invasive capacity of breast CSCs (Iliopoulos et al. 2010). This interaction between miR-200 and Suz12 is highly conserved, and probably is an ancient regulatory mechanism to control stem cell growth.

Future Problems, Future Prospects

Now we have reviewed the most relevant results from the scientific literature in the last 3–4 years and, although several new potential anti-CSC targets have been identified, the aftertaste is still somewhat sour: no real breakthroughs have been made. Furthermore, research in the field of anti-CSC-targeted therapy is starting to look disappointingly similar to the basic cancer research of the last 20 years: many small incremental findings about minor details of molecular mechanisms, usually in very specific experimental contexts. All very interesting and new, but all of it too far from any realistic therapeutic application. If we really want the CSC theory to make a difference on the way we treat cancer, we really have to start accepting all the implications of the theory, rather than keeping on doing the same kind of research and just adding the "CSC" label to it.

From our point of view, new research tools and approaches are required that take the CSC hypothesis into account from a bottom-up perspective. This means that the CSC theory should be the starting point for the experimental design, rather than the top-bottom approaches used today, in which old systems and models are still used, just pretending that they are equally useful under the constrains of the CSC theory as they were before.

One of the first requirements for the design of a feasible anti-CSC drug is to be able to validate its lack of toxicity against normal stem cells (Jordan 2008). This requires a solid and rational screening design with well-defined controls. Recently, a multi-step *in vitro* and *in vivo* approach has been designed to identify compounds that can target AML CSCs and at the same time, counter-selecting the compounds against normal HSCs to screen for undesired toxicity. 4000 compounds were screened on leukemia cell lines derived from human experimental leukemogenesis models (McDermott et al. 2011). This high-throughput screening gave 80 hits, of which 10 compounds were less toxic to HSCs. Among them one compound (kinetic riboside (KR)) was characterized further and it showed comparable efficacy to standard therapies against blast cells from 63 primary leukemias (McDermott et al. 2011). Also recently, the sesquiterpene lactone parthenolide (PTL) has been identified as an agent able to selectively target leukemic CSCs without damaging HSC, due to an overdependence on NF-kB signaling in the tumoral versus the normal cells (Guzman et al. 2002, Guzman et al. 2005, Guzman et al. 2007). Using these results as a starting point, Hassane et al. (Hassane et al. 2008) cleverly designed an screening in which the transcriptional profile of cells in response to PTL was determined by microarray analysis, and afterwards used to identify other compounds producing a similar transcriptional response. In this way, one does not need to understand the pathways involved in the response, but can just focus on the response itself. Using this approach they identified 2 new agents, celastrol and 4- hydroxy-2-nonenal, that effectively eradicate AML CSCs (Hassane et al. 2008). Based on the results obtained in this last study, other structurally related compounds were tested, and it was found that Δ12-prostaglandin J3, an omega-3 fatty acid–derived metabolite, also selectively kills leukemia stem cells in mice infected with Friend erythroleukemia virus or expressing BCR-ABL (Hegde et al. 2011). This kind of approach has also been applied to non-hematological tumors; in the case of breast cancer, a chemical screening allowed to discover compounds with selective toxicity for breast CSCs. Among them, salinomycin reduced CSC with 100 times more efficiency that the commonly used chemotherapeutic agent paclitaxel (Gupta et al. 2009). Treatment of mice with salinomycin inhibits tumor growth *in vivo* and induces global gene expression changes suggestive of the loss of breast CSCs.

All these screenings must, for obvious throughput reasons, be performed *in vitro*. However, their *in vivo* validation in a suitable model is an essential requisite to move them forward into trials. Indeed, we have mentioned some of the problems related to the need of avoiding toxicity to normal stem cells. But these are not the only ones, there can be many other unexpected difficulties. For example, stem cells might not be the only important cells requiring a particular molecular route. Indeed, there are other types of cells that can present stem cell-like features at certain point of their life, and therefore an unspecific therapy might have unwanted effects on them. For instance, in the mouse immune system, both memory T and memory B cells share a transcriptional self-renewal profile with HSCs (Luckey et al. 2006), a property that has also been found in humans (Gattinoni et al. 2011), and B cells in the germinal center have been shown to divide asymmetrically, a property highly related to stem cell self-renewal (Barnett et al. 2012). Another problem might be related to the fact that maybe more than one type of stem cells contribute to generate a given tissue, so testing *in vitro* the toxicity against a certain stem cell type might not account for the full effect that a drug might have *in vivo* on tissue renewal. For example, the mammary gland contains different types of stem cells that contribute to its development and maintenance (Van Keymeulen et al. 2011). Also, in the small intestine, a new population of stem cells has recently been identified, showing that Bmi1-expressing stem cells form a reserve stem cell pool for the replenishment, under stress conditions, of the Lgr5-expressing cells that work under nonpathological circumstances (Tian et al. 2011). Finally, the niche will undoubtedly play an essential role in the biology of CSCs as it does for normal stem cells, and this aspect will have to be taken into account in the evaluation of any new strategy (Beck et al. 2011, Gobaa et al. 2011).

Therefore, the development of *in vivo* models that can accurately reproduce CSC-driven tumors similarly as how they happen in humans is not just an essential step for our understanding of the biology of CSCs and for the identification of new anti-CSCs targets and biomarkers but, furthermore, improved tumor models will be required as the definitive system in which new therapies can be tested before they can be translated into humans (Sanchez-Garcia et al. 2007, Vicente-Duenas et al. 2009, Castellanos et al. 2010, Vicente-Duenas et al. 2010). Since, as of today, most models are still based in theoretical concepts that ignore the CSC theory, the generation of CSC-based animal models will be the main bottleneck in the coming years for the discovery, development and translation of new real anti-CSC targets that can help the patients.

Acknowledgements

Research in C.C. lab was partially supported by FEDER, Fondo de Investigaciones Sanitarias (PI080164), CSIC P.I.E. 200920I055 and 201120E060, and from an Institutional grant from the "Fundación Ramón Areces". Research in I.S.G. group was partially supported by FEDER and by MICINN (SAF2009-08803 to I.S.G.), by Junta de Castilla y León (REF. CSI007A11-2 and Proyecto Biomedicina 2009–2010), by MEC OncoBIO Consolider-Ingenio 2010 (Ref. CSD2007-0017), by NIH grant (R01 CA109335-04A1), by Sandra Ibarra Foundation, by Group of Excellence Grant (GR15) from Junta de Castilla y Leon and by Proyecto en Red de Investigación en Celulas Madre Tumorales en Cancer de Mama, supported by Obra Social

Kutxa y Conserjería de Sanidad de la Junta de Castilla y Leon. All Spanish funding is co-sponsored by the European Union FEDER program. I.S.G. is an API lab of the EuroSyStem project. E.C.S. is recipient of a JAE-predoc fellowship from CSIC. I.R.C. is recipient of a FPU fellowship from MICINN. E.C.S. is recipient of a "Residencia de Estudiantes" fellowship.

References

Abollo-Jimenez, F., R. Jimenez, and C. Cobaleda. 2010. Physiological cellular reprogramming and cancer. Semin Cancer Biol 20: 98–106.

Anderson, K., C. Lutz, F.W. van Delft, C.M. Bateman, Y. Guo, S.M. Colman, H. Kempski, A.V. Moorman, I. Titley, J. Swansbury, L. Kearney, T. Enver, and M. Greaves. 2011. Genetic variegation of clonal architecture and propagating cells in leukaemia. Nature 469: 356–361.

Angers, S., and R.T. Moon. 2009. Proximal events in Wnt signal transduction. Nat Rev Mol Cell Biol 10: 468–477.

Barker, N. 2008. The canonical Wnt/beta-catenin signalling pathway. Methods Mol Biol 468: 5–15.

Barker, N., and H. Clevers. 2006. Mining the Wnt pathway for cancer therapeutics. Nat Rev Drug Discov 5: 997–1014.

Barnett, B.E., M.L. Ciocca, R. Goenka, L.G. Barnett, J. Wu, T.M. Laufer, J.K. Burkhardt, M.P. Cancro, and S.L. Reiner. 2012. Asymmetric B Cell Division in the Germinal Center Reaction. Science 335: 342–344.

Baumann, M., M. Krause, and R. Hill. 2008. Exploring the role of cancer stem cells in radioresistance. Nat Rev Cancer 8: 545–554.

Beck, B., G. Driessens, S. Goossens, K.K. Youssef, A. Kuchnio, A. Caauwe, P.A. Sotiropoulou, S. Loges, G. Lapouge, A. Candi, G. Mascre, B. Drogat, S. Dekoninck, J.J. Haigh, P. Carmeliet, and C. Blanpain. 2011. A vascular niche and a VEGF-Nrp1 loop regulate the initiation and stemness of skin tumours. Nature 478: 399–403.

Ben-Porath, I., M.W. Thomson, V.J. Carey, R. Ge, G. W. Bell, A. Regev, and R.A. Weinberg. 2008. An embryonic stem cell-like gene expression signature in poorly differentiated aggressive human tumors. Nat Genet 40: 499–507.

Bertucci, F., N. Cervera, and D. Birnbaum. 2007. A gene signature in breast cancer. N Engl J Med 356: 1887-8; author reply 1887–1888.

Bonnet, D., and J.E. Dick. 1997. Human acute myeloid leukemia is organized as a hierarchy that originates from a primitive hematopoietic cell. Nat Med 3: 730–737.

Brennan, S.K., Q. Wang, R. Tressler, C. Harley, N. Go, E. Bassett, C.A. Huff, R.J. Jones, and W. Matsui. 2010. Telomerase inhibition targets clonogenic multiple myeloma cells through telomere length-dependent and independent mechanisms. PLoS One 5.

Castellanos, A., C. Vicente-Duenas, E. Campos-Sanchez, J.J. Cruz, F.J. Garcia-Criado, M.B. Garcia-Cenador, P.A. Lazo, J. Perez-Losada, and I. Sanchez-Garcia. 2010. Cancer as a reprogramming-like disease: implications in tumor development and treatment. Semin Cancer Biol 20: 93–97.

Clevers, H. 2011. The cancer stem cell: premises, promises and challenges. Nat Med 17: 313–319.

Cobaleda, C., and I. Sanchez-Garcia. 2009. B-cell acute lymphoblastic leukaemia: towards understanding its cellular origin. Bioessays 31: 600–609.

Corbin, A.S., A. Agarwal, M. Loriaux, J. Cortes, M.W. Deininger, and B.J. Druker. 2011. Human chronic myeloid leukemia stem cells are insensitive to imatinib despite inhibition of BCR-ABL activity. J Clin Invest 121: 396–409.

Chabner, B.A., and T.G. Roberts, Jr. 2005. Timeline: Chemotherapy and the war on cancer. Nat Rev Cancer 5: 65–72.

Chalhoub, N., and S.J. Baker. 2009. PTEN and the PI3-kinase pathway in cancer. Annu Rev Pathol 4: 127–150.

Cheli, Y., S. Giuliano, T. Botton, S. Rocchi, V. Hofman, P. Hofman, P. Bahadoran, C. Bertolotto, and R. Ballotti. 2011. Mitf is the key molecular switch between mouse or human melanoma initiating cells and their differentiated progeny. Oncogene 30: 2307–2318.

Chen, Y., Y. Hu, S. Michaels, D. Segal, D. Brown, and S. Li. 2009. Inhibitory effects of omacetaxine on leukemic stem cells and BCR-ABL-induced chronic myeloid leukemia and acute lymphoblastic leukemia in mice. Leukemia 23: 1446–1454.

Chen, Y., Y. Hu, H. Zhang, C. Peng, and S. Li. 2009. Loss of the Alox5 gene impairs leukemia stem cells and prevents chronic myeloid leukemia. Nat Genet 41: 783–792.

Chen, Y., D. Li, and S. Li. 2009. The Alox5 gene is a novel therapeutic target in cancer stem cells of chronic myeloid leukemia. Cell Cycle 8: 3488–3492.

de Sousa, E.M.F., S. Colak, J. Buikhuisen, J. Koster, K. Cameron, J.H. de Jong, J.B. Tuynman, P.R. Prasetyanti, E. Fessler, S.P. van den Bergh, H. Rodermond, E. Dekker, C.M. van der Loos, S.T. Pals, M.J. van de Vijver, R. Versteeg, D.J. Richel, L. Vermeulen, and J.P. Medema. 2011. Methylation of cancer-stem-cell-associated wnt target genes predicts poor prognosis in colorectal cancer patients. Cell Stem Cell 9: 476–485.

Dick, J.E. 2008. Stem cell concepts renew cancer research. Blood 112: 4793–4807.

Diehn, M., R.W. Cho, N.A. Lobo, T. Kalisky, M.J. Dorie, A.N. Kulp, D. Qian, J.S. Lam, L.E. Ailles, M. Wong, B. Joshua, M.J. Kaplan, I. Wapnir, F.M. Dirbas, G. Somlo, C. Garberoglio, B. Paz, J. Shen, S.K. Lau, S.R. Quake, J.M. Brown, I.L. Weissman, and M.F. Clarke. 2009. Association of reactive oxygen species levels and radioresistance in cancer stem cells. Nature 458: 780–783.

Dierks, C., R. Beigi, G.R. Guo, K. Zirlik, M.R. Stegert, P. Manley, C. Trussell, A. Schmitt-Graeff, K. Landwerlin, H. Veelken, and M. Warmuth. 2008. Expansion of Bcr-Abl-positive leukemic stem cells is dependent on Hedgehog pathway activation. Cancer Cell 14: 238–249.

Ding, L., T.J. Ley, D.E. Larson, C.A. Miller, D.C. Koboldt, J.S. Welch, J.K. Ritchey, M.A. Young, T. Lamprecht, M.D. McLellan, J.F. McMichael, J.W. Wallis, C. Lu, D. Shen, C.C. Harris, D.J. Dooling, R.S. Fulton, L.L. Fulton, K. Chen, H. Schmidt, J. Kalicki-Veizer, V.J. Magrini, L. Cook, S.D. McGrath, T.L. Vickery, M.C. Wendl, S. Heath, M.A. Watson, D.C. Link, M.H. Tomasson, W.D. Shannon, J.E. Payton, S. Kulkarni, P. Westervelt, M.J. Walter, T.A. Graubert, E.R. Mardis, R.K. Wilson, and J.F. Dipersio. 2012. Clonal evolution in relapsed acute myeloid leukaemia revealed by whole-genome sequencing. Nature 481: 157–163.

Donahue, T.R., and D.W. Dawson. 2011. Nodal/Activin signaling: a novel target for pancreatic cancer stem cell therapy. Cell Stem Cell 9: 383–384.

Ellisen, L.W., J. Bird, D.C. West, A.L. Soreng, T.C. Reynolds, S.D. Smith, and J. Sklar. 1991. TAN-1, the human homolog of the Drosophila notch gene, is broken by chromosomal translocations in T lymphoblastic neoplasms. Cell 66: 649–661.

Eppert, K., K. Takenaka, E.R. Lechman, L. Waldron, B. Nilsson, P. van Galen, K.H. Metzeler, A. Poeppl, V. Ling, J. Beyene, A.J. Canty, J.S. Danska, S.K. Bohlander, C. Buske, M.D. Minden, T.R. Golub, I. Jurisica, B.L. Ebert, and J.E. Dick. 2011. Stem cell gene expression programs influence clinical outcome in human leukemia. Nat Med 17: 1086–1093.

Erler, J.T., K.L. Bennewith, T.R. Cox, G. Lang, D. Bird, A. Koong, Q.T. Le, and A.J. Giaccia. 2009. Hypoxia-induced lysyl oxidase is a critical mediator of bone marrow cell recruitment to form the premetastatic niche. Cancer Cell 15: 35–44.

Essers, M.A., S. Offner, W.E. Blanco-Bose, Z. Waibler, U. Kalinke, M.A. Duchosal, and A. Trumpp. 2009. IFNalpha activates dormant haematopoietic stem cells *in vivo*. Nature 458: 904–908.

Etzioni, R., N. Urban, S. Ramsey, M. McIntosh, S. Schwartz, B. Reid, J. Radich, G. Anderson, and L. Hartwell. 2003. The case for early detection. Nat Rev Cancer 3: 243–252.

Fialkow, P.J., R.J. Jacobson, and T. Papayannopoulou. 1977. Chronic myelocytic leukemia: clonal origin in a stem cell common to the granulocyte, erythrocyte, platelet and monocyte/macrophage. Am J Med 63: 125–130.

Frangioni, J.V. 2008. New technologies for human cancer imaging. J Clin Oncol 26: 4012–4021.

Gao, J., S. Graves, U. Koch, S. Liu, V. Jankovic, S. Buonamici, A. El Andaloussi, S.D. Nimer, B.L. Kee, R. Taichman, F. Radtke, and I. Aifantis. 2009. Hedgehog signaling is dispensable for adult hematopoietic stem cell function. Cell Stem Cell 4: 548–558.

Gattinoni, L., E. Lugli, Y. Ji, Z. Pos, C.M. Paulos, M.F. Quigley, J.R. Almeida, E. Gostick, Z. Yu, C. Carpenito, E. Wang, D.C. Douek, D.A. Price, C.H. June, F.M. Marincola, M. Roederer, and N.P. Restifo. 2011. A human memory T cell subset with stem cell-like properties. Nat Med 17:1290–1297.

Gentles, A.J., S.K. Plevritis, R. Majeti, and A.A. Alizadeh. 2010. Association of a leukemic stem cell gene expression signature with clinical outcomes in acute myeloid leukemia. JAMA 304: 2706–2715.

Gobaa, S., S. Hoehnel, M. Roccio, A. Negro, S. Kobel, and M.P. Lutolf. 2011. Artificial niche microarrays for probing single stem cell fate in high throughput. Nat Methods 8:949–955.

Goff, D., and C. Jamieson. 2010. Cycling toward elimination of leukemic stem cells. Cell Stem Cell 6: 296–297.

Goldstein, A.S., J. Huang, C. Guo, I.P. Garraway, and O.N. Witte. 2010. Identification of a cell of origin for human prostate cancer. Science 329: 568–571.

Goss, P.E., and A.F. Chambers. 2010. Does tumour dormancy offer a therapeutic target? Nat Rev Cancer 10: 871–877.

Greaves, M. 2010. Cancer stem cells: back to Darwin? Semin Cancer Biol 20: 65–70.

Greaves, M. 2011. Cancer stem cells renew their impact. Nat Med 17: 1046–1048.

Gregory, M.A., T.L. Phang, P. Neviani, F. Alvarez-Calderon, C.A. Eide, T. O'Hare, V. Zaberezhnyy, R.T. Williams, B.J. Druker, D. Perrotti, and J. Degregori. 2010. Wnt/Ca2+/NFAT signaling maintains survival of Ph+ leukemia cells upon inhibition of Bcr-Abl. Cancer Cell 18: 74–87.

Gupta, P.B., C.M. Fillmore, G. Jiang, S.D. Shapira, K. Tao, C. Kuperwasser, and E.S. Lander. 2011. Stochastic state transitions give rise to phenotypic equilibrium in populations of cancer cells. Cell 146: 633–644.

Gupta, P.B., T.T. Onder, G. Jiang, K. Tao, C. Kuperwasser, R.A. Weinberg, and E.S. Lander. 2009. Identification of selective inhibitors of cancer stem cells by high-throughput screening. Cell 138: 645–659.

Guzman, M.L., X. Li, C.A. Corbett, R.M. Rossi, T. Bushnell, J.L. Liesveld, J. Hebert, F. Young, and C.T. Jordan. 2007. Rapid and selective death of leukemia stem and progenitor cells induced by the compound 4-benzyl, 2-methyl, 1,2,4-thiadiazolidine, 3,5 dione (TDZD-8). Blood 110: 4436–4444.

Guzman, M.L., R.M. Rossi, L. Karnischky, X. Li, D.R. Peterson, D.S. Howard, and C.T. Jordan. 2005. The sesquiterpene lactone parthenolide induces apoptosis of human acute myelogenous leukemia stem and progenitor cells. Blood 105: 4163–4169.

Guzman, M.L., C.F. Swiderski, D.S. Howard, B.A. Grimes, R.M. Rossi, S.J. Szilvassy, and C.T. Jordan. 2002. Preferential induction of apoptosis for primary human leukemic stem cells. Proc Natl Acad Sci USA 99: 16220–16225.

Hamburger, A., and S.E. Salmon. 1977. Primary bioassay of human myeloma stem cells. J Clin Invest 60: 846–854.

Han, Y.C., C.Y. Park, G. Bhagat, J. Zhang, Y. Wang, J.B. Fan, M. Liu, Y. Zou, I.L. Weissman, and H. Gu. 2010. microRNA-29a induces aberrant self-renewal capacity in hematopoietic progenitors, biased myeloid development, and acute myeloid leukemia. J Exp Med 207: 475–489.

Hassane, D.C., M.L. Guzman, C. Corbett, X. Li, R. Abboud, F. Young, J.L. Liesveld, M. Carroll, and C.T. Jordan. 2008. Discovery of agents that eradicate leukemia stem cells using an in silico screen of public gene expression data. Blood 111: 5654–5662.

Hegde, S., N. Kaushal, K.C. Ravindra, C. Chiaro, K.T. Hafer, U.H. Gandhi, J.T. Thompson, J.P. van den Heuvel, M.J. Kennett, P. Hankey, R.F. Paulson, and K.S. Prabhu. 2011. {Delta}12-prostaglandin J3, an omega-3 fatty acid-derived metabolite, selectively ablates leukemia stem cells in mice. Blood 118: 6909–6919.

Hermann, P.C., S. Bhaskar, M. Cioffi, and C. Heeschen. 2010. Cancer stem cells in solid tumors. Semin Cancer Biol 20: 77–84.

Heuser, M., H. Yun, T. Berg, E. Yung, B. Argiropoulos, F. Kuchenbauer, G. Park, I. Hamwi, L. Palmqvist, C.K. Lai, M. Leung, G. Lin, A. Chaturvedi, B.K. Thakur, M. Iwasaki, M. Bilenky, N. Thiessen, G. Robertson, M. Hirst, D. Kent, N.K. Wilson, B. Gottgens, C. Eaves, M.L. Cleary, M. Marra, A. Ganser, and R.K. Humphries. 2011. Cell of origin in AML: susceptibility to MN1-induced transformation is regulated by the MEIS1/AbdB-like HOX protein complex. Cancer Cell 20: 39–52.

Hoey, T., W.C. Yen, F. Axelrod, J. Basi, L. Donigian, S. Dylla, M. Fitch-Bruhns, S. Lazetic, I.K. Park, A. Sato, S. Satyal, X. Wang, M.F. Clarke, J. Lewicki, and A. Gurney. 2009. DLL4 blockade inhibits tumor growth and reduces tumor-initiating cell frequency. Cell Stem Cell 5: 168–177.

Hofmann, I., E.H. Stover, D.E. Cullen, J. Mao, K.J. Morgan, B.H. Lee, M.G. Kharas, P.G. Miller, M.G. Cornejo, R. Okabe, S.A. Armstrong, N. Ghilardi, S. Gould, F.J. de Sauvage, A.P. McMahon, and D.G. Gilliland. 2009. Hedgehog signaling is dispensable for adult murine hematopoietic stem cell function and hematopoiesis. Cell Stem Cell 4: 559–567.

Huang, S. 2011. On the intrinsic inevitability of cancer: from foetal to fatal attraction. Semin Cancer Biol 21: 183–199.

Hurtz, C., K. Hatzi, L. Cerchietti, M. Braig, E. Park, Y.M. Kim, S. Herzog, P. Ramezani-Rad, H. Jumaa, M.C. Muller, W.K. Hofmann, A. Hochhaus, B.H. Ye, A. Agarwal, B.J. Druker, N.P. Shah, A.M. Melnick, and M. Muschen. 2011. BCL6-mediated repression of p53 is critical for leukemia stem cell survival in chronic myeloid leukemia. J Exp Med 11:2163–2174.

Husemann, Y., J.B. Geigl, F. Schubert, P. Musiani, M. Meyer, E. Burghart, G. Forni, R. Eils, T. Fehm, G. Riethmuller, and C.A. Klein. 2008. Systemic spread is an early step in breast cancer. Cancer Cell 13: 58–68.

Iliopoulos, D., M. Lindahl-Allen, C. Polytarchou, H.A. Hirsch, P.N. Tsichlis, and K. Struhl. 2010. Loss of miR-200 inhibition of Suz12 leads to polycomb-mediated repression required for the formation and maintenance of cancer stem cells. Mol Cell 39: 761–772.

Ishikawa, F., S. Yoshida, Y. Saito, A. Hijikata, H. Kitamura, S. Tanaka, R. Nakamura, T. Tanaka, H. Tomiyama, N. Saito, M. Fukata, T. Miyamoto, B. Lyons, K. Ohshima, N. Uchida, S. Taniguchi, O. Ohara, K. Akashi, M. Harada, and L.D. Shultz. 2007. Chemotherapy-resistant human AML stem cells home to and engraft within the bone-marrow endosteal region. Nat Biotechnol 25: 1315–1321.

Ishimoto, T., O. Nagano, T. Yae, M. Tamada, T. Motohara, H. Oshima, M. Oshima, T. Ikeda, R. Asaba, H. Yagi, T. Masuko, T. Shimizu, T. Ishikawa, K. Kai, E. Takahashi, Y. Imamura, Y. Baba, M. Ohmura, M. Suematsu, H. Baba, and H. Saya. 2011. CD44 variant regulates redox status in cancer cells by stabilizing the xCT subunit of system xc(–) and thereby promotes tumor growth. Cancer Cell 19: 387–400.

Ito, K., R. Bernardi, and P.P. Pandolfi. 2009. A novel signaling network as a critical rheostat for the biology and maintenance of the normal stem cell and the cancer-initiating cell. Curr Opin Genet Dev 19: 51–59.

Jaiswal, S., C.H. Jamieson, W.W. Pang, C.Y. Park, M.P. Chao, R. Majeti, D. Traver, N. van Rooijen, and I.L. Weissman. 2009. CD47 is upregulated on circulating hematopoietic stem cells and leukemia cells to avoid phagocytosis. Cell 138: 271–285.

Jin, L., E.M. Lee, H.S. Ramshaw, S.J. Busfield, A.G. Peoppl, L. Wilkinson, M.A. Guthridge, D. Thomas, E.F. Barry, A. Boyd, D.P. Gearing, G. Vairo, A.F. Lopez, J.E. Dick, and R.B.

Lock. 2009. Monoclonal antibody-mediated targeting of CD123, IL-3 receptor alpha chain, eliminates human acute myeloid leukemic stem cells. Cell Stem Cell 5: 31–42.

Jordan, C.T. 2008. Can we finally target the leukemic stem cells? Best Pract Res Clin Haematol 21: 615–620.

Kasinski, A.L., and F.J. Slack. 2011. Epigenetics and genetics. MicroRNAs en route to the clinic: progress in validating and targeting microRNAs for cancer therapy. Nat Rev Cancer 11: 849–864.

Kikushige, Y., T. Shima, S. Takayanagi, S. Urata, T. Miyamoto, H. Iwasaki, K. Takenaka, T. Teshima, T. Tanaka, Y. Inagaki, and K. Akashi. 2010. TIM-3 is a promising target to selectively kill acute myeloid leukemia stem cells. Cell Stem Cell 7: 708–717.

Koch, U., M. Krause, and M. Baumann. 2010. Cancer stem cells at the crossroads of current cancer therapy failures—radiation oncology perspective. Semin Cancer Biol 20: 116–124.

Kuster, L., R. Grausenburger, G. Fuka, U. Kaindl, G. Krapf, A. Inthal, G. Mann, M. Kauer, J. Rainer, R. Kofler, A. Hall, M. Metzler, L.H. Meyer, C. Meyer, J. Harbott, R. Marschalek, S. Strehl, O.A. Haas, and R. Panzer-Grumayer. 2011. ETV6/RUNX1-positive relapses evolve from an ancestral clone and frequently acquire deletions of genes implicated in glucocorticoid signaling. Blood 117: 2658–2667.

Lane, S.W., and D.G. Gilliland. 2010. Leukemia stem cells. Semin Cancer Biol 20: 71–76.

Lathia, J.D., J. Gallagher, J.M. Heddleston, J. Wang, C.E. Eyler, J. Macswords, Q. Wu, A. Vasanji, R.E. McLendon, A.B. Hjelmeland, and J.N. Rich. 2010. Integrin alpha 6 regulates glioblastoma stem cells. Cell Stem Cell 6: 421–432.

le Viseur, C., M. Hotfilder, S. Bomken, K. Wilson, S. Rottgers, A. Schrauder, A. Rosemann, J. Irving, R.W. Stam, L.D. Shultz, J. Harbott, H. Jurgens, M. Schrappe, R. Pieters, and J. Vormoor. 2008. In childhood acute lymphoblastic leukemia, blasts at different stages of immunophenotypic maturation have stem cell properties. Cancer Cell 14: 47–58.

Leder, K., E.C. Holland, and F. Michor. 2010. The therapeutic implications of plasticity of the cancer stem cell phenotype. PLoS One 5: e14366.

Lee, H.E., J.H. Kim, Y.J. Kim, S.Y. Choi, S.W. Kim, E. Kang, I.Y. Chung, I.A. Kim, E.J. Kim, Y. Choi, H.S. Ryu, and S.Y. Park. 2011. An increase in cancer stem cell population after primary systemic therapy is a poor prognostic factor in breast cancer. Br J Cancer 104: 1730–1738.

Lee, Y.S., and A. Dutta. 2009. MicroRNAs in cancer. Annu Rev Pathol 4: 199–227.

Lessard, J., and G. Sauvageau. 2003. Bmi-1 determines the proliferative capacity of normal and leukaemic stem cells. Nature 423: 255–260.

Li, Y., A. Li, M. Glas, B. Lal, M. Ying, Y. Sang, S. Xia, D. Trageser, H. Guerrero-Cazares, C.G. Eberhart, A. Quinones-Hinojosa, B. Scheffler, and J. Laterra. 2011. c-Met signaling induces a reprogramming network and supports the glioblastoma stem-like phenotype. Proc Natl Acad Sci USA 108: 9951–9956.

Li, Z., S. Bao, Q. Wu, H. Wang, C. Eyler, S. Sathornsumetee, Q. Shi, Y. Cao, J. Lathia, R.E. McLendon, A.B. Hjelmeland, and J.N. Rich. 2009. Hypoxia-inducible factors regulate tumorigenic capacity of glioma stem cells. Cancer Cell 15: 501–513.

Liu, C., J.C. Sage, M.R. Miller, R.G. Verhaak, S. Hippenmeyer, H. Vogel, O. Foreman, R.T. Bronson, A. Nishiyama, L. Luo, and H. Zong. 2011. Mosaic analysis with double markers reveals tumor cell of origin in glioma. Cell 146: 209–221.

Liu, H., A.W. Chi, K.L. Arnett, M.Y. Chiang, L. Xu, O. Shestova, H. Wang, Y.M. Li, A. Bhandoola, J.C. Aster, S.C. Blacklow, and W.S. Pear. 2010. Notch dimerization is required for leukemogenesis and T-cell development. Genes Dev 24: 2395–2407.

Lobry, C., P. Oh, and I. Aifantis. 2011. Oncogenic and tumor suppressor functions of Notch in cancer: it's NOTCH what you think. J Exp Med 208: 1931–1935.

Lonardo, E., P.C. Hermann, M.T. Mueller, S. Huber, A. Balic, I. Miranda-Lorenzo, S. Zagorac, S. Alcala, I. Rodriguez-Arabaolaza, J.C. Ramirez, R. Torres-Ruiz, E. Garcia, M. Hidalgo, D.A. Cebrian, R. Heuchel, M. Lohr, F. Berger, P. Bartenstein, A. Aicher, and C. Heeschen.

2011. Nodal/Activin signaling drives self-renewal and tumorigenicity of pancreatic cancer stem cells and provides a target for combined drug therapy. Cell Stem Cell 9: 433–446.
Luckey, C.J., D. Bhattacharya, A.W. Goldrath, I.L. Weissman, C. Benoist, and D. Mathis. 2006. Memory T and memory B cells share a transcriptional program of self-renewal with long-term hematopoietic stem cells. Proc Natl Acad Sci USA 103: 3304–3309.
Luis, T.C., B.A. Naber, P.P. Roozen, M.H. Brugman, E.F. de Haas, M. Ghazvini, W.E. Fibbe, J.J. van Dongen, R. Fodde, and F.J. Staal. 2011. Canonical wnt signaling regulates hematopoiesis in a dosage-dependent fashion. Cell Stem Cell 9: 345–356.
Majeti, R., M.P. Chao, A.A. Alizadeh, W.W. Pang, S. Jaiswal, K.D. Gibbs, Jr., N. van Rooijen, and I.L. Weissman. 2009. CD47 is an adverse prognostic factor and therapeutic antibody target on human acute myeloid leukemia stem cells. Cell 138: 286–299.
Malanchi, I., A. Santamaria-Martinez, E. Susanto, H. Peng, H.A. Lehr, J.F. Delaloye, and J. Huelsken. 2012. Interactions between cancer stem cells and their niche govern metastatic colonization. Nature 481: 85–89.
Marder, J. 2010. Cancer research. Childhood's cures haunted by adulthood's 'late effects'. Science 328: 1474–1475.
Marshall, E. 2011. Cancer research and the $90 billion metaphor. Science 331: 1540–1541.
McDermott, S.P., K. Eppert, F. Notta, M. Isaac, A. Datti, R. Al-Awar, J. Wrana, M.D. Minden, and J.E. Dick. 2011. A small molecule screening strategy with validation on human leukemia stem cells uncovers the therapeutic efficacy of kinetin riboside. Blood doi: 10.1182/blood-2011-01-330019
Merlos-Suarez, A., F.M. Barriga, P. Jung, M. Iglesias, M.V. Cespedes, D. Rossell, M. Sevillano, X. Hernando-Momblona, V. da Silva-Diz, P. Munoz, H. Clevers, E. Sancho, R. Mangues, and E. Batlle. 2011. The intestinal stem cell signature identifies colorectal cancer stem cells and predicts disease relapse. Cell Stem Cell 8: 511–524.
Meyer, L.H., S.M. Eckhoff, M. Queudeville, J.M. Kraus, M. Giordan, J. Stursberg, A. Zangrando, E. Vendramini, A. Moricke, M. Zimmermann, A. Schrauder, G. Lahr, K. Holzmann, M. Schrappe, G. Basso, K. Stahnke, H.A. Kestler, G. Te Kronnie, and K.M. Debatin. 2011. Early relapse in all is identified by time to leukemia in NOD/SCID mice and is characterized by a gene signature involving survival pathways. Cancer Cell 19: 206–217.
Molyneux, G., F.C. Geyer, F.A. Magnay, A. McCarthy, H. Kendrick, R. Natrajan, A. Mackay, A. Grigoriadis, A. Tutt, A. Ashworth, J.S. Reis-Filho, and M.J. Smalley. 2010. BRCA1 basal-like breast cancers originate from luminal epithelial progenitors and not from basal stem cells. Cell Stem Cell 7: 403–417.
Mullighan, C.G., L.A. Phillips, X. Su, J. Ma, C.B. Miller, S.A. Shurtleff, and J.R. Downing. 2008. Genomic analysis of the clonal origins of relapsed acute lymphoblastic leukemia. Science 322: 1377–1380.
Naka, K., T. Hoshii, T. Muraguchi, Y. Tadokoro, T. Ooshio, Y. Kondo, S. Nakao, N. Motoyama, and A. Hirao. 2010. TGF-beta-FOXO signalling maintains leukaemia-initiating cells in chronic myeloid leukaemia. Nature 463: 676–680.
Nalls, D., S.N. Tang, M. Rodova, R.K. Srivastava, and S. Shankar. 2011. Targeting Epigenetic Regulation of miR-34a for Treatment of Pancreatic Cancer by Inhibition of Pancreatic Cancer Stem Cells. PLoS One 6: e24099.
Ng, J.M., and T. Curran. 2011. The Hedgehog's tale: developing strategies for targeting cancer. Nat Rev Cancer 11: 493–501.
Notta, F., C.G. Mullighan, J.C. Wang, A. Poeppl, S. Doulatov, L.A. Phillips, J. Ma, M.D. Minden, J.R. Downing, and J.E. Dick. 2011. Evolution of human BCR-ABL1 lymphoblastic leukaemia-initiating cells. Nature 469: 362–367.
Park, I.K., D. Qian, M. Kiel, M.W. Becker, M. Pihalja, I.L. Weissman, S.J. Morrison, and M.F. Clarke. 2003. Bmi-1 is required for maintenance of adult self-renewing haematopoietic stem cells. Nature 423: 302–305.
Park, K.S., L.G. Martelotto, M. Peifer, M.L. Sos, A.N. Karnezis, M.R. Mahjoub, K. Bernard, J.F. Conklin, A. Szczepny, J. Yuan, R. Guo, B. Ospina, J. Falzon, S. Bennett, T.J. Brown,

A. Markovic, W.L. Devereux, C.A. Ocasio, J.K. Chen, T. Stearns, R.K. Thomas, M. Dorsch, S. Buonamici, D.N. Watkins, C.D. Peacock, and J. Sage. 2011. A crucial requirement for Hedgehog signaling in small cell lung cancer. Nat Med 7: 1504–1508.

Pece, S., D. Tosoni, S. Confalonieri, G. Mazzarol, M. Vecchi, S. Ronzoni, L. Bernard, G. Viale, P.G. Pelicci, and P.P. Di Fiore. 2010. Biological and molecular heterogeneity of breast cancers correlates with their cancer stem cell content. Cell 140: 62–73.

Penuelas, S., J. Anido, R.M. Prieto-Sanchez, G. Folch, I. Barba, I. Cuartas, D. Garcia-Dorado, M.A. Poca, J. Sahuquillo, J. Baselga, and J. Seoane. 2009. TGF-beta increases glioma-initiating cell self-renewal through the induction of LIF in human glioblastoma. Cancer Cell 15: 315–327.

Persson, A.I., C. Petritsch, F.J. Swartling, M. Itsara, F.J. Sim, R. Auvergne, D.D. Goldenberg, S.R. Vandenberg, K.N. Nguyen, S. Yakovenko, J. Ayers-Ringler, A. Nishiyama, W.B. Stallcup, M.S. Berger, G. Bergers, T.R. McKnight, S.A. Goldman, and W.A. Weiss. 2010. Non-stem cell origin for oligodendroglioma. Cancer Cell 18: 669–682.

Pietersen, A.M., and M. van Lohuizen. 2008. Stem cell regulation by polycomb repressors: postponing commitment. Curr Opin Cell Biol 20: 201–207.

Pui, C.H., L.L. Robison, and A.T. Look. 2008. Acute lymphoblastic leukaemia. Lancet 371: 1030–1043.

Quintana, E., M. Shackleton, H.R. Foster, D.R. Fullen, M.S. Sabel, T.M. Johnson, and S.J. Morrison. 2010. Phenotypic heterogeneity among tumorigenic melanoma cells from patients that is reversible and not hierarchically organized. Cancer Cell 18: 510–523.

Ranganathan, P., K.L. Weaver, and A.J. Capobianco. 2011. Notch signalling in solid tumours: a little bit of everything but not all the time. Nat Rev Cancer 11: 338–351.

Reynaud, D., E. Pietras, K. Barry-Holson, A. Mir, M. Binnewies, M. Jeanne, O. Sala-Torra, J.P. Radich, and E. Passegue. 2011. IL-6 controls leukemic multipotent progenitor cell fate and contributes to chronic myelogenous leukemia development. Cancer Cell 20: 661–673.

Ritchie, D.S., and M.J. Smyth. 2009. A new therapeutic target for leukemia comes to the surface. Cell 138: 226–8.

Roesch, A., M. Fukunaga-Kalabis, E.C. Schmidt, S.E. Zabierowski, P.A. Brafford, A. Vultur, D. Basu, P. Gimotty, T. Vogt, and M. Herlyn. 2010. A temporarily distinct subpopulation of slow-cycling melanoma cells is required for continuous tumor growth. Cell 141: 583–594.

Sagrera, A., J. Perez-Losada, M. Perez-Caro, R. Jimenez, I. Sanchez-Garcia, and C. Cobaleda 2009. Elimination of Cancer Stem Cells. Stem Cell Biology in Health and Disease. T. Dittmar and K. S. Zänker, Springer Science + Bussiness Media: 356–384.

Saito, Y., N. Uchida, S. Tanaka, N. Suzuki, M. Tomizawa-Murasawa, A. Sone, Y. Najima, S. Takagi, Y. Aoki, A. Wake, S. Taniguchi, L.D. Shultz, and F. Ishikawa. 2010. Induction of cell cycle entry eliminates human leukemia stem cells in a mouse model of AML. Nat Biotechnol 28: 275–280.

Sanchez-Garcia, I. 2009. The crossroads of oncogenesis and metastasis. N Engl J Med 360: 297–299.

Sanchez-Garcia, I. 2010. Getting to the stem of cancer. Semin Cancer Biol 20: 63–64.

Sanchez-Garcia, I., C. Vicente-Duenas, and C. Cobaleda. 2007. The theoretical basis of cancer-stem-cell-based therapeutics of cancer: can it be put into practice? Bioessays 29: 1269–1280.

Schurch, C., C. Riether, M.S. Matter, A. Tzankov, and A.F. Ochsenbein. 2012. CD27 signaling on chronic myelogenous leukemia stem cells activates Wnt target genes and promotes disease progression. J Clin Invest doi: 10.1172/JCI45977.

Shackleton, M. 2010. Normal stem cells and cancer stem cells: similar and different. Semin Cancer Biol 20: 85–92.

Shah, N., and S. Sukumar. 2010. The Hox genes and their roles in oncogenesis. Nat Rev Cancer 10: 361–371.

Sharma, S.V., D.Y. Lee, B. Li, M.P. Quinlan, F. Takahashi, S. Maheswaran, U. McDermott, N. Azizian, L. Zou, M.A. Fischbach, K.K. Wong, K. Brandstetter, B. Wittner, S. Ramaswamy,

M. Classon, and J. Settleman. 2010. A chromatin-mediated reversible drug-tolerant state in cancer cell subpopulations. Cell 141: 69–80.

Shats, I., M.L. Gatza, J.T. Chang, S. Mori, J. Wang, J. Rich, and J.R. Nevins. 2011. Using a stem cell-based signature to guide therapeutic selection in cancer. Cancer Res 71: 1772–1780.

Shimono, Y., M. Zabala, R.W. Cho, N. Lobo, P. Dalerba, D. Qian, M. Diehn, H. Liu, S.P. Panula, E. Chiao, F.M. Dirbas, G. Somlo, R.A. Pera, K. Lao, and M.F. Clarke. 2009. Downregulation of miRNA-200c links breast cancer stem cells with normal stem cells. Cell 138: 592–603.

Smith, L.L., J. Yeung, B.B. Zeisig, N. Popov, I. Huijbers, J. Barnes, A.J. Wilson, E. Taskesen, R. Delwel, J. Gil, M. Van Lohuizen, and C.W. So. 2011. Functional crosstalk between Bmi1 and MLL/Hoxa9 axis in establishment of normal hematopoietic and leukemic stem cells. Cell Stem Cell 8: 649–662.

Somervaille, T.C., C.J. Matheny, G.J. Spencer, M. Iwasaki, J.L. Rinn, D.M. Witten, H.Y. Chang, S.A. Shurtleff, J.R. Downing, and M.L. Cleary. 2009. Hierarchical maintenance of MLL myeloid leukemia stem cells employs a transcriptional program shared with embryonic rather than adult stem cells. Cell Stem Cell 4: 129–140.

Tatarek, J., K. Cullion, T. Ashworth, R. Gerstein, J.C. Aster, and M.A. Kelliher. 2011. Notch1 inhibition targets the leukemia-initiating cells in a Tal1/Lmo2 mouse model of T-ALL. Blood 118: 1579–1590.

Tian, H., B. Biehs, S. Warming, K.G. Leong, L. Rangell, O.D. Klein, and F.J. de Sauvage. 2011. A reserve stem cell population in small intestine renders Lgr5-positive cells dispensable. Nature 478: 255–259.

Uhr, J.W., and K. Pantel. 2011. Controversies in clinical cancer dormancy. Proc Natl Acad Sci USA 108: 12396–12400.

van Delft, F.W., S. Horsley, S. Colman, K. Anderson, C. Bateman, H. Kempski, J. Zuna, C. Eckert, V. Saha, L. Kearney, A. Ford, and M. Greaves. 2011. Clonal origins of relapse in ETV6-RUNX1 acute lymphoblastic leukemia. Blood 117: 6247–6254.

Van Keymeulen, A., A.S. Rocha, M. Ousset, B. Beck, G. Bouvencourt, J. Rock, N. Sharma, S. Dekoninck, and C. Blanpain. 2011. Distinct stem cells contribute to mammary gland development and maintenance. Nature 49: 189–193.

Varjosalo, M., and J. Taipale. 2008. Hedgehog: functions and mechanisms. Genes Dev 22: 2454–2472.

Veeman, M.T., J.D. Axelrod, and R.T. Moon. 2003. A second canon. Functions and mechanisms of beta-catenin-independent Wnt signaling. Dev Cell 5: 367–377.

Vicente-Duenas, C., C. Cobaleda, J. Perez-Losada, and I. Sanchez-Garcia. 2010. The evolution of cancer modeling: the shadow of stem cells. Dis Model Mech 3: 149–155.

Vicente-Duenas, C., M. Perez-Caro, F. Abollo-Jimenez, C. Cobaleda, and I. Sanchez-Garcia. 2009. Stem-cell driven cancer: "hands-off" regulation of cancer development. Cell Cycle 8: 1314–1318.

Visvader, J.E. 2011. Cells of origin in cancer. Nature 469: 314–322.

Wang, Y., A.V. Krivtsov, A.U. Sinha, T.E. North, W. Goessling, Z. Feng, L.I. Zon, and S.A. Armstrong. 2010. The Wnt/beta-catenin pathway is required for the development of leukemia stem cells in AML. Science 327: 1650–1653.

Wang, Y., Y. Liu, S.N. Malek, and P. Zheng. 2011. Targeting HIF1alpha eliminates cancer stem cells in hematological malignancies. Cell Stem Cell 8: 399–411.

Warr, M.R., E.M. Pietras, and E. Passegue. 2011. Mechanisms controlling hematopoietic stem cell functions during normal hematopoiesis and hematological malignancies. Wiley Interdiscip Rev Syst Biol Med 3: 681–701.

Weng, A.P., A.A. Ferrando, W. Lee, J.P. t. Morris, L.B. Silverman, C. Sanchez-Irizarry, S.C. Blacklow, A.T. Look, and J.C. Aster. 2004. Activating mutations of NOTCH1 in human T cell acute lymphoblastic leukemia. Science 306: 269–271.

Wingo, P.A., C.J. Cardinez, S.H. Landis, R.T. Greenlee, L.A. Ries, R.N. Anderson, and M.J. Thun. 2003. Long-term trends in cancer mortality in the United States, 1930–1998. Cancer 97: 3133–3275.

Yachida, S., S. Jones, I. Bozic, T. Antal, R. Leary, B. Fu, M. Kamiyama, R.H. Hruban, J.R. Eshleman, M.A. Nowak, V.E. Velculescu, K.W. Kinzler, B. Vogelstein, and C.A. Iacobuzio-Donahue. 2010. Distant metastasis occurs late during the genetic evolution of pancreatic cancer. Nature 467: 1114–1117.

Yeung, J., M.T. Esposito, A. Gandillet, B.B. Zeisig, E. Griessinger, D. Bonnet, and C.W. So. 2010. beta-Catenin mediates the establishment and drug resistance of MLL leukemic stem cells. Cancer Cell 18: 606–618.

Yuan, J., M. Takeuchi, M. Negishi, H. Oguro, H. Ichikawa, and A. Iwama. 2011. Bmi1 is essential for leukemic reprogramming of myeloid progenitor cells. Leukemia 25: 1335–1343.

Zhang, B., A.C. Strauss, S. Chu, M. Li, Y. Ho, K.D. Shiang, D.S. Snyder, C.S. Huettner, L. Shultz, T. Holyoake, and R. Bhatia. 2010. Effective targeting of quiescent chronic myelogenous leukemia stem cells by histone deacetylase inhibitors in combination with imatinib mesylate. Cancer Cell 17: 427–442.

Zhang, N., P. Wei, A. Gong, W.T. Chiu, H.T. Lee, H. Colman, H. Huang, J. Xue, M. Liu, Y. Wang, R. Sawaya, K. Xie, W.K. Yung, R.H. Medema, X. He, and S. Huang. 2011. FoxM1 Promotes beta-Catenin Nuclear Localization and Controls Wnt Target-Gene Expression and Glioma Tumorigenesis. Cancer Cell 20: 427–442.

Zhang, W.C., N. Shyh-Chang, H. Yang, A. Rai, S. Umashankar, S. Ma, B.S. Soh, L.L. Sun, B.C. Tai, M.E. Nga, K.K. Bhakoo, S.R. Jayapal, M. Nichane, Q. Yu, D.A. Ahmed, C. Tan, W.P. Sing, J. Tam, A. Thiruganananam, M.S. Noghabi, Y. Huei Pang, H.S. Ang, P. Robson, P. Kaldis, R.A. Soo, S. Swarup, E.H. Lim, and B. Lim. 2012. Glycine Decarboxylase Activity Drives Non-Small Cell Lung Cancer Tumor-Initiating Cells and Tumorigenesis. Cell 48: 259–272.

Zhao, C., A. Chen, C.H. Jamieson, M. Fereshteh, A. Abrahamsson, J. Blum, H.Y. Kwon, J. Kim, J.P. Chute, D. Rizzieri, M. Munchhof, T. VanArsdale, P.A. Beachy, and T. Reya. 2009. Hedgehog signalling is essential for maintenance of cancer stem cells in myeloid leukaemia. Nature 458: 776–769.

Zoller, M. 2011. CD44: can a cancer-initiating cell profit from an abundantly expressed molecule? Nat Rev Cancer 11: 254–267.

Chapter 8

Novel Cancer Therapies by Molecular Targeting Distinct Growth Factor Pathways and Drug Resistance-Associated Molecules in Cancer- and Metastasis-Initiating Cells

Murielle Mimeault[a,*] and Surinder K. Batra[b,*]

Introduction

Major advances in stem cell biology have revealed that the accumulation of genetic and/or epigenetic alterations in tissue-resident adult stem/progenitor cells or their early progenies endowed with a high self-renewal capacity may result in their malignant transformation into cancer stem/progenitor cells also designated as cancer- or tumor-initiating cells (Beachy et al. 2004, Liu et al. 2005, Kim et al. 2005, Rizo et al. 2006, Li and Neaves 2006, Vaish 2007, Nijhof et al. 2007, Shiras et al. 2007, Mimeault et al. 2007b, Gumucio et al. 2008, Mimeault and Batra 2010b, Sugiarto et al. 2011, Mimeault and Batra 2011b, Mimeault et al. 2012). Moreover, it has been shown that cancer progression is generally associated with the occurrence of distinct molecular events in cancer stem/progenitor cells and their progenies that may contribute to their acquisition of a more malignant

Department of Biochemistry and Molecular Biology, Eppley Institute for Research in Cancer and Allied Diseases, University of Nebraska Medical Center, Omaha, Nebraska 68198-5870, USA.
[a]E-mail: mmimeault@unmc.edu
[b]E-mail: sbatra@unmc.edu
*Corresponding authors

behavior. Typically, the inactivation of different tumor suppressor gene products combined with a stimulation of diverse oncogenic cascades initiated by different growth factors are involved in the progression of cancer to locally invasive and metastatic disease states (Beachy et al. 2004, Bell and Van Zant 2004, Larue and Bellacosa 2005, Li and Neaves 2006, Liu et al. 2006b, Nicolis 2007, Sengupta et al. 2007, Chen et al. 2007, Clement et al. 2007, Wang et al. 2007a, Marusyk and DeGregori 2008, Mimeault and Batra 2009, Mimeault and Batra 2010b, Motohara et al. 2011, Mimeault and Batra 2011b, Mimeault and Batra 2012). More specifically, the acquisition of a migratory phenotype by highly tumorigenic cancer stem/progenitor cells during epithelial-mesenchymal transition (EMT) process may lead to their invasion, dissemination and metastases at distant tissues or organs (Thiery 2002, Gotzmann et al. 2004, Derycke and Bracke 2004, Larue and Bellacosa 2005, Brabletz et al. 2005, Kuphal et al. 2005, Tso et al. 2006, Onoue et al. 2006, Mimeault and Batra 2007b, Mimeault et al. 2007b, Mani et al. 2008, Kabashima et al. 2009, Kurrey et al. 2009, Mimeault and Batra 2010b, Bao et al. 2011, Cao et al. 2011, Gulhati et al. 2011, Mimeault and Batra 2011b, Mimeault et al. 2012, Lamouille et al. 2012, Rhim et al. 2012, Zhang et al. 2012, Wang et al. 2012a). Additionally, the changes in the local microenvironment of cancer stem/progenitor cells, including in hypoxic zones and the release of soluble factors by host activated stromal cells at primary and secondary neoplasms, may also promote tumor growth (Lyden et al. 2001, Neiva et al. 2005, Li and Neaves 2006, Mimeault and Batra 2007b, Mimeault et al. 2007b, Olivotto and Dello 2008, Das et al. 2008, Gerlee and Anderson 2008, Mohyeldin et al. 2010, Box et al. 2010, Kuphal et al. 2010, Mimeault and Batra 2010b, Mathieu et al. 2011, Ma et al. 2011, Persano et al. 2011, Hashimoto et al. 2011, Shiozawa et al. 2011b, Mimeault and Batra 2011b, Mimeault et al. 2012).

To support the postulate that cancer stem/progenitor cells play a critical role in cancer development, subpopulations of immature cancer cells with stem cell-like properties have been identified in most of the human cancers (Hemmati et al. 2003, Al-Hajj et al. 2003, Singh et al. 2004, Yuan et al. 2004, Galli et al. 2004, Kim et al. 2005, Ponti et al. 2005, Bapat et al. 2005, Maitland et al. 2006, Hermann et al. 2007, Ricci-Vitiani et al. 2007, Prince et al. 2007, Mitsutake et al. 2007, Mimeault et al. 2007b, Yu et al. 2008, Wright et al. 2008, Fillmore and Kuperwasser 2008, Eramo et al. 2008, Yang et al. 2008, She et al. 2008, Chiba et al. 2008, Shi et al. 2008, Sung et al. 2008, Friel et al. 2008, Zhang et al. 2008a, Huang et al. 2009, Mimeault and Batra 2011b, Mimeault et al. 2012, Marsden et al. 2012). Cancer stem/progenitor cells expressed several stem cell-like markers such as telomerase, aldehyde dehydrogenase (ADLH), CD133, CD44, ATP binding-cassette (ABC) multidrug transporters and/or CXC chemokine receptor-4 (CXCR4) but lacked terminally differentiation marker expression. It has also been shown that highly tumorigenic cancer stem/progenitor cells were able to give rise *in vitro* and *in vivo* to the total mass of differentiated cancer cells expressing phenotypes of original patients' tumors (Hemmati et al. 2003, Al-Hajj et al. 2003, Singh et al. 2004, Yuan et al. 2004, Galli et al. 2004, Ponti et al. 2005, Kim et al. 2005, Bapat et al. 2005, Frank et al. 2005, Maitland et al. 2006, Hermann et al. 2007, Ricci-Vitiani et al. 2007, Prince et al. 2007, Yu et al. 2008, Wright et al. 2008, Fillmore and Kuperwasser 2008, Eramo et al. 2008, Yang et al. 2008, She et al. 2008, Chiba et al. 2008, Shi et al. 2008, Sung et al. 2008, Friel et al. 2008, Zhang et al. 2008a, Huang et al. 2009, Marsden et al. 2012). Importantly, the results from numerous recent studies have also revealed that cancer stem/progenitor cells may be more resistant than their differentiated progenies to current clinical cancer therapies (Hirschmann-Jax et al. 2004, Frank et al. 2005, Salmaggi et al. 2006, Bao et al. 2006, Haraguchi et al. 2006, Maitland et al. 2006, Liu et al. 2006a, Rich 2007, Hermann et al. 2007, Todaro et al. 2007, Wang et al. 2007a, Mimeault et al. 2007b, Ma et al. 2008, Fillmore and Kuperwasser 2008, Wright et al. 2008, She et al. 2008, Chiba et al. 2008, Shi et al. 2008, Sung et al. 2008, Friel et al. 2008, Loebinger et al. 2008, Zhou et al. 2008, Zhang et al. 2008a, Zhang et al. 2008b, Huang et al. 2009, Kurrey et al. 2009, Alvero et al. 2009, Mimeault et al. 2010a, Mimeault et al. 2010b, Mimeault and Batra 2011b, Mimeault et al. 2012, Hamada et al. 2012, Steg et al. 2012, Wang et al. 2012a). Consequently, the persistence of cancer- and metastasis-initiating cells with stem cell-like features at the primary and secondary malignant neoplasms after treatment initiation may be responsible for tumor re-growth and disease relapse. On this issue, we review here the intrinsic properties of cancer stem/progenitor cells as well as the molecular oncogenic events that frequently occur in these immature cancer cells and their local tumor microenvironment during the early and late stages of cancer progression and metastases at distant sites. The emphasis is on the deregulated growth factor cascades and drug resistance-associated molecules in cancer stem/progenitor cells that can contribute to their resistance to current cancer therapies and disease relapse. Of great clinical interest: we also report potential therapeutic targets in cancer- and metastasis-initiating cells and their progenies as well as their local microenvironment that may be exploited for improving the current anti-hormonal, radiation and chemotherapeutic treatments against aggressive, recurrent and lethal cancers.

Critical Functions of Cancer Stem/Progenitor Cells in Primary Cancer Development and Metastases

A growing body of evidence suggests that cancer development may derive from the clonal expansion and aberrant differentiation from cancer stem/

progenitor cells and/or their early progenies endowed with a high self-renewal capacity and aberrant differentiation potential (Beachy et al. 2004, Kim et al. 2005, Rizo et al. 2006, Li and Neaves 2006, Vaish 2007, Nijhof et al. 2007, Mimeault et al. 2007b, Gumucio et al. 2008, Mimeault and Batra 2010b, Sugiarto et al. 2011, Mimeault and Batra 2011b, Mimeault et al. 2012). Recent investigations have led to the identification and isolation of different subpopulations of cancer stem/progenitor cells in most of the common human cancers and established cancer cell lines (Mimeault et al. 2007b, Mimeault and Batra 2010b). The cancer types harboring a subset of cancer stem/progenitor cells include leukemias, melanomas, brain tumors and a variety of epithelial cancers such as skin, head and neck, thyroid, lung, cervical, renal, hepatic, esophageal, gastrointestinal, colon, bladder, pancreatic, prostatic, mammary and ovarian cancers (Hemmati et al. 2003, Al-Hajj et al. 2003, Singh et al. 2004, Yuan et al. 2004, Galli et al. 2004, Kim et al. 2005, Ponti et al. 2005, Bapat et al. 2005, Maitland et al. 2006, Hermann et al. 2007, Ricci-Vitiani et al. 2007, Prince et al. 2007, Mitsutake et al. 2007, Mimeault et al. 2007b, Yu et al. 2008, Wright et al. 2008, Fillmore and Kuperwasser 2008, Eramo et al. 2008, Yang et al. 2008, She et al. 2008, Chiba et al. 2008, Shi et al. 2008, Sung et al. 2008, Friel et al. 2008, Zhang et al. 2008a, Huang et al. 2009, Mimeault and Batra 2011b, Mimeault et al. 2012, Marsden et al. 2012). It has been observed that a complex signaling network involving an aberrant expression and/or activation of different growth factor signaling cascades in cancer cells may lead to cancer progression and metastases at distant tissues and organs (Beachy et al. 2004, Tso et al. 2006, Bao et al. 2006, Haraguchi et al. 2006, Liu et al. 2006a, Katoh 2007, Nicolis 2007, Sato et al. 2007, Sengupta et al. 2007, Mimeault et al. 2007b, Glinsky 2008, Mimeault and Batra 2010b, Mimeault and Batra 2011b, Shiozawa et al. 2011b, Mimeault et al. 2012). The common transforming events include the sustained activation of telomerase and diverse growth factor cascades such as hedgehog, epidermal growth factor (EGF)-EGFR receptor, Wnt/β-catenin, Notch, hyaluronan/CD44, transforming growth factor-β (TGF-β)/TGF-βR receptors, stromal cell-derived factor-1 (SDF-1)-CXCR4 and/or polycomb group (PcG) protein signaling pathways (Beachy et al. 2004, Kucia et al. 2005, Chen et al. 2006, Tso et al. 2006, Haraguchi et al. 2006, Liu et al. 2006a, Liu et al. 2006b, Katoh 2007, Clement et al. 2007, Nicolis 2007, Sato et al. 2007, Sengupta et al. 2007, Mimeault et al. 2007b, Bourguignon et al. 2008, Glinsky 2008, Bourguignon et al. 2009, Moriyama et al. 2010, Mimeault and Batra 2010b, Bao et al. 2011, Huynh et al. 2011, Mimeault and Batra 2011b, Mimeault et al. 2012, Cheng et al. 2012, Eichhorn et al. 2012, Wang et al. 2012a, Wang et al. 2012b). These tumorigenic pathways may induce the sustained stimulation of different downstream oncogenic signaling elements such as Ras/mitogen-activated protein kinases (MAPKs), phosphatidylinositol 3'-kinase (PI3K)/Akt/molecular target of rapamycin (mTOR), Myc, nuclear factor-kappaB (NF-κB) and B lymphoma Mo-MLV insertion region 1 homolog (Bmi-1) in cancer stem/progenitor cells and their progenies that contribute to their sustained growth, survival, invasion, metastases and treatment resistance (DePinho 2000, Schwarze et al. 2002, Jarrard et al. 2002, Bell and Van Zant 2004, Shay and Wright 2005, Shin et al. 2006, Maurelli et al. 2006, Sato et al. 2007, Deng and Chang 2007, Shiras et al. 2007, Vaish 2007, Zhou et al. 2007, Marusyk and DeGregori 2007, Mimeault et al. 2007b, Birnie et al. 2008, Ma et al. 2008, Pacifico et al. 2008, Marusyk and DeGregori 2008, Chiba et al. 2008, Mimeault et al. 2012, Venugopal et al. 2012, Wang et al. 2012b).

Critical Functions of Cancer Stem/Progenitor Cells in the Resistance to Current Cancer Therapies

Recent investigations have also indicated that cancer stem/progenitor cells may be intrinsically resistant to radiation therapy and different chemotherapeutic drugs and/or acquire an enhanced multidrug resistance (MDR) phenotype with cancer development (Hirschmann-Jax et al. 2004, Dean et al. 2005, Frank et al. 2005, Salmaggi et al. 2006, Bao et al. 2006, Liu et al. 2006a, Ginestier et al. 2007, Ho et al. 2007, de Jonge-Peeters et al. 2007, Kang and Kang 2007, Mimeault et al. 2007b, Dylla et al. 2008, Mimeault et al. 2008a, Mimeault and Batra 2008c, Mimeault et al. 2010a, Mimeault and Batra 2010a, Mimeault et al. 2010b, Mimeault and Batra 2011a, Hamada et al. 2012, Mimeault et al. 2012, Steg et al. 2012, Wang et al. 2012a). Several intrinsic properties of cancer stem/progenitor cells common with their normal counterpart, tissue-resident adult stem/progenitor cells, may contribute to their resistance to current cancer therapies and disease relapse. The phenotypic features associated with the treatment resistance of primitive cancer stem/progenitor cells include their slow division, high expression levels of diverse anti-apoptotic factors such as Bcl-2, survivin and NF-κB, ABC multidrug transporters and DNA repair and intracellular detoxifying enzymes (ALDH and O6-methylguanine DNA methyltransferase "MGMT") (Gottesman et al. 2002, Hirschmann-Jax et al. 2004, Dean et al. 2005, Frank et al. 2005, Bao et al. 2006, Salmaggi et al. 2006, Liu et al. 2006a, Ho et al. 2007, Kang and Kang 2007, Rich 2007, Xu et al. 2007, Johannessen et al. 2008, Matsui et al. 2008, Dylla et al. 2008, Shervington and Lu 2008, Chen et al. 2008, Widera et al. 2008, Birnie et al. 2008, Douville et al. 2009, Mimeault et al. 2010a, Mimeault and Batra 2010a, Mimeault et al. 2010b, Hamada et al. 2012, Mimeault et al. 2012). More particularly, the poorly differentiated cancer stem/progenitor cells at primary and secondary tumors may exist under a quiescent and less metabolically active state, and consequently they may be more resistant than their differentiated progenies to irradiation and cytotoxic drugs targeting proliferating cancer cells. Moreover, the most of cancer stem/progenitor cells, including side population (SP), also express high levels of ABC-multidrug transporters such as multidrug resistance 1 (MDR1/ABCB1) encoding P-glycoprotein (P-gp), breast cancer resistance protein-1 (BCRP-1/ABCB2) and/or multidrug resistance-associated proteins (MRPs) (Gerk and Vore 2002, Han and Zhang 2004, Mao and Unadkat 2005, Ozben 2006, Krishnamurthy and Schuetz 2006, Deeley et al. 2006, Wang et al. 2007a, Mimeault et al. 2007b, Loebinger et al. 2008, Shervington and Lu 2008, Shi et al. 2008, Bourguignon et al. 2008, Das et al. 2008, She et al. 2008, Mimeault et al. 2008a, Mimeault et al. 2010a, Mimeault et al. 2010b, Mimeault and Batra 2011b, Hamada et al. 2012, Mimeault et al. 2012). The ABC multidrug transporters may protect these immature cancer cells from cytotoxic effects induced by diverse structurally and functionally non-related chemotherapeutic drugs and in this manner contribute to their MDR phenotype. Specifically, the ABC transporters localized at the plasmic membrane can activately pump the intracellular cytotoxic agents outside of the cells through the expanse of the hydrolysis of ATP molecules. Certain ABC transporter types such as ABCA3 and ABCB2 that are found in endolysosomal compartment in cancer cells including cancer stem/progenitor cells may also mediate a sequestration of intracellular toxic compounds, including the chemotherapeutic drugs, and thereby promote the MDR phenomenon (Boonstra et al. 2004, Mack et al. 2006, Kaminski et al. 2006). On the other hand, high expression levels of DNA-mismatch repair and detoxifying enzymes in cancer stem/progenitor cells may be also responsible in part for their resistance to some chemotherapeutic drugs or ionizing radiation (Dean et al. 2005, Cai et al. 2005, Rabik et al. 2006, Liu et al. 2006a, Juillerat and Juillerat-Jeanneret 2007, Johannessen et al. 2008). For instance, a high expression level of MGMT in cancer stem/progenitor cells may protect them against the cytotoxic effects induced by methylating and chloroethylating agents [carmustine, 1,3-bis (2-chloroethyl)-1-nitrosourea or BCNU, methyl methanesulfonate (MMS) and temozolomide (TMZ)] by repairing alkyl adducts at the O6-position of guanine in DNA (Dean et al. 2005, Cai et al. 2005, Rabik et al. 2006, Liu et al. 2006a, Juillerat and Juillerat-Jeanneret 2007, Johannessen et al. 2008). Furthermore, the cancer-initiating cells found in melanoma, retinoblastoma, brain, breast, prostate, pancreatic and colorectal cancers also can express a high level of ALDH activity that may participate in the detoxification of a variety of compounds such as the active metabolite of an alkylating agent, cyclophosphamide and treatment resistance (Seigel et al. 2005, Feldmann et al. 2007, Bar et al. 2007, Dylla et al. 2008, Wright et al. 2008, Douville et al. 2009, van den Hoogen et al. 2010, Boonyaratanakornkit et al. 2010, Ishizawa et al. 2010, Stuelten et al. 2010, van den Hoogen et al. 2011).

In addition, the acquisition of stem cell-like features and more malignant behavior by cancer stem/progenitor cells and their progenies during cancer progression, and more particularly through the EMT process generally occurs in numerous aggressive cancers such as melanomas and brain, skin, prostate, ovarian, mammary, hepatic, gastrointestinal, pancreatic and colorectal carcinomas (Fig. 1) (Thiery 2002, Larue and Bellacosa 2005, Brabletz et al. 2005, Kuphal et al. 2005, Tso et al. 2006, Onoue et al. 2006, Mimeault and Batra 2007b, Mimeault et al. 2007b, Mani et al. 2008, Kabashima et al. 2009, Kurrey et al. 2009, Mimeault and Batra 2010b, Bao et al. 2011, Cao et al. 2011, Gulhati et al. 2011, Mimeault and Batra 2011b, Mimeault et al. 2012, Lamouille et al. 2012, Rhim et al. 2012, Zhang et al. 2012).

The gene products often altered in the cancer-initiating cells and their progenies during the EMT program include a decreased expression

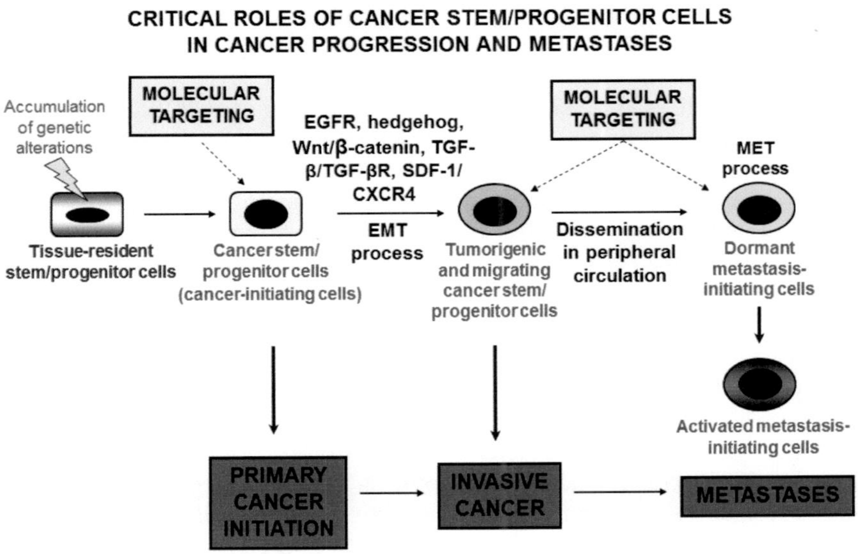

Fig. 1. Proposed models of the critical functions of cancer stem/progenitor cells for epithelial cancer initiation, progression and metastases. The scheme shows the malignant transformation of tissue-resident adult stem/progenitor cells into tumorigenic cancer stem/progenitor cells also designated cancer-initiating cells that may be induced through the genetic and/or epigenetic alterations in these immature cells and the changes in their local microenvironments. The cancer stem/progenitor cells endowed with a high self-renewal ability and aberrant differentiation potential may give rise to the bulk mass of differentiated cancer cells forming primary cancer. The acquisition of migratory phenotype by tumorigenic cancer stem/progenitor cells during the epithelial-mesenchymal transition (EMT) process induced through the sustained activation of diverse growth factors may lead to their invasion, dissemination *via* the peripheral circulation and metastases at distant tissues and organs under specific environmental conditions. Therefore, the molecular targeting of cancer- and metastasis-initiating cells, which are resistant to current cancer therapies, is of major importance to prevent the tumor re-growth and disease relapse after treatment initiation.

of E-cadherin concomitant with an up-regulation of different oncogenic signaling elements such as N-cadherin, vimentin, snail, slug, twist, β-catenin, CXCR4 and anti-apoptotic factors organs (Thiery 2002, Gotzmann et al. 2004, Derycke and Bracke 2004, Larue and Bellacosa 2005, Brabletz et al. 2005, Kuphal et al. 2005, Tso et al. 2006, Onoue et al. 2006, Mimeault and Batra 2007b, Mimeault et al. 2007b, Mani et al. 2008, Kabashima et al. 2009, Kurrey et al. 2009, Mimeault and Batra 2010b, Bao et al. 2011, Cao et al. 2011, Gulhati et al. 2011, Mimeault and Batra 2011b, Mimeault et al. 2012, Lamouille et al. 2012, Rhim et al. 2012, Zhang et al. 2012, Wang et al. 2012a). These EMT process-associated molecules in turn may confer a more invasive phenotype and enhanced treatment resistance to cancer-initiating cells during primary cancer progression (Fig. 1) (Gotzmann et al. 2004, Brabletz et al. 2005, Kuphal et al. 2005, Tso et al. 2006, Onoue et al. 2006, Mimeault and Batra 2007b, Mimeault et al. 2007b, Mani et al. 2008, Kabashima et al. 2009, Kurrey et al. 2009, Mimeault and Batra 2010b, Bao et al. 2011, Cao et al. 2011, Gulhati et al. 2011, Mimeault and Batra 2011b, Mimeault et al. 2012, Lamouille et al. 2012, Rhim et al. 2012, Zhang et al. 2012, Wang et al. 2012a). Moreover, diverse alterations in the oncogenic signaling elements such as hypoxia-inducible factors (HIFs), vascular endothelial growth factor (VEGF)-VEGFR receptor system, matrix metalloproteases (MMPs) and cyclooxygenase-2 (COX-2) may occur under hypoxic conditions and promote the cancer progression to locally invasive and metastatic states (Olivotto and Dello 2008, Gerlee and Anderson 2008, Das et al. 2008, Kondoh 2008, Mohyeldin et al. 2010, Kuphal et al. 2010, Mathieu et al. 2011, Ma et al. 2011, Frank et al. 2011, Hashimoto et al. 2011, Shiozawa et al. 2011b).

In addition, the cancer development is often accompanied by an enhanced glycolysis in cancer cells including cancer stem/progenitor cells (Olivotto and Dello 2008, Gerlee and Anderson 2008, Das et al. 2008, Kondoh 2008). This phenomenon known as Warburg effect may contribute to the resistance of cancer cells to oxidative stress as well as their survival in intratumoral hypoxic conditions (Olivotto and Dello 2008, Gerlee and Anderson 2008, Das et al. 2008, Kondoh 2008). Hence, the molecular targeting of these different oncogenic signaling elements and drug resistance-associated molecules in cancer stem/progenitor cells and their progenies may constitute promising strategies for overcoming treatment resistance and disease recurrence.

Novel Cancer Therapies by Targeting Cancer Stem/ Progenitor Cells and Their Local Tumor Microenvironment

The cancer progression to locally advanced, invasive and metastatic disease states that are resistant to current cancer therapies remains one of the leading causes of disease recurrence and cancer-related deaths (Mimeault and

Batra 2010b, Siegel et al. 2012, Mimeault and Batra 2012). Consequently, the molecular targeting of distinct oncogenic products in cancer- and metastasis-initiating cells offers great promise for developing new therapeutic options against aggressive and recurrent cancers (Mimeault and Batra 2008c, Mimeault and Batra 2010c, Mimeault and Batra 2010d, Mimeault and Batra 2012). In this regard, we are report here recent advances in the development of novel targeted strategies to eradicate cancer stem/progenitor cells and their progenies, and which could be used for improving the current treatments against locally invasive, metastatic and recurrent cancers.

Molecular targeting of cancer- and metastasis-initiating cells and their progenies

The molecular targeting of distinct deregulated signaling elements that mediate the transforming events occurring in highly tumorigenic cancer- and metastasis-initiating cells represents a new promising approach to improve the current cancer therapies (Mimeault and Batra 2007a, Mimeault and Batra 2008c, Mimeault and Batra 2010c, Mimeault and Batra 2010d). Among the potential molecular targets often altered in cancer stem/progenitor cells and their progenies during cancer progression, there are diverse developmental signaling pathways. These deregulated pathways include hedgehog, EGFR, Wnt/β-catenin, Notch, hyaluronan (HA)/CD44, interleukin-4 (IL-4)/IL-4 receptor α (IL-4Rα), TGF-β/TGF-βRs, Bmi-1, stem cell factor (SCF)/KIT receptor, extracellular matrix (ECM)/integrin and/or SDF-1/CXCR4 signaling elements (Figs. 1 and 2) (Ghatak et al. 2002, Beachy et al. 2004, Karhadkar et al. 2004, Misra et al. 2006, Rubin and de Sauvage 2006, Chen et al. 2006, Todaro et al. 2007, Feldmann et al. 2007, Sims-Mourtada et al. 2007, Villanueva et al. 2007, Mimeault and Batra 2007a, Mimeault et al. 2007b, Mimeault and Batra 2007b, Bourguignon et al. 2008, Mimeault and Batra 2008a, Mimeault et al. 2008b, Mimeault and Batra 2008b, Mimeault and Batra 2008c, Kabashima et al. 2009, Mimeault and Batra 2010c, van der Horst et al. 2011, Huynh et al. 2011, Venugopal et al. 2012, Cheng et al. 2012, Eichhorn et al. 2012, Wang et al. 2012c).

It has been shown that the blockage of these tumorigenic pathways by using specific inhibitor or antagonist, monoclonal antibody (mAb) or antisense oligonucleotide (As) led to growth inhibition, apoptotic cell death and/or a reduction of invasiveness or metastatic spread of cancer-initiating cells and their progenies *in vitro* or in animal models *in vivo* (Table 1, Figs. 1 and 2) (Ghatak et al. 2002, Beachy et al. 2004, Karhadkar et al. 2004, Misra et al. 2006, Rubin and de Sauvage 2006, Chen et al. 2006, Todaro et al. 2007, Feldmann et al. 2007, Sims-Mourtada et al. 2007, Villanueva et al. 2007, Mimeault and Batra 2007a, Mimeault et al. 2007b, Mimeault and Batra 2007b, Bourguignon et al. 2008, Mimeault and Batra 2008a, Mimeault et al.

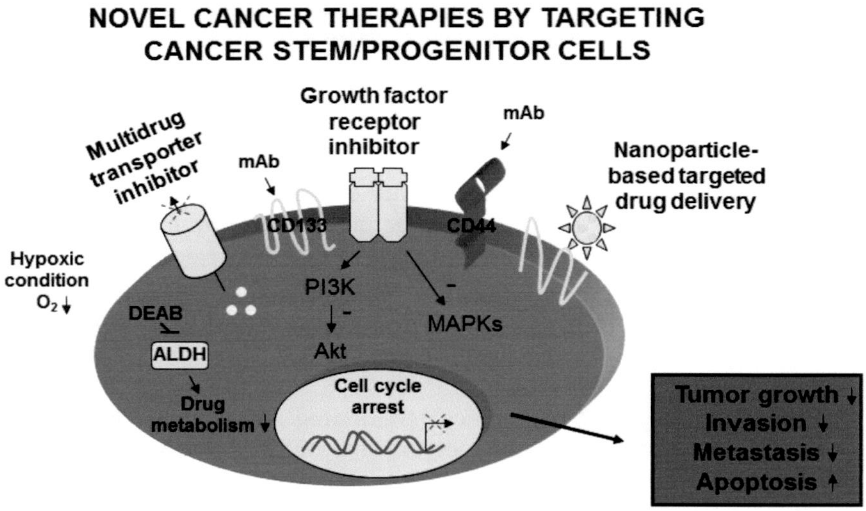

Fig. 2. New cancer therapies by molecular targeting distinct signaling pathways involved in aggressive behavior and multidrug resistance phenotype of cancer stem/progenitor cells. The inhibitory effects induced by distinct pharmacological agents such as the selective inhibitors of growth factor receptors or monoclonal antibody (mAb) directed against CD133 or CD44 stem cell-like surface marker, which may lead in an inhibition of PI3K/Akt and MAPK pathways in cancer stem/progenitor cells are indicated. The targeting of ABC multidrug transporters or aldehyde dehydrogenase (ALDH) by diethylaminobenzaldehyde (DEAB), which may result in the intracellular accumulation of chemotherapeutic drugs, is also shown. Moreover, the possibility to target the EMT process- and drug resistance-associated molecules as well as the signaling elements involved in the metabolic process under hypoxic condition are also indicated. The potential targeted delivery of drugs into cancer cells by using novel nanotechnology-based approaches is also illustrated. Hence, multitargeted strategies are likely to be effective at inhibiting tumor growth, invasion, metastases and overcoming treatment resistance.

Color image of this figure appears in the color plate section at the end of the book.

2008b, Mimeault and Batra 2008b, Mimeault and Batra 2008c, Mimeault et al. 2010a, Mimeault and Batra 2010a, Mimeault and Batra 2010c, Huynh et al. 2011, Mimeault et al. 2012, Cheng et al. 2012, Eichhorn et al. 2012).

For instance, the inhibition of smoothened (SMO) hedgehog signaling element by cyclopamine has been observed to eliminate the total cancer cell mass including tumor-initiating cells and improve the cytotoxic and anti-metastatic effects induced by current chemotherapeutic drugs on gliomasphere cells as well as prostatic and pancreatic cancer cell lines *in vitro* and/or *in vivo* (Karhadkar et al. 2004, Clement et al. 2007, Feldmann et al. 2007). The results from our recent work have also revealed that the molecular targeting of EGFR, SMO co-receptor of hedgehog, NF-κB, pAkt or macrophage inhibitory cytokine-1 (MIC-1) was effective to promote the apoptotic effect induced by current chemotherapeutic drug docetaxel

Table 1. Potential therapeutic targets in cancer- and metastasis-initiating cells and their progenies.

Targeted deregulated element	Name of inhibitory agent
CD133 stem cell-like marker	mAb
Growth factor signaling inhibitor:	
EGFR (erbB1) antibody	mAb-C225, IMC-C225
EGFR-TKI	Gefitinib, erlotinib, AG1478, EKB-569
anti-EGF antibody	ABX-EGF
erbB1/erbB2/erbB3/erbB4-TKI	CI1033
HA/CD44	small HA oligosaccharides, anti-CD44 mAb
Hedgehog	SMO inhibitor (cyclopamine), anti-SHH antibody
Wnt/β-catenin	anti-Wnt antibody, WIF-1
Notch	γ-secretase inhibitor DAPT, RO4929097, MK-0752, GSI-18
TGF-β/TGF-βR	anti-TGF-βR antibody, SB-431542
SDF-1/CXCR4	anti-SDF-1 or anti-CXCR4 antibody CXCR4 antagonist (TC14012, TN14003 or AMD3100)
VEGF	anti-VEGF antibody (bevacizumab)
VEGF-R	anti-VEGFR antibody (DC101)
VEGF-R/EGFR	ZD6474
ECM component/integrin	anti-integrin antibody, α_v-integrin antagonist GLPG0187
Intracellular signaling inhibitor:	
Telomerase	G-quadruplex interacting agents (RHPS4), inhibitors of RT and telomerase phosphorylation or assembly As-TERT or -TR, telomerase template antagonist
BMI-1	Vorinostat, azacitidine decitabine
PI3K	LY294002, rapamycin, CCL-779
NF-κB	IkBα inhibitor, sulfasalazine, bortezomib (PS-341), salinosporamides A (NPI-0052), parthenolide
FTI	BMS-214662
COX-2	NS-396, etodolax, celecoxib, rofecoxib
Drug resistance signaling element inhibitor:	
Bcl2	ABT-737
Sphingosine kinase-1	N,N-dimethylsphingosine (DMS), F12509A, B-5354c
Glucosylceramide synthase	D,L-*threo*-PDMP, PPMP, tamoxifen
Ceramidase	N-oleoylethanolamide (OE), B13, LCL204, D-e-MAPP
MGMT	O6-benzylguanine (O6-BG)
ALDH	Disulfiram, diethylaminobenzaldehyde (DEAB)
MDR-1/ABCB1/P-gp	MS-209, Zosuquidar, gefitinib, CI1033, tamoxifen, cyclopamine
MRP1/ABCC1	MS-209
BCRP-1/ABCG2	Gefitinib, CI1033, tamoxifen derivatives, cyclopamine

on $CD133^+$ SP and $CD133^-$ non-SP cells from highly invasive WPE1-NB26 prostate cancer cells (Mimeault et al. 2010a, Mimeault et al. 2012). In the same way, the molecular targeting of IL-4 cytokine in tumorigenic $CD133^+$ cancer stem/progenitor cells from human colon carcinoma by using IL-4Rα antagonist or anti-IL-4 neutralizing antibody sensitized these tumor-initiating cells to the anti-tumor effects induced by standard chemotherapeutic drugs (Todaro et al. 2007). Also, it has been observed that the inhibition of HA/CD44 tumorigenic pathway using anti-CD44 mAb induced the apoptotic death of $CD90^+$\$CD44^+$ progenitor cells from hepatocellular carcinoma cell lines and prevented the formation of metastases in the liver and lungs derived from these immature cancer cells *in vivo* (Yang et al. 2008). Furthermore, the blockage of the SDF-1/CXCR4 axis by using a mAb directed against SDF-1 or CXCR4 or a selective CXCR4 antagonist, such as 14-mer peptide (TN14003) or AMD 3100, has given promising results to counteract the migration of cancer stem/progenitor cells to distant tissues and organs and metastases (Rubin et al. 2003, Liang et al. 2004, Kulbe et al. 2004, Balkwill 2004, Kang et al. 2005, Kucia et al. 2005, Hart et al. 2005, Neiva et al. 2005, Engl et al. 2006, Onoue et al. 2006, Singh et al. 2010, Mimeault and Batra 2010a, Shiozawa et al. 2011a).

Targeting EMT process-associated molecules

The epithelial-mesenchymal transition (EMT) process, which occurs during normal embryogenesis, is often re-activated during the progression of numerous aggressive and metastatic epithelial cancers such as lung, colorectal, pancreatic, prostate, breast and ovarian cancers (Beachy et al. 2004, Dean et al. 2005, Strizzi et al. 2005, Hu and Xing 2005, Mimeault and Batra 2006, Feldmann et al. 2007, Chen et al. 2007, Wang et al. 2007b, Mani et al. 2008, Mimeault et al. 2008a, Mimeault and Batra 2008b, Mimeault et al. 2008b, Alvero et al. 2009, Kabashima et al. 2009, Wang et al. 2009, Mimeault and Batra 2010c, Cao et al. 2011, Zhang et al. 2012, Rhim et al. 2012, Wang et al. 2012a). More specifically, a loss of epithelial characteristics and cell–cell contacts combined with the acquisition of stem cell-like and mesenchymal features and migratory properties by cancer cells during the EMT program is typically associated with their invasion from the primary tumor and metastatic spread at near lymph nodes and distant tissues (Beachy et al. 2004, Dean et al. 2005, Strizzi et al. 2005, Hu and Xing 2005, Mimeault and Batra 2006, Feldmann et al. 2007, Chen et al. 2007, Mimeault et al. 2007a, Mimeault and Batra 2007b, Wang et al. 2007b, Mani et al. 2008, Mimeault et al. 2008a, Mimeault and Batra 2008b, Mimeault et al. 2008b, Alvero et al. 2009, Kabashima et al. 2009, Wang et al. 2009, Mimeault and Batra 2010c, Cao et al. 2011, Bao et al. 2011, Zhang et al. 2012, Rhim et al. 2012, Wang et al. 2012a). Consequently, the targeting of distinct gene products that are frequently altered during the EMT process in cancer-initiating cells and their progenies represent another potential approach to prevent their invasion and metastatic spread (Table 1, Figs. 1 and 2) (Beachy et al. 2004, Dean et al. 2005, Strizzi et al. 2005, Hu and Xing 2005, Mimeault and Batra 2006, Feldmann et al. 2007, Chen et al. 2007, Mimeault and Batra 2007b, Wang et al. 2007b, Mani et al. 2008, Mimeault and Batra 2008b, Mimeault et al. 2008b, Alvero et al. 2009, Kabashima et al. 2009, Wang et al. 2009, Mimeault and Batra 2010c, Cao et al. 2011, Bao et al. 2011, Zhang et al. 2012, Rhim et al. 2012, Wang et al. 2012a). The gene products that are often altered through the EMT program include different growth factor signaling elements such as TGF-β/TGF-βR, EGFR and its ligands including EGF, hedgehog, Notch and platelet-derived growth factor (PDGF)/PDGFR receptors that may cooperate to induce a fully EMT program in cancer cells (Mimeault and Batra 2007b, Kurrey et al. 2009, Wang et al. 2009, Kabashima et al. 2009, Mimeault and Batra 2010b, Cao et al. 2011, Bao et al. 2011, Wang et al. 2012a). The activation of these growth factor pathways may lead to the sustained stimulation of the downstream signaling elements such as PI3K/Akt/mTOR, NF-κB, Cripto-1, tenascin C, snail, slug and twist that are involved in the sustained growth, enhanced survival, invasion, metastases and/or drug resistance of cancer cells (Figs. 1 and 2) (Mimeault and Batra 2007b, Kurrey et al. 2009, Wang et al. 2009, Kabashima et al. 2009, Mimeault and Batra 2010b, Cao et al. 2011, Bao et al. 2011, Wang et al. 2012a). More specifically, it has been reported that the activation of TGF-β/TGF-βR pathway may induce the EMT program and cooperate with other growth factors such as EGF/EGFR system in the malignant transformation of cancer stem/progenitor cells and their progenies, invasion and metastases (Mimeault and Batra 2007b, Kabashima et al. 2009, Mimeault and Batra 2010b, Gulhati et al. 2011, van den Hoogen et al. 2011, Lamouille et al. 2012, Wang et al. 2012a). For instance, the activation of TGF-β/TGF-βR axis may mediate, in both an autocrine and a manner, the stimulation of multiple intracellular signaling elements [MAPKs, PI3K/Akt/mTOR complex 1 and 2 (mTORC1 and mTORC2) and small GTPases, RhoA and Rac-1, tissue transglutaminase (TG2) and ALDH] and enhanced expression of the EMT process-associated molecules in cancer-initiating cells and their progenies (Mimeault and Batra 2007b, Gulhati et al. 2011, van den Hoogen et al. 2011, Cao et al. 2011, Lamouille et al. 2012). These TGF-β signaling effectors in turn may induce a cytoskeletal remodeling and gain of migratory and invasive capacity by cancer cells (Kabashima et al. 2009, Gulhati et al. 2011, van den Hoogen et al. 2011, Cao et al. 2011, Lamouille et al. 2012). Importantly, it has been shown that a specific inhibition or inactivation of TGF-β/TGF-βR or its downstream signaling effectors such as mTORC2, TG2 and ALDH was effective at inhibiting the EMT process in different cancer epithelial cell lines and preventing their invasion and dissemination *in vitro* and *in vivo* (van den Hoogen et al. 2011, Cao et al. 2011, Lamouille et al. 2012). Of particular interest, the activation of TGF-β pathway, which acts as a key player in prostate cancer invasiveness and bone metastasis, strongly induced ALDH7A1 activity in human prostate cancer PC3M-Pro4 cells (van den Hoogen et al. 2011). In addition, the knockdown of ALDH7A1 expression resulted in a significant decrease of the number of integrin-$\alpha 2^{high}$/integrin-αv^{high}/$CD44^{+}$ stem/progenitor cell subpopulation detected in PC3M-Pro4 cell line, inhibited their clonogenic and migratory ability *in vitro* and intra-bone growth and metastasis *in vivo* (van den Hoogen et al. 2011).

Targeting drug resistance-associated molecules

Other potential molecular targets in cancer stem/progenitor cells and their progenies also comprise the gene products that are involved in the drug resistance to current cancer therapies such as different anti-apoptotic factors (Bcl-2 and survivin), altered ceramide signaling elements (sphingosine kinase-1, glucosylceramide synthase and ceramidase), ALDH, MGMT and ABC multidrug transporters (Table 1, Figs. 1 and 2) (Beachy et al. 2004, Dean et al. 2005, Strizzi et al. 2005, Hu and Xing 2005, Mimeault and Batra 2006, Feldmann et al. 2007, Chen et al. 2007, Mimeault et al. 2007a, Mimeault and Batra 2007b, Wang et al. 2007b, Loebinger et al. 2008, Schatton et al. 2008, Rappa et al. 2008, Mimeault et al. 2008a, Mimeault and Batra 2008b, Mimeault et al. 2008b, Mimeault and Batra 2010c, Hamada et al. 2012). For instance, the treatment of the SP cell fraction from human oral squamous cell carcinoma H357 cells with an ABC multidrug transporter inhibitor, verapamil has been reported to inhibit the efflux of mitoxantrone and restore the growth-inhibitory effect induced by mitoxantrone on these cancer stem cell-like cells (Loebinger et al. 2008). Moreover, the treatment of melanoma-initiating cell xenograft-derived tumors with a mAb directed against ABCB5 transporter also induced the growth inhibitory effect *in vivo* (Schatton et al. 2008). Importantly, the molecular targeting of CD133 stem cell-like surface marker with anti-CD133 mAb was also effective at inducing the cytotoxic effects on FEMX-1 melanoma cells and reducing their metastatic spread in mouse model *in vivo* (Rappa et al. 2008). On the other hand, the induction of the differentiation of cancer stem/progenitor cells by using agents such as retinoic acid and its synthetic analogues, interferons (IFNs) or histone deacetylase inhibitor, also may represent a promising adjuvant therapeutic strategy (Sell 2004, Sano et al. 2005, Massard et al. 2006, Perabo and Muller 2006, Moserle et al. 2008, Ying et al. 2011, Whitworth et al. 2012). For instance, it has been reported that the IFN-α treatment caused a dramatic reduction in the verapamil-sensitive SP cell fraction from diverse ovarian cancer cell lines (Moserle et al. 2008).

Molecular targeting of the local microenvironment of cancer- and metastasis-initiating cells

In view of the fact that the local microenvironment of cancer stem/progenitor cells and tumor angiogenic process play key roles in cancer progression and metastases, well efforts are being made to counteract cancer progression by targeting the tumor host cells such as stromal cells and endothelial cells. In particular, the targeting of the myofibroblasts and immune cells that support the malignant transformation of cancer stem/progenitor cells as well as the use of anti-angiogenic agents may constitute an adjuvant treatment for counteracting cancer progression to metastatic and lethal disease states (Davidoff et al. 2001, Rafii et al. 2002, Byrne et al. 2005, Ganss 2006, Rosell et al. 2006, Onoue et al. 2006, Moreira et al. 2007, Calabrese et al. 2007, Folkins et al. 2007, Yang and Wechsler-Reya 2007, Mimeault et al. 2007a, Mimeault et al. 2007b, Mimeault et al. 2008a, Mimeault and Batra 2008b, Mimeault and Batra 2010c). More specifically, the combined use of the cytotoxic agents targeting cancer stem/progenitor cells plus a selective inhibitor of the angiogenic process, such as cyclooxygenase-1 or -2 (COX-1 or -2), NF-κB and/or VEGF-VEFGR, may represent the effective strategies to prevent disease relapse (Table 1) (Lyden et al. 2001, Sun et al. 2004, Miller and Sorensen 2005, Arora and Scholar 2005, Vosseler et al. 2005, Zhong and Bowen 2006, Meric et al. 2006, Aggarwal et al. 2006, Kobayashi and Lin 2006, Grosch et al. 2006, Cairns et al. 2006, Lu et al. 2006, van Beijnum et al. 2006, Frank et al. 2011). As a matter of fact, it has been observed that the treatment of the mice-bearing orthotopic U87 glioma cell xenografts with an anti-VEGF monoclonal antibody called bevacizumab noticeably reduced the microvasculature density and tumor growth. This anti-carcinogenic effect of bevacizumab was also accompanied by a decrease in the number of vessel-associated self-renewing $CD133^+$/$nestin^+$ tumor-initiating cells detected in tumors (Calabrese et al. 2007).

Conclusions and Future Directions

Recent advances in basic and clinical oncology have revealed that the tumorigenic and migrating cancer stem/progenitor cells can provide critical functions in primary cancer initiation and progression, metastases, treatment resistance and disease relapse. Consequently, the molecular targeting of tumor- and metastasis-initiating cells and their local microenvironment may represent new potential strategies for improving the efficacy of current cancer treatments. Future investigations are still essential to more precisely determine the specific biomarkers and altered gene products occurring in cancer- and metastasis-initiating cells and the changes in their niches during the primary cancer development and metastases at distant sites. These studies should lead to the identification of new biomarkers and molecular therapeutic targets that could be exploited to develop new diagnostic and prognostic methods and therapeutic approaches for treating and even curing patients diagnosed with locally advanced, invasive, metastatic and recurrent cancers.

Acknowledgements

The authors on this work are supported by the grants from the National Institutes of Health (EDRN UO1, CA111294, and CA138791) and the U.S. Department of Defense (PC074489).

References

Aggarwal, B.B., S. Shishodia, S.K. Sandur, M.K. Pandey, and G. Sethi. 2006. Inflammation and cancer: How hot is the link? Biochem Pharmacol 72: 1605–1621.

Al-Hajj, M., M.S. Wicha, A. ito-Hernandez, S.J. Morrison, and M.F. Clarke. 2003. Prospective identification of tumorigenic breast cancer cells. Proc Natl Acad Sci USA 100: 3983–3988.

Alvero, A.B., R. Chen, H.H. Fu, M. Montagna, P.E. Schwartz, T. Rutherford, D.A. Silasi, K.D. Steffensen, M. Waldstrom, I. Visintin, and G. Mor. 2009. Molecular phenotyping of human ovarian cancer stem cells unravels the mechanisms for repair and chemoresistance. Cell Cycle 8: 158–166.

Arora, A., and E.M. Scholar. 2005. Role of tyrosine kinase inhibitors in cancer therapy. J Pharmacol Exp Ther 315: 971–979.

Balkwill, F. 2004. Cancer and the chemokine network. Nat Rev Cancer 4: 540–550.

Bao, B., Z. Wang, S. Ali, D. Kong, Y. Li, A. Ahmad, S. Banerjee, A.S. Azmi, L. Miele, and F.H. Sarkar. 2011. Notch-1 induces epithelial-mesenchymal transition consistent with cancer stem cell phenotype in pancreatic cancer cells. Cancer Lett 307: 26–36.

Bao, S., Q. Wu, R.E. McLendon, Y. Hao, Q. Shi, A.B. Hjelmeland, M.W. Dewhirst, D.D. Bigner, and J.N. Rich. 2006. Glioma stem cells promote radioresistance by preferential activation of the DNA damage response. Nature 444: 756–760.

Bapat, S.A., A.M. Mali, C.B. Koppikar, and N.K. Kurrey. 2005. Stem and progenitor-like cells contribute to the aggressive behavior of human epithelial ovarian cancer. Cancer Res 65: 3025–3029.

Bar, E.E., A. Chaudhry, A. Lin, X. Fan, K. Schreck, W. Matsui, S. Piccirillo, A.L. Vescovi, F. Dimeco, A. Olivi, and C.G. Eberhart. 2007. Cyclopamine-mediated hedgehog pathway inhibition depletes stem-like cancer cells in glioblastoma. Stem Cells 25: 2524–2533.

Beachy, P.A., S.S. Karhadkar, and D.M. Berman. 2004. Tissue repair and stem cell renewal in carcinogenesis. Nature 432: 324–331.

Bell, D.R., and G. Van Zant. 2004. Stem cells, aging, and cancer: inevitabilities and outcomes. Oncogene 23: 7290–7296.

Birnie, R., S.D. Bryce, C. Roome, V. Dussupt, A. Droop, S.H. Lang, P.A. Berry, C.F. Hyde, J.L. Lewis, M.J. Stower, N.J. Maitland, and A.T. Collins. 2008. Gene expression profiling of human prostate cancer stem cells reveals a pro-inflammatory phenotype and the importance of extracellular matrix interactions. Genome Biol 9: R83.

Boonstra, R., H. Timmer-Bosscha, J. van Echten-Arends, D.M. van der Kolk, A. van den Berg, B. de Jong, K.D. Tew, S. Poppema, and E.G. de Vries. 2004. Mitoxantrone resistance in a small cell lung cancer cell line is associated with ABCA2 upregulation. Br J Cancer 90: 2411–2417.

Boonyaratanakornkit, J.B., L. Yue, L.R. Strachan, K.J. Scalapino, P.E. LeBoit, Y. Lu, S.P. Leong, J.E. Smith, and R. Ghadially. 2010. Selection of tumorigenic melanoma cells using ALDH. J Invest Dermatol 130: 2799–2808.

Bourguignon, L.Y., K. Peyrollier, W. Xia, and E. Gilad. 2008. Hyaluronan-CD44 interaction activates stem cell marker Nanog, Stat-3-mediated MDR1 gene expression, and ankyrin-regulated multidrug efflux in breast and ovarian tumor cells. J Biol Chem 283: 17635–17651.

Bourguignon, L.Y., C.C. Spevak, G. Wong, W. Xia, and E. Gilad. 2009. Hyaluronan-CD44 interaction with PKC-epsilon promotes oncogenic signaling by the stem cell marker, Nanog and the production of microRNA-21 leading to downregulation of the tumor suppressor protein, PDCD4, anti-apoptosis and chemotherapy resistance in breast tumor cells. J Biol Chem 294: 26533–26546.

Box, N.F., E.C. Torchia, and D.R. Roop. 2010. Are stem cell niches shared for skin cancers? Pigment Cell Melanoma Res 23: 517–520.

Brabletz, T., A. Jung, S. Spaderna, F. Hlubek, and T. Kirchner. 2005. Opinion: migrating cancer stem cells - an integrated concept of malignant tumour progression. Nat Rev Cancer 5: 744–749.

Byrne, A.M., D.J. Bouchier-Hayes, and J.H. Harmey. 2005. Angiogenic and cell survival functions of vascular endothelial growth factor (VEGF). J Cell Mol Med 9: 777–794.

Cai, S., Y. Xu, R.J. Cooper, M.J. Ferkowicz, J.R. Hartwell, K.E. Pollok, and M.R. Kelley. 2005. Mitochondrial targeting of human O6-methylguanine DNA methyltransferase protects against cell killing by chemotherapeutic alkylating agents. Cancer Res 65: 3319–3327.

Cairns, R., I. Papandreou, and N. Denko. 2006. Overcoming physiologic barriers to cancer treatment by molecularly targeting the tumor microenvironment. Mol Cancer Res 4: 61–70.

Calabrese, C., H. Poppleton, M. Kocak, T.L. Hogg, C. Fuller, B. Hamner, E.Y. Oh, M.W. Gaber, D. Finklestein, M. Allen, A. Frank, I.T. Bayazitov, S.S. Zakharenko, A. Gajjar, A. Davidoff, and R.J. Gilbertson. 2007. A perivascular niche for brain tumor stem cells. Cancer Cell 11: 69–82.

Cao, L., M. Shao, J. Schilder, T. Guise, K.S. Mohammad, and D. Matei. 2011. Tissue transglutaminase links TGF-beta, epithelial to mesenchymal transition and a stem cell phenotype in ovarian cancer. Oncogene (in press).

Chen, B.Y., J.Y. Liu, H.H. Chang, C.P. Chang, W.Y. Lo, W.H. Kuo, C.R. Yang, and D.P. Lin. 2007. Hedgehog is involved in prostate basal cell hyperplasia formation and its progressing towards tumorigenesis. Biochem. Biophys Res Commun 357: 1084–1089.

Chen, J.S., F.S. Pardo, J. Wang-Rodriguez, T.S. Chu, J.P. Lopez, J. Aguilera, X. Altuna, R.A. Weisman, and W.M. Ongkeko. 2006. EGFR regulates the side population in head and neck squamous cell carcinoma. Laryngoscope 116: 401–406.

Chen, Y.C., H.S. Hsu, Y.W. Chen, T.H. Tsai, C.K. How, C.Y. Wang, S.C. Hung, Y.L. Chang, M.L. Tsai, Y.Y. Lee, H.H. Ku, and S.H. Chiou. 2008. Oct-4 expression maintained cancer stem-like properties in lung cancer-derived CD133-positive cells. PLoS. ONE 3: e2637.

Cheng, W., T. Liu, X. Wan, Y. Gao, and H. Wang. 2012. MicroRNA-199a targets CD44 to suppress the tumorigenicity and multidrug resistance of ovarian cancer-initiating cells. FEBS J (In press).

Chiba, T., S. Miyagi, A. Saraya, R. Aoki, A. Seki, Y. Morita, Y. Yonemitsu, O. Yokosuka, H. Taniguchi, H. Nakauchi, and A. Iwama. 2008. The polycomb gene product BMI1 contributes to the maintenance of tumor-initiating side population cells in hepatocellular carcinoma. Cancer Res 68: 7742–7749.

Clement, V., P. Sanchez, N. de Tribolet, I. Radovanovic, and A. Altaba. 2007. HEDGEHOG-GLI1 signaling regulates human glioma growth, cancer stem cell self-renewal, and tumorigenicity. Curr Biol 17: 165–172.

Das, B., R. Tsuchida, D. Malkin, G. Koren, S. Baruchel, and H. Yeger. 2008. Hypoxia enhances tumor stemness by increasing the invasive and tumorigenic side population fraction. Stem Cells 26: 1818–1830.

Davidoff, A.M., C.Y. Ng, P. Brown, M.A. Leary, W.W. Spurbeck, J. Zhou, E. Horwitz, E.F. Vanin, and A.W. Nienhuis. 2001. Bone marrow-derived cells contribute to tumor neovasculature and, when modified to express an angiogenesis inhibitor, can restrict tumor growth in mice. Clin Cancer Res 7: 2870–2879.

de Jonge-Peeters, S.D., F. Kuipers, E.G. de Vries, and E. Vellenga. 2007. ABC transporter expression in hematopoietic stem cells and the role in AML drug resistance. Crit Rev Oncol Hematol 62: 214–226.

Dean, M., T. Fojo, and S. Bates. 2005. Tumour stem cells and drug resistance. Nat Rev Cancer 5: 275–284.

Deeley, R.G., C. Westlake, and S.P. Cole. 2006. Transmembrane transport of endo- and xenobiotics by mammalian ATP-binding cassette multidrug resistance proteins. Physiol Rev 86: 849–899.

Deng, Y., and S. Chang. 2007. Role of telomeres and telomerase in genomic instability, senescence and cancer. Lab Invest 87: 1071–1076.

DePinho, R.A. 2000. The age of cancer. Nature 408: 248–254.

Derycke, L.D., and M.E. Bracke 2004. N-cadherin in the spotlight of cell-cell adhesion, differentiation, embryogenesis, invasion and signalling. Int. J Dev Biol 48: 463–476.

Douville, J., R. Beaulieu, and D. Balicki. 2009. ALDH1 as a functional marker of cancer stem and progenitor cells. Stem Cells Dev 18: 17–25.

Dylla, S.J., L. Beviglia, I.K. Park, C. Chartier, J. Raval, L. Ngan, K. Pickell, J. Aguilar, S. Lazetic, S. Smith-Berdan, M.F. Clarke, T. Hoey, J. Lewicki, and A.L. Gurney. 2008. Colorectal cancer stem cells are enriched in xenogeneic tumors following chemotherapy. PLoS ONE 3: e2428.

Eichhorn, P.J., L. Rodon, A. Gonzalez-Junca, A. Dirac, M. Gili, E. Martinez-Saez, C. Aura, I. Barba, V. Peg, A. Prat, I. Cuartas, J. Jimenez, D. Garcia-Dorado, J. Sahuquillo, R. Bernards, J. Baselga, and J. Seoane. 2012. USP15 stabilizes TGF-beta receptor I and promotes oncogenesis through the activation of TGF-beta signaling in glioblastoma. Nat Med 18: 429–435.

Engl, T., B. Relja, D. Marian, C. Blumenberg, I. Muller, W.D. Beecken, J. Jones, E.M. Ringel, J. Bereiter-Hahn, D. Jonas, and R.A. Blaheta. 2006. CXCR4 chemokine receptor mediates prostate tumor cell adhesion through alpha5 and beta3 integrins. Neoplasia 8: 290–301.

Eramo, A., F. Lotti, G. Sette, E. Pilozzi, M. Biffoni, A. Di Virgilio, C. Conticello, L. Ruco, C. Peschle, and R. De Maria. 2008. Identification and expansion of the tumorigenic lung cancer stem cell population. Cell Death. Differ 15: 504–514.

Feldmann, G., S. Dhara, V. Fendrich, D. Bedja, R. Beaty, M. Mullendore, C. Karikari, H. Alvarez, C. Iacobuzio-Donahue, A. Jimeno, K.L. Gabrielson, W. Matsui, and A. Maitra, 2007. Blockade of hedgehog signaling inhibits pancreatic cancer invasion and metastases: a new paradigm for combination therapy in solid cancers. Cancer Res 67: 2187–2196.

Fillmore, C.M., and C. Kuperwasser. 2008. Human breast cancer cell lines contain stem-like cells that self-renew, give rise to phenotypically diverse progeny and survive chemotherapy. Breast Cancer Res 10: R25.

Folkins, C., S. Man, P. Xu, Y. Shaked, D.J. Hicklin, and R.S. Kerbel. 2007. Anticancer therapies combining antiangiogenic and tumor cell cytotoxic effects reduce the tumor stem-like cell fraction in glioma xenograft tumors. Cancer Res 67: 3560–3564.

Frank, N.Y., A. Margaryan, Y. Huang, T. Schatton, A.M. Waaga-Gasser, M. Gasser, M.H. Sayegh, W. Sadee, and M.H. Frank. 2005. ABCB5-mediated doxorubicin transport and chemoresistance in human malignant melanoma. Cancer Res 65: 4320–4333.

Frank, N.Y., T. Schatton, S. Kim, Q. Zhan, B.J. Wilson, J. Ma, K.R. Saab, V. Osherov, H.R. Widlund, M. Gasser, A.M. Waaga-Gasser, T.S. Kupper, G.F. Murphy, and M.H. Frank. 2011. VEGFR-1 expressed by malignant melanoma-initiating cells is required for tumor growth. Cancer Res 71: 1474–1485.

Friel, A.M., P.A. Sergent, C. Patnaude, P.P. Szotek, E. Oliva, D.T. Scadden, M.V. Seiden, R. Foster, and B.R. Rueda. 2008. Functional analyses of the cancer stem cell-like properties of human endometrial tumor initiating cells. Cell Cycle 7: 242–249.

Galli, R., E. Binda, U. Orfanelli, B. Cipelletti, A. Gritti, S. De Vitis, R. Fiocco, C. Foroni, F. Dimeco, and A. Vescovi. 2004. Isolation and characterization of tumorigenic, stem-like neural precursors from human glioblastoma. Cancer Res 64: 7011–7021.

Ganss, R. 2006. Tumor stroma fosters neovascularization by recruitment of progenitor cells into the tumor bed. J Cell Mol Med 10: 857–865.

Gerk, P.M., and M. Vore. 2002. Regulation of expression of the multidrug resistance-associated protein 2 (MRP2) and its role in drug disposition. J Pharmacol Exp Ther 302: 407–415.

Gerlee, P., and A.R. Anderson. 2008. A hybrid cellular automaton model of clonal evolution in cancer: the emergence of the glycolytic phenotype. J Theor Biol 250: 705–722.

Ghatak, S., S. Misra, and B.P. Toole. 2002. Hyaluronan oligosaccharides inhibit anchorage-independent growth of tumor cells by suppressing the phosphoinositide 3-kinase/Akt cell survival pathway. J Biol Chem 277: 38013–38020.

Ginestier, C., M.H. Hur, E. Charafe-Jauffret, F. Monville, J. Dutcher, M. Brown, J. Jacquemier, P. Viens, C.G. Kleer, S. Liu, A. Schott, D. Hayes, D. Birnbaum, M.S. Wicha, and G. Dontu. 2007. ALDH1 is a marker of normal and malignant human mammary stem cells and a predictor of poor clinical outcome. Cell Stem Cell 1: 555–567.

Glinsky, G.V. 2008. "Stemness" genomics law governs clinical behavior of human cancer: implications for decision making in disease management. J Clin Oncol 26: 2846–2853.

Gottesman, M.M., T. Fojo, and S.E. Bates 2002. Multidrug resistance in cancer: role of ATP-dependent transporters. Nat Rev Cancer 2: 48–58.

Gotzmann, J., M. Mikula, A. Eger, R. Schulte-Hermann, R. Foisner, H. Beug, and W. Mikulits. 2004. Molecular aspects of epithelial cell plasticity: implications for local tumor invasion and metastasis. Mutat Res 566: 9–20.

Grosch, S., T.J. Maier, S. Schiffmann, and G. Geisslinger. 2006. Cyclooxygenase-2 (COX-2)-independent anticarcinogenic effects of selective COX-2 inhibitors. J Natl Cancer Inst 98: 736–747.

Gulhati, P., K.A. Bowen, J. Liu, P.D. Stevens, P.G. Rychahou, M. Chen, E.Y. Lee, H.L. Weiss, K.L. O'Connor, T. Gao, and B.M. Evers. 2011. mTORC1 and mTORC2 regulate EMT, motility, and metastasis of colorectal cancer via RhoA and Rac1 signaling pathways. Cancer Res 71: 3246–3256.

Gumucio, D.L., S. Fagoonee, X.T. Qiao, M. Liebert, J.L. Merchant, F. Altruda, M. Rizzetto, and R. Pellicano. 2008. Tissue stem cells and cancer stem cells: potential implications for gastric cancer. Panminerva Med 50: 65–71.

Hamada, S., K. Satoh, M. Hirota, A. Kanno, J. Umino, H. Ito, A. Masamune, K. Kikuta, K. Kume, and T. Shimosegawa. 2012. The homeobox gene MSX2 determines chemosensitivity of pancreatic cancer cells *via* the regulation of transporter gene ABCG2. J Cell Physiol 227: 729–738.

Han, B., and J.T. Zhang. 2004. Multidrug resistance in cancer chemotherapy and xenobiotic protection mediated by the half ATP-binding cassette transporter ABCG2. Curr Med Chem Anticancer Agents 4: 31–42.

Haraguchi, N., T. Utsunomiya, H. Inoue, F. Tanaka, K. Mimori, G.F. Barnard, and M. Mori. 2006. Characterization of a side population of cancer cells from human gastrointestinal system. Stem Cells 24: 506–513.

Hart, C.A., M. Brown, S. Bagley, M. Sharrard, and N.W. Clarke. 2005. Invasive characteristics of human prostatic epithelial cells: understanding the metastatic process. Br J Cancer 92: 503–512.

Hashimoto, O., K. Shimizu, S. Semba, S. Chiba, Y. Ku, H. Yokozaki, and Y. Hori. 2011. Hypoxia induces tumor aggressiveness and the expansion of CD133-positive cells in a hypoxia-inducible factor-1alpha-dependent manner in pancreatic cancer cells. Pathobiology 78: 181–192.

Hemmati, H.D., I. Nakano, J.A. Lazareff, M. Masterman-Smith, D.H. Geschwind, M. Bronner-Fraser, and H.I. Kornblum. 2003. Cancerous stem cells can arise from pediatric brain tumors. Proc. Natl Acad Sci USA 100: 15178–15183.

Hermann, P.C., S.L. Huber, T. Herrler, A. Aicher, J.W. Ellwart, M. Guba, C.J. Bruns, and C. Heeschen. 2007. Distinct populations of cancer stem cells determine tumor growth and metastatic activity in human pancreatic cancer. Cell Stem Cell 1: 313–323.

Hirschmann-Jax, C., A.E. Foster, G.G. Wulf, J.G. Nuchtern, T.W. Jax, U. Gobel, M.A. Goodell, and M.K. Brenner. 2004. A distinct "side population" of cells with high drug efflux capacity in human tumor cells. Proc Natl Acad Sci USA 101: 14228–14233.

Ho, M.M., A.V. Ng, S. Lam, and J.Y. Hung. 2007. Side population in human lung cancer cell lines and tumors is enriched with stem-like cancer cells. Cancer Res 67: 4827–4833.

Hu, X.F., and P.X. Xing. 2005. Cripto as a target for cancer immunotherapy. Expert Opin Ther Targets 9: 383–394.

Huang, D., Q. Gao, L. Guo, C. Zhang, J. Wei, H. Li, W.J. Jing, X. Han, Y. Shi, and H.L. Shih. 2009. Isolation and identification of cancer stem-like cells in esophageal carcinoma cell lines. Stem Cells Dev 18: 465–473.

Huynh, C., L. Poliseno, M.F. Segura, R. Medicherla, A. Haimovic, S. Menendez, S. Shang, A. Pavlick, Y. Shao, F. Darvishian, J.F. Boylan, I. Osman, and E. Hernando. 2011. The novel gamma Secretase inhibitor RO4929097 reduces the tumor initiating potential of melanoma. PLoS ONE 6: e25264.

Ishizawa, K., Z.A. Rasheed, R. Karisch, Q. Wang, J. Kowalski, E. Susky, K. Pereira, C. Karamboulas, N. Moghal, N.V. Rajeshkumar, M. Hidalgo, M. Tsao, L. Ailles, T.K. Waddell, A. Maitra, B.G. Neel, and W. Matsui. 2010. Tumor-initiating cells are rare in many human tumors. Cell Stem Cell 7: 279–282.

Jarrard, D.F., J. Modder, P. Fadden, V. Fu, L. Sebree, D. Heisey, S.R. Schwarze, and A. Friedl. 2002. Alterations in the p16/pRb cell cycle checkpoint occur commonly in primary and metastatic human prostate cancer. Cancer Lett 185: 191–199.

Johannessen, T.C., R. Bjerkvig, and B.B. Tysnes. 2008. DNA repair and cancer stem-like cells - Potential partners in glioma drug resistance? Cancer Treat Rev 34: 558–567.

Juillerat, A., and L. Juillerat-Jeanneret. 2007. S-alkylthiolation of O6-methylguanine-DNA-methyltransferase (MGMT) to sensitize cancer cells to anticancer therapy. Expert Opin Ther Targets 11: 349–361.

Kabashima, A., H. Higuchi, H. Takaishi, Y. Matsuzaki, S. Suzuki, M. Izumiya, H. Iizuka, G. Sakai, S. Hozawa, T. Azuma, and T. Hibi. 2009. Side population of pancreatic cancer cells predominates in TGF-beta-mediated epithelial to mesenchymal transition and invasion. Int J Cancer 124: 2771–2779.

Kaminski, W.E., A. Piehler, and J.J. Wenzel. 2006. ABC A-subfamily transporters: structure, function and disease. Biochim Biophys Acta 1762: 510–524.

Kang, H., G. Watkins, A. Douglas-Jones, R.E. Mansel, and W.G. Jiang. 2005. The elevated level of CXCR4 is correlated with nodal metastasis of human breast cancer. Breast 14: 360–367.

Kang, M.K., and S.K. Kang. 2007. Tumorigenesis of chemotherapeutic drug-resistant cancer stem-like cells in brain glioma. Stem Cells Dev 16: 837–847.

Karhadkar, S.S., G.S. Bova, N. Abdallah, S. Dhara, D. Gardner, A. Maitra, J.T. Isaacs, D.M. Berman, and P.A. Beachy. 2004. Hedgehog signalling in prostate regeneration, neoplasia and metastasis. Nature 431: 707–712.

Katoh, M. 2007. Networking of WNT, FGF, Notch, BMP, and Hedgehog signaling pathways during carcinogenesis. Stem Cell Rev 3: 30–38.

Kim, C.F., E.L. Jackson, A.E. Woolfenden, S. Lawrence, I. Babar, S. Vogel, D. Crowley, R.T. Bronson, and T. Jacks. 2005. Identification of bronchioalveolar stem cells in normal lung and lung cancer. Cell 121: 823–835.

Kobayashi, H. and P.C. Lin. 2006. Antiangiogenic and radiotherapy for cancer treatment. Histol Histopathol 21: 1125–1134.

Kondoh, H. 2008. Cellular life span and the Warburg effect. Exp Cell Res 314: 1923–1928.

Krishnamurthy, P., and J.D. Schuetz. 2006. Role of ABCG2/BCRP in biology and medicine. Annu Rev Pharmacol Toxicol 46: 381–410.
Kucia, M., R. Reca, K. Miekus, J. Wanzeck, W. Wojakowski, A. Janowska-Wieczorek, J. Ratajczak, and M.Z. Ratajczak. 2005. Trafficking of normal stem cells and metastasis of cancer stem cells involve similar mechanisms: pivotal role of the SDF-1-CXCR4 axis. Stem Cells 23: 879–894.
Kulbe, H., N.R. Levinson, F. Balkwill, and J.L. Wilson. 2004. The chemokine network in cancer—much more than directing cell movement. Int J Dev Biol 48: 489–496.
Kuphal, S., H.G. Palm, I. Poser, and A.K. Bosserhoff. 2005. Snail-regulated genes in malignant melanoma. Melanoma Res 15: 305–313.
Kuphal, S., A. Winklmeier, C. Warnecke, and A.K. Bosserhoff. 2010. Constitutive HIF-1 activity in malignant melanoma. Eur J Cancer 46: 1159–1169.
Kurrey, N.K., S.P. Jalgaonkar, A.V. Joglekar, A.D. Ghanate, P.D. Chaskar, R.Y. Doiphode, and S.A. Bapat. 2009. Snail and slug mediate radioresistance and chemoresistance by antagonizing p53-mediated apoptosis and acquiring a stem-like phenotype in ovarian cancer cells. Stem Cells 27: 2059–2068.
Lamouille, S., E. Connolly, J.W. Smyth, R.J. Akhurst, and R. Derynck. 2012. TGF-beta-induced activation of mTOR complex 2 drives epithelial-mesenchymal transition and cell invasion. J Cell Sci (In press)
Larue, L., and A. Bellacosa. 2005. Epithelial-mesenchymal transition in development and cancer: role of phosphatidylinositol 3′ kinase/AKT pathways. Oncogene 24: 7443–7454.
Li, L., and W.B. Neaves. 2006. Normal stem cells and cancer stem cells: the niche matters. Cancer Res 66: 4553–4557.
Liang, Z., T. Wu, H. Lou, X. Yu, R.S. Taichman, S.K. Lau, S. Nie, J. Umbreit, and H. Shim. 2004. Inhibition of breast cancer metastasis by selective synthetic polypeptide against CXCR4. Cancer Res 64: 4302–4308.
Liu, G., X. Yuan, Z. Zeng, P. Tunici, H. Ng, I.R. Abdulkadir, L. Lu, D. Irvin, K.L. Black, and J.S. Yu. 2006a. Analysis of gene expression and chemoresistance of CD133+ cancer stem cells in glioblastoma. Mol Cancer 5: 67.
Liu, S., G. Dontu, I.D. Mantle, S. Patel, N.S. Ahn, K.W. Jackson, P. Suri, and M.S. Wicha. 2006b. Hedgehog signaling and Bmi-1 regulate self-renewal of normal and malignant human mammary stem cells. Cancer Res 66: 6063–6071.
Liu, S., G. Dontu, and M.S. Wicha. 2005. Mammary stem cells, self-renewal pathways, and carcinogenesis. Breast Cancer Res 7: 86–95.
Loebinger, M.R., A. Giangreco, K.R. Groot, L. Prichard, K. Allen, C. Simpson, L. Bazley, N. Navani, S. Tibrewal, D. Davies, and S.M. Janes. 2008. Squamous cell cancers contain a side population of stem-like cells that are made chemosensitive by ABC transporter blockade. Br J Cancer 98: 380–387.
Lu, H., W. Ouyang, and C. Huang. 2006. Inflammation, a key event in cancer development. Mol Cancer Res 4: 221–233.
Lyden, D., K. Hattori, S. Dias, C. Costa, P. Blaikie, L. Butros, A. Chadburn, B. Heissig, W. Marks, L. Witte, Y. Wu, D. Hicklin, Z. Zhu, N.R. Hackett, R.G. Crystal, M.A. Moore, K.A. Hajjar, K. Manova, R. Benezra, and S. Rafii 2001. Impaired recruitment of bone-marrow-derived endothelial and hematopoietic precursor cells blocks tumor angiogenesis and growth. Nat Med 7: 1194–1201.
Ma, S., T.K. Lee, B.J. Zheng, K.W. Chan, and X.Y. Guan. 2008. CD133+ HCC cancer stem cells confer chemoresistance by preferential expression of the Akt/PKB survival pathway. Oncogene 27: 1749–1758.
Ma, Y., D. Liang, J. Liu, K. Axcrona, G. Kvalheim, T. Stokke, J.M. Nesland, and Z. Suo. 2011. Prostate cancer cell lines under hypoxia exhibit greater stem-like properties. PLoS ONE 6: e29170.
Mack, J.T., V. Beljanski, K.D. Tew, and D.M. Townsend. 2006. The ATP-binding cassette transporter ABCA2 as a mediator of intracellular trafficking. Biomed Pharmacother 60: 587–592.

Maitland, N.J., S.D. Bryce, M.J. Stower, and A.T. Collins. 2006. Prostate cancer stem cells: a target for new therapies. Ernst. Schering. Found Symp Proc 5: 155–179.

Mani, S.A., W. Guo, M.J. Liao, E.N. Eaton, A. Ayyanan, A.Y. Zhou, M. Brooks, F. Reinhard, C.C. Zhang, M. Shipitsin, L.L. Campbell, K. Polyak, C. Brisken, J. Yang, and R.A. Weinberg. 2008. The epithelial-mesenchymal transition generates cells with properties of stem cells. Cell 133: 704–715.

Mao, Q., and J.D. Unadkat. 2005. Role of the breast cancer resistance protein (ABCG2) in drug transport. AAPS J 7: E118–E133.

Marsden, C.G., M.J. Wright, L. Carrier, K. Moroz, R. Pochampally, and B.G. Rowan. 2012. "A novel *in vivo* model for the study of human breast cancer metastasis using primary breast tumor-initiating cells from patient biopsies". BMC Cancer 12: 10.

Marusyk, A., and J. DeGregori. 2007. Replicational stress selects for p53 mutation. Cell Cycle 6: 2148–2151.

Marusyk, A., and J. DeGregori. 2008. Declining cellular fitness with age promotes cancer initiation by selecting for adaptive oncogenic mutations. Biochim Biophys Acta 1785: 1–11.

Massard, C., E. Deutsch, and J.C. Soria. 2006. Tumour stem cell-targeted treatment: elimination or differentiation. Ann Oncol 17: 1620–1624.

Mathieu, J., Z. Zhang, W. Zhou, A.J. Wang, J.M. Heddleston, C.M. Pinna, A. Hubaud, B. Stadler, M. Choi, M. Bar, M. Tewari, A. Liu, R. Vessella, R. Rostomily, D. Born, M. Horwitz, C. Ware, C.A. Blau, M.A. Cleary, J.N. Rich, and H. Ruohola-Baker. 2011. HIF induces human embryonic stem cell markers in cancer cells. Cancer Res 71: 4640–4652.

Matsui, W., Q. Wang, J.P. Barber, S. Brennan, B.D. Smith, I. Borrello, I. McNiece, L. Lin, R.F. Ambinder, C. Peacock, D.N. Watkins, C.A. Huff, and R.J. Jones. 2008. Clonogenic multiple myeloma progenitors, stem cell properties, and drug resistance. Cancer Res 68: 190–197.

Maurelli, R., G. Zambruno, L. Guerra, C. Abbruzzese, G. Dimri, M. Gellini, S. Bondanza, and E. Dellambra. 2006. Inactivation of p16INK4a (inhibitor of cyclin-dependent kinase 4A) immortalizes primary human keratinocytes by maintaining cells in the stem cell compartment. FASEB J 20: 1516–1518.

Meric, J.B., S. Rottey, K. Olaussen, J.C. Soria, D. Khayat, O. Rixe, and J.P. Spano. 2006. Cyclooxygenase-2 as a target for anticancer drug development. Crit Rev Oncol Hematol 59: 51–64.

Miller, J.C., and A.G. Sorensen. 2005. Imaging biomarkers predictive of disease/therapy outcome: ischemic stroke and drug development. Prog Drug Res 62: 319–356.

Mimeault, M., R. Hauke, and S.K. Batra. 2007a. Stem cells—A revolution in therapeutics—Recent advances on the stem cell biology and their therapeutic applications in regenerative medicine and cancer therapies. Clin Pharmacol Ther 82: 252–264.

Mimeault, M., R. Hauke, P.P. Mehta, and S.K. Batra. 2007b. Recent advances on cancer stem/progenitor cell research: therapeutic implications for overcoming resistance to the most aggressive cancers. J Mol Cell Med 11: 981–1011.

Mimeault, M., and S.K. Batra. 2006. Recent advances on the significance of stem cells in tissue regeneration and cancer therapies. Stem Cells 24: 2319–2345.

Mimeault, M., and S.K. Batra. 2007a. Functions of tumorigenic and migrating cancer progenitor cells in cancer progression and metastasis and their therapeutic implications. Cancer Metastasis Rev 26: 203–214.

Mimeault, M., and S.K. Batra. 2007b. Interplay of distinct growth factors during epithelial-mesenchymal transition of cancer progenitor cells and molecular targeting as novel cancer therapies. Ann Oncol 18: 1605–1619.

Mimeault, M., and S.K. Batra. 2008a. Recent advances on the development of novel anti-cancer drugs targeting cancer stem/progenitor cells. Drug Develop Res 69: 415–430.

Mimeault, M., and S.K. Batra. 2008b. Recent progress on normal and malignant pancreatic stem/progenitor cell research: therapeutic implications for the treatment of type 1 or 2 diabetes mellitus and aggressive pancreatic cancer. Gut 57: 1456–1468.

Mimeault, M., and S.K. Batra. 2008c. Targeting of cancer stem/progenitor cells plus stem cell-based therapies: the ultimate hope for treating and curing aggressive and recurrent cancers. Panminerva Med 50: 3–18.

Mimeault, M., and S.K. Batra. 2009. Recent insights into the molecular mechanisms involved in aging and the malignant transformation of adult stem/progenitor cells and their therapeutic implications. Ageing Res Rev 8: 94–112.

Mimeault, M., and S.K. Batra. 2010a. Frequent deregulations in the hedgehog signaling network and cross-talks with the epidermal growth factor receptor pathway involved in cancer progression and targeted therapies. Pharmacol Rev 62: 497–524.

Mimeault, M., and S.K. Batra. 2010b. New advances on critical implications of tumor- and metastasis-initiating cells in cancer progression, treatment resistance and disease recurrence. Histol Histopathol 25: 1057–1073.

Mimeault, M., and S.K. Batra. 2010c. New promising drug targets in cancer- and metastasis-initiating cells. Drug Discov Today 15: 354–364.

Mimeault, M., and S.K. Batra. 2010d. Novel therapies against aggressive and recurrent epithelial cancers by molecular targeting tumor- and metastasis-initiating cells and their progenies. Anticancer Agents Med Chem 10: 137–151.

Mimeault, M., and Batra, S.K. 2011a. Animal models of prostate carcinogenesis underlining the critical implication of prostatic stem progenitor cells. Biochim. Biophys. Acta 1816: 25–37.

Mimeault, M., and S.K. Batra. 2011b. Complex oncogenic signaling networks regulate brain tumor-initiating cells and their progenies: pivotal roles of wild-type EGFR, EGFRvIII mutant and hedgehog cascades and novel multitargeted therapies. Brain Pathol 21: 479–500.

Mimeault, M., and S.K. Batra. 2012. Novel biomarkers and therapeutic targets for optimizing the therapeutic management of melanomas. World J Clin Oncol 3: 32–42.

Mimeault, M., R. Hauke, and S.K. Batra. 2008a. Recent advances on the molecular mechanisms involved in the drug resistance of cancer cells and novel targeting therapies. Clin Pharmacol Ther 83: 673–691.

Mimeault, M., S.L. Johansson, and S.K. Batra. 2012. Pathobiological implications of the expression of EGFR, pAkt, NF-kB and MIC-1 in prostate cancer stem cells and their progenies. PLoS ONE 7: e31919.

Mimeault, M., S.L. Johansson, J.P. Henichart, P. Depreux, and S.K. Batra. 2010a. Cytotoxic effects induced by docetaxel, gefitinib, and cyclopamine on side population and non-side population cell fractions from human invasive prostate cancer cells. Mol Cancer Ther 9: 617–630.

Mimeault, M., S.L. Johansson, S. Senapati, N. Momi, S. Chakraborty, and S.K. Batra. 2010b. MUC4 down-regulation reverses chemoresistance of pancreatic cancer stem/progenitor cells and their progenies. Cancer Lett 295: 69–84.

Mimeault, M., P.P. Mehta, R. Hauke, and S.K. Batra. 2008b. Functions of normal and malignant prostatic stem/progenitor cells in tissue regeneration and cancer progression and novel targeting therapies. Endocr Rev 29: 234–252.

Misra, S., B.P. Toole, and S. Ghatak. 2006. Hyaluronan constitutively regulates activation of multiple receptor tyrosine kinases in epithelial and carcinoma cells. J Biol Chem 281: 34936–34941.

Mitsutake, N., A. Iwao, K. Nagai, H. Namba, A. Ohtsuru, V. Saenko, and S. Yamashita. 2007. Characterization of side population in thyroid cancer cell lines: cancer stem-like cells are enriched partly but not exclusively. Endocrinology 148: 1797–1803.

Mohyeldin, A., T. Garzon-Muvdi, and A. Quinones-Hinojosa. 2010. Oxygen in stem cell biology: a critical component of the stem cell niche. Cell Stem Cell 7: 150–161.

Moreira, I.S., P.A. Fernandes, and M.J. Ramos. 2007. Vascular endothelial growth factor (VEGF) inhibition—a critical review. Anticancer Agents Med Chem 7: 223–245.

Moriyama, T., K. Ohuchida, K. Mizumoto, L. Cui, N. Ikenaga, N. Sato, and M. Tanaka. 2010. Enhanced cell migration and invasion of CD133+ pancreatic cancer cells cocultured with pancreatic stromal cells. Cancer 116: 3357–3368.

Moserle, L., S. Indraccolo, M. Ghisi, C. Frasson, E. Fortunato, S. Canevari, S. Miotti, V. Tosello, R. Zamarchi, A. Corradin, S. Minuzzo, E. Rossi, G. Basso, and A. Amadori. 2008. The side population of ovarian cancer cells is a primary target of IFN-alpha antitumor effects. Cancer Res 68: 5658–5668.

Motohara, T., S. Masuko, T. Ishimoto, T. Yae, N. Onishi, T. Muraguchi, A. Hirao, Y. Matsuzaki, H. Tashiro, H. Katabuchi, H. Saya, and O. Nagano. 2011. Transient depletion of p53 followed by transduction of c-Myc and K-Ras converts ovarian stem-like cells into tumor-initiating cells. Carcinogenesis 32: 1597–1606.

Neiva, K., Y.X. Sun, and R.S. Taichman. 2005. The role of osteoblasts in regulating hematopoietic stem cell activity and tumor metastasis. Braz J Med Biol Res 38: 1449–1454.

Nicolis, S.K. 2007. Cancer stem cells and "stemness" genes in neuro-oncology. Neurobiol Dis 25: 217–229.

Nijhof, J.G., C. van Pelt, A.A. Mulder, D.L. Mitchell, L.H. Mullenders, and F.R. de Gruijl. 2007. Epidermal stem and progenitor cells in murine epidermis accumulate UV damage despite NER proficiency. Carcinogenesis 28: 792–800.

Olivotto, M., and S.P. Dello. 2008. Environmental restrictions within tumor ecosystems select for a convergent, hypoxia-resistant phenotype of cancer stem cells. Cell Cycle 7: 176–187.

Onoue, T., D. Uchida, N.M. Begum, Y. Tomizuka, H. Yoshida, and M. Sato. 2006. Epithelial-mesenchymal transition induced by the stromal cell-derived factor-1/CXCR4 system in oral squamous cell carcinoma cells. Int J Oncol 29: 1133–1138.

Ozben, T. 2006. Mechanisms and strategies to overcome multiple drug resistance in cancer. FEBS Lett 580: 2903–2909.

Pacifico, A., L.H. Goldberg, K. Peris, S. Chimenti, G. Leone, and H.N. Ananthaswamy. 2008. Loss of CDKN2A and p14ARF expression occurs frequently in human nonmelanoma skin cancers. Br J Dermatol 158: 291–297.

Perabo, F.G., and S.C. Muller. 2006. New agents for treatment of advanced transitional cell carcinoma. Ann Oncol 18: 835–843.

Persano, L., E. Rampazzo, P.A. Della, F. Pistollato, and G. Basso. 2011. The three-layer concentric model of glioblastoma: cancer stem cells, microenvironmental regulation, and therapeutic implications. Scientific World Journal 11: 1829–1841.

Ponti, D., A. Costa, N. Zaffaroni, G. Pratesi, G. Petrangolini, D. Coradini, S. Pilotti, M.A. Pierotti, and M.G. Daidone. 2005. Isolation and *in vitro* propagation of tumorigenic breast cancer cells with stem/progenitor cell properties. Cancer Res 65: 5506–5511.

Prince, M.E., R. Sivanandan, A. Kaczorowski, G.T. Wolf, M.J. Kaplan, P. Dalerba, I.L. Weissman, M.F. Clarke, and L.E. Ailles. 2007. Identification of a subpopulation of cells with cancer stem cell properties in head and neck squamous cell carcinoma. Proc Natl Acad Sci USA 104: 973–978.

Rabik, C.A., M.C. Njoku, and M.E. Dolan. 2006. Inactivation of O6-alkylguanine DNA alkyltransferase as a means to enhance chemotherapy. Cancer Treat Rev 32: 261–276.

Rafii, S., Lyden, D., R. Benezra, K. Hattori, and B. Heissig. 2002. Vascular and haematopoietic stem cells: novel targets for anti-angiogenesis therapy? Nat Rev Cancer 2: 826–835.

Rappa, G., O. Fodstad, and A. Lorico. 2008. The stem cell-associated antigen CD133 (Prominin-1) is a molecular therapeutic target for metastatic melanoma. Stem Cells 26: 3008–3017.

Rhim, A.D., E.T. Mirek, N.M. Aiello, A. Maitra, J.M. Bailey, F. McAllister, M. Reichert, G.L. Beatty, A.K. Rustgi, R.H. Vonderheide, S.D. Leach, and B.Z. Stanger. 2012. EMT and dissemination precede pancreatic tumor formation. Cell 148: 349–361.

Ricci-Vitiani, L., D.G. Lombardi, E. Pilozzi, M. Biffoni, M. Todaro, C. Peschle, and R. De Maria. 2007. Identification and expansion of human colon-cancer-initiating cells. Nature 445: 111–115.

Rich, J.N. 2007. Cancer stem cells in radiation resistance. Cancer Res 67: 8980–8984.

Rizo, A., E. Vellenga, G. de Haan, and J.J. Schuringa. 2006. Signaling pathways in self-renewing hematopoietic and leukemic stem cells: do all stem cells need a niche? Hum Mol Genet 15: R210–R219.

Rosell, R., F. Cecere, F. Cognetti, M. Cuello, J.M. Sanchez, M. Taron, N. Reguart, and D. Jablons. 2006. Future directions in the second-line treatment of non-small cell lung cancer. Semin Oncol 33: S45–S51.

Rubin, J.B., A.L. Kung, R.S. Klein, J.A. Chan, Y. Sun, K. Schmidt, M.W. Kieran, A.D. Luster, and R.A. Segal. 2003. A small-molecule antagonist of CXCR4 inhibits intracranial growth of primary brain tumors. Proc Natl Acad Sci USA 100: 13513–13518.

Rubin, L.L., and F.J. de Sauvage. 2006. Targeting the Hedgehog pathway in cancer. Nat Rev Drug Discov 5: 1026–1033.

Salmaggi, A., A. Boiardi, M. Gelati, A. Russo, C. Calatozzolo, E. Ciusani, F.L. Sciacca, A. Ottolina, E.A. Parati, C. La Porta, G. Alessandri, C. Marras, D. Croci, and M. De Rossi. 2006. Glioblastoma-derived tumorospheres identify a population of tumor stem-like cells with angiogenic potential and enhanced multidrug resistance phenotype. Glia 54: 850–860.

Sano, T., M. Kagawa, M. Okuno, N. Ishibashi, M. Hashimoto, M. Yamamoto, R. Suzuki, H. Kohno, R. Matsushima-Nishiwaki, Y. Takano, H. Tsurumi, S. Kojima, S.L. Friedman, H. Moriwaki, and T. Tanaka. 2005. Prevention of rat hepatocarcinogenesis by acyclic retinoid is accompanied by reduction in emergence of both TGF-alpha-expressing oval-like cells and activated hepatic stellate cells. Nutr Cancer 51: 197–206.

Sato, M., D.S. Shames, A.F. Gazdar, and J.D. Minna. 2007. A translational view of the molecular pathogenesis of lung cancer. J Thorac Oncol 2: 327–343.

Schatton, T., G.F. Murphy, N.Y. Frank, K. Yamaura, A.M. Waaga-Gasser, M. Gasser, Q. Zhan, S. Jordan, L.M. Duncan, C. Weishaupt, R.C. Fuhlbrigge, T.S. Kupper, M.H. Sayegh, and M.H. Frank. 2008. Identification of cells initiating human melanomas. Nature 451: 345–349.

Schwarze, S.R., S.E. DePrimo, L.M. Grabert, V.X. Fu, J.D. Brooks, and D.F. Jarrard. 2002. Novel pathways associated with bypassing cellular senescence in human prostate epithelial cells. J Biol Chem 277: 14877–14883.

Seigel, G.M., L.M. Campbell, M. Narayan, and F. Gonzalez-Fernandez. 2005. Cancer stem cell characteristics in retinoblastoma. Mol Vis 11: 729–737.

Sell, S. 2004. Stem cell origin of cancer and differentiation therapy. Crit Rev Oncol Hematol 51: 1–28.

Sengupta, A., D. Banerjee, S. Chandra, S.K. Banerji, R. Ghosh, R. Roy, and S. Banerjee. 2007. Deregulation and cross talk among Sonic hedgehog, Wnt, Hox and Notch signaling in chronic myeloid leukemia progression. Leukemia 21: 949–955.

Shay, J.W., and W.E. Wright. 2005. Senescence and immortalization: role of telomeres and telomerase. Carcinogenesis 26: 867–874.

She, J.J., P.G. Zhang, Z.M. Wang, W.M. Gan, and X.M. Che. 2008. Identification of side population cells from bladder cancer cells by dyecycle violet staining. Cancer Biol Ther 7: 1663–1668.

Shervington, A., and C. Lu. 2008. Expression of multidrug resistance genes in normal and cancer stem cells. Cancer Invest 26: 535–542.

Shi, G.M., Y. Xu, J. Fan, J. Zhou, X.R. Yang, S.J. Qiu, Y. Liao, W.Z. Wu, Y. Ji, A.W. Ke, Z.B. Ding, Y.Z. He, B. Wu, G.H. Yang, W.Z. Qin, W. Zhang, J. Zhu, Z.H. Min, and Z.Q. Wu. 2008. Identification of side population cells in human hepatocellular carcinoma cell lines with stepwise metastatic potentials. J Cancer Res Clin Oncol 134: 1155–1163.

Shin, J.S., A. Hong, M.J. Solomon, and C.S. Lee. 2006. The role of telomeres and telomerase in the pathology of human cancer and aging. Pathology 38: 103–113.

Shiozawa, Y., E.A. Pedersen, A.M. Havens, Y. Jung, A. Mishra, J. Joseph, J.K. Kim, L.R. Patel, C. Ying, A.M. Ziegler, M.J. Pienta, J. Song, J. Wang, R.D. Loberg, P.H. Krebsbach, K.J. Pienta, and R.S. Taichman. 2011a. Human prostate cancer metastases target the

hematopoietic stem cell niche to establish footholds in mouse bone marrow. J Clin Invest 121: 1298–1312.
Shiozawa, Y., K.J. Pienta, and R.S. Taichman. 2011b. Hematopoietic stem cell niche is a potential therapeutic target for bone metastatic tumors. Clin Cancer Res 17: 5553–5558.
Shiras, A., S.T. Chettiar, V. Shepal, G. Rajendran, G.R. Prasad, and P. Shastry. 2007. Spontaneous transformation of human adult nontumorigenic stem cells to cancer stem cells is driven by genomic instability in a human model of glioblastoma. Stem Cells 25: 1478–1489.
Siegel, R., D. Naishadham, and A. Jemal. 2012. Cancer statistics, 2012. CA Cancer J Clin 62: 10–29.
Sims-Mourtada, J., J.G. Izzo, J. Ajani, and K.S. Chao. 2007. Sonic Hedgehog promotes multiple drug resistance by regulation of drug transport. Oncogene 26: 5674–5679.
Singh, S., S.K. Srivastava, A. Bhardwaj, L.B. Owen, and A.P. Singh. 2010. CXCL12-CXCR4 signalling axis confers gemcitabine resistance to pancreatic cancer cells: a novel target for therapy. Br J Cancer 103: 1671–1679.
Singh, S.K., C. Hawkins, I.D. Clarke, J.A. Squire, J. Bayani, T. Hide, R.M. Henkelman, M.D. Cusimano, and P.B. Dirks. 2004. Identification of human brain tumour initiating cells. Nature 432: 396–401.
Steg, A.D., K.S. Bevis, A.A. Katre, A. Ziebarth, Z.C. Dobbin, R.D. Alvarez, K. Zhang, M. Conner, and C.N. Landen. 2012. Stem cell pathways contribute to clinical chemoresistance in ovarian cancer. Clin Cancer Res 18: 869–881.
Strizzi, L., C. Bianco, N. Normanno, and D. Salomon. 2005. Cripto-1: a multifunctional modulator during embryogenesis and oncogenesis. Oncogene 24: 5731–5741.
Stuelten, C.H., S.D. Mertins, J.I. Busch, M. Gowens, D.A. Scudiero, M.W. Burkett, K.M. Hite, M. Alley, M. Hollingshead, R.H. Shoemaker, and J.E. Niederhuber. 2010. Complex display of putative tumor stem cell markers in the NCI60 tumor cell line panel. Stem Cells 28: 649–660.
Sugiarto, S., A.I. Persson, E.G. Munoz, M. Waldhuber, C. Lamagna, N. Andor, P. Hanecker, J. Ayers-Ringler, J. Phillips, J. Siu, D.A. Lim, S. Vandenberg, W. Stallcup, M.S. Berger, G. Bergers, W.A. Weiss, and C. Petritsch. 2011. Asymmetry-defective oligodendrocyte progenitors are glioma precursors. Cancer Cell 20: 328–340.
Sun, S.Y., N. Hail, Jr. and R. Lotan. 2004. Apoptosis as a novel target for cancer chemoprevention. J. Natl. Cancer Inst 96: 662–672.
Sung, J.M., H.J. Cho, H. Yi, C.H. Lee, H.S. Kim, D.K. Kim, A.M. bd El-Aty, J.S. Kim, C.P. Landowski, M.A. Hediger, and H.C. Shin. 2008. Characterization of a stem cell population in lung cancer A549 cells. Biochem Biophys Res Commun 371: 163–167.
Thiery, J.P. 2002. Epithelial-mesenchymal transitions in tumour progression. Nat Rev Cancer 2: 442–454.
Todaro, M., M.P. Alea, A.B. Di Stefano, P. Cammareri, L. Vermeulen, F. Iovino, C. Tripodo, A. Russo, G. Gulotta, J.-P. Medema, and G. Stassi. 2007. Colon cancer stem cells dictate tumor growth and resist cell death by production of interleukin-4. Cell Stem Cell 1: 389–402.
Tso, C.L., P. Shintaku, J. Chen, Q. Liu, J. Liu, Z. Chen, K. Yoshimoto, P.S. Mischel, T.F. Cloughesy, L.M. Liau, and S.F. Nelson. 2006. Primary glioblastomas express mesenchymal stem-like properties. Mol Cancer Res 4: 607–619.
Vaish, M. 2007. Mismatch repair deficiencies transforming stem cells into cancer stem cells and therapeutic implications. Mol Cancer 6: 26.
van Beijnum, J., R.P. Dings, E. van der Linden, B.M. Zwaans, F.C. Ramaekers, K.H. Mayo, and A.W. Griffioen. 2006. Gene expression of tumor angiogenesis dissected; specific targeting of colon cancer angiogenic vasculature. Blood 108: 2339–2348.
van den Hoogen, C., G. van der Horst, H. Cheung, J.T. Buijs, J.M. Lippitt, N. Guzman-Ramirez, F.C. Hamdy, C.L. Eaton, G.N. Thalmann, M.G. Cecchini, R.C. Pelger, and G. van der Pluijm. 2010. High aldehyde dehydrogenase activity identifies tumor-initiating and metastasis-initiating cells in human prostate cancer. Cancer Res 70: 5163–5173.

van den Hoogen, C., G. van der Horst, H. Cheung, J.T. Buijs, R.C. Pelger, and G. van der Pluijm. 2011. The aldehyde dehydrogenase enzyme 7A1 is functionally involved in prostate cancer bone metastasis. Clin Exp Metastasis 28: 615–625.

van der Horst, G., C. van den Hoogen, J.T. Buijs, H. Cheung, H. Bloys, R.C. Pelger, G. Lorenzon, B. Heckmann, J. Feyen, P. Pujuguet, R. Blanque, P. Clement-Lacroix, and G. van der Pluijm 2011. Targeting of alpha(v)-integrins in stem/progenitor cells and supportive microenvironment impairs bone metastasis in human prostate cancer. Neoplasia 13: 516–525.

Venugopal, C., N. Li, X. Wang, B. Manoranjan, C. Hawkins, T. Gunnarsson, R. Hollenberg, P. Klurfan, N. Murty, J. Kwiecien, F. Farrokhyar, J.P. Provias, C. Wynder, and S.K. Singh. 2012. Bmi1 marks intermediate precursors during differentiation of human brain tumor initiating cells. Stem Cell Res 8: 141–153.

Villanueva, A., P. Newell, D.Y. Chiang, S.L. Friedman, and J.M. Llovet. 2007. Genomics and signaling pathways in hepatocellular carcinoma. Semin Liver Dis 27: 55–76.

Vosseler, S., N. Mirancea, P. Bohlen, M.M. Mueller, and N.E. Fusenig. 2005. Angiogenesis inhibition by vascular endothelial growth factor receptor-2 blockade reduces stromal matrix metalloproteinase expression, normalizes stromal tissue, and reverts epithelial tumor phenotype in surface heterotransplants. Cancer Res 65: 1294–1305.

Wang, J., L.P. Guo, L.Z. Chen, Y.X. Zeng, and S.H. Lu. 2007a. Identification of cancer stem cell-like side population cells in human nasopharyngeal carcinoma cell line. Cancer Res 67: 3716–3724.

Wang, L., R. Mezencev, N.J. Bowen, L.V. Matyunina, and J.F. McDonald. 2012a. Isolation and characterization of stem-like cells from a human ovarian cancer cell line. Mol Cell Biochem 363: 257–268.

Wang, X., K. Belguise, N. Kersual, K.H. Kirsch, N.D. Mineva, F. Galtier, D. Chalbos, and G.E. Sonenshein. 2007b. Oestrogen signalling inhibits invasive phenotype by repressing RelB and its target BCL2. Nat Cell Biol 9: 470–478.

Wang, X., C. Venugopal, B. Manoranjan, N. McFarlane, E. O'Farrell, S. Nolte, T. Gunnarsson, R. Hollenberg, J. Kwiecien, P. Northcott, M.D. Taylor, C. Hawkins, and S.K. Singh. 2012b. Sonic hedgehog regulates Bmi1 in human medulloblastoma brain tumor-initiating cells. Oncogene 31: 187–199.

Wang, Z., Y. Li, D. Kong, S. Banerjee, A. Ahmad, A.S. Azmi, S. Ali, J.L. Abbruzzese, G.E. Gallick, and F.H. Sarkar. 2009. Acquisition of epithelial-mesenchymal transition phenotype of gemcitabine-resistant pancreatic cancer cells is linked with activation of the notch signaling pathway. Cancer Res 69: 2400–2407.

Whitworth, J.M., A.I. Londono-Joshi, J.C. Sellers, P.J. Oliver, D.D. Muccio, V.R. Atigadda, J.M. Straughn, Jr. and D.J. Buchsbaum. 2012. The impact of novel retinoids in combination with platinum chemotherapy on ovarian cancer stem cells. Gynecol Oncol 125: 226–230.

Widera, D., A. Kaus, C. Kaltschmidt, and B. Kaltschmidt. 2008. Neural stem cells, inflammation and NF-kappaB: basic principle of maintenance and repair or origin of brain tumours? J Cell Mol Med 12: 459–470.

Wright, M.H., A.M. Calcagno, C.D. Salcido, M.D. Carlson, S.V. Ambudkar, and L. Varticovski. 2008. Brca1 breast tumors contain distinct CD44+/CD24– and CD133+ cells with cancer stem cell characteristics. Breast Cancer Res 10: R10.

Xu, J.X., E. Morii, Y. Liu, N. Nakamichi, J. Ikeda, H. Kimura, and K. Aozasa. 2007. High tolerance to apoptotic stimuli induced by serum depletion and ceramide in side-population cells: high expression of CD55 as a novel character for side-population. Exp Cell Res 313: 1877–1885.

Yang, Z.F., D.W. Ho, M.N. Ng, C.K. Lau, W.C. Yu, P. Ngai, P.W. Chu, C.T. Lam, R.T. Poon, and S.T. Fan. 2008. Significance of CD90(+) cancer stem cells in human liver cancer. Cancer Cell 13: 153–166.

Yang, Z.J., and R.J. Wechsler-Reya. 2007. Hit 'em where they live: targeting the cancer stem cell niche. Cancer Cell 11: 3–5.

Ying, M., S. Wang, Y. Sang, P. Sun, B. Lal, C.R. Goodwin, H. Guerrero-Cazares, A. Quinones-Hinojosa, J. Laterra, and S. Xia. 2011. Regulation of glioblastoma stem cells by retinoic acid: role for Notch pathway inhibition. Oncogene 30: 3454–3467.

Yu, S.C., Y.F. Ping, L. Yi, Z.H. Zhou, J.H. Chen, X.H. Yao, L. Gao, J.M. Wang, and X.W. Bian. 2008. Isolation and characterization of cancer stem cells from a human glioblastoma cell line U87. Cancer Lett 265: 124–134.

Yuan, X., J. Curtin, Y. Xiong, G. Liu, S. Waschsmann-Hogiu, D.L. Farkas, K.L. Black, and J.S. Yu. 2004. Isolation of cancer stem cells from adult glioblastoma multiforme. Oncogene 23: 9392–9400.

Zhang, S., C. Balch, M.W. Chan, H.C. Lai, D. Matei, J.M. Schilder, P.S. Yan, T.H. Huang, and K.P. Nephew. 2008a. Identification and characterization of ovarian cancer-initiating cells from primary human tumors. Cancer Res 68: 4311–4320.

Zhang, X., R. Komaki, L. Wang, B. Fang, and J.Y. Chang. 2008b. Treatment of radioresistant stem-like esophageal cancer cells by an apoptotic gene-armed, telomerase-specific oncolytic adenovirus. Clin Cancer Res 14: 2813–2823.

Zhang, Y., J. Wei, H. Wang, X. Xue, Y. An, D. Tang, Z. Yuan, F. Wang, J. Wu, J. Zhang, and Y. Miao. 2012. Epithelial mesenchymal transition correlates with CD24+CD44+ and CD133+ cells in pancreatic cancer. Oncol Rep 27: 1599–1605.

Zhong, H., and J.P. Bowen. 2006. Antiangiogenesis drug design: multiple pathways targeting tumor vasculature. Curr Med Chem 13: 849–862.

Zhou, J., C.Y. Wang, T. Liu, B. Wu, F. Zhou, J.X. Xiong, H.S. Wu, J. Tao, G. Zhao, M. Yang, and S.M. Gou. 2008. Persistence of side population cells with high drug efflux capacity in pancreatic cancer. World J Gastroenterol 14: 925–930.

Zhou, Z., A. Flesken-Nikitin, and A.Y. Nikitin. 2007. Prostate cancer associated with p53 and Rb deficiency arises from the stem/progenitor cell-enriched proximal region of prostatic ducts. Cancer Res 67: 5683–5690.

Chapter 9

Salinomycin, a Candidate Drug for the Elimination of Cancer Stem Cells

Cord Naujokat[1,*] and Stefan Laufer[2]

Introduction

The challenge of targeting cancer stem cells and their progeny

The experimental demonstration of cancer stem cells (CSCs) in a variety of human malignant tumors, including cancers of the blood, breast, brain, bone, skin, liver, lung, bladder, ovary, prostate, colon, pancreas and head and neck has led to the conceptual hypothesis that tumors, like physiologic proliferative tissues, can be hierarchically organized and propagated by limited numbers of dedicated stem cells (Reya et al. 2001, Pardal et al. 2003, Dick 2008, Frank et al. 2010, Hermann et al. 2010, Clevers 2011, Nguyen et al. 2012). According to a consensus definition, these CSCs are cells within a tumor that possess the capacity to self-renew and to give rise to the heterogeneous lineages of cancer cells that comprise the tumor (Clarke et al. 2006). CSCs can therefore only be defined experimentally by their ability to recapitulate the generation of a continuously growing tumor in serial xenotransplantation settings (Clarke et al. 2006).

Unfortunately, CSCs possess numerous intrinsic mechanisms of resistance to conventional chemotherapeutic drugs and radiation therapy, including expression of ATP-binding cassette (ABC) drug pumps such

[1]Institute of Immunology, University of Heidelberg, Im Neuenheimer Feld 305, 69120 Heidelberg, Germany.
E-mail: cord.naujokat@med.uni-heidelberg.de
[2]Department of Pharmaceutical/Medical Chemistry, Eberhard-Karls-University Tübingen, Auf der Morgenstelle 8, 72076 Tübingen, Germany.
E-mail: Stefan.Laufer@uni-tuebingen.de
*Corresponding author

as ABCG2/BCRP and P-glycoprotein/MDR1 (Dean et al. 2005, An and Ongkeko 2009, Dean 2009, Calcagno et al. 2010, Elliot et al. 2010, Fuchs et al. 2010, Moitra et al. 2011), amplified checkpoint activation and efficient DNA and oxydative damage repair (Bao et al. 2006, Phillips et al. 2006, Diehn et al. 2009, Ropolo et al. 2009, Viale et al. 2009, Pajonk et al. 2010), metabolic alterations with a preference for hypoxia (Zhou et al. 2011), activation of Wnt/β-catenin signaling (Chen et al. 2007, Woodward et al. 2007, Teng et al. 2010, Yeung et al. 2010, Takebe et al. 2011a, Janikova and Skarda 2012), activation of the hedgehog and notch signaling pathways (Zhao et al. 2009, Kobune et al. 2009, Wang et al. 2009, 2010a, Kelleher 2011, Takebe et al. 2011a, Ray et al. 2011, Janikova and Skarda 2012), expression and activation of the Akt/PKB and ATR/CHK1 survival pathways (Ma et al. 2008, Korkaya et al. 2009, Gallmeier et al. 2011), aberrant PI3K/Akt/mTor-mediated signaling and loss of phosphatase and tensin homolog (PTEN) (Hambardzumyan et al. 2008, Hill and Wu 2009, Martelli et al. 2011), amplified activity of aldehyde dehydrogenase 1 (ALDH1) (Pearce et al. 2005, Ginestier et al. 2007, Dylla et al. 2008, Awad et al. 2010, Charafe-Jauffret et al. 2010), constitutive activation of NF-κB (Guzman et al. 2002, Zhou et al. 2008, Liu et al. 2010), acquisition of epithelial-mesenchymal transition (EMT) (Creighton et al. 2009, Kim et al. 2009, Wang et al. 2009, Chen et al. 2011), protection from apoptosis by autocrine production of interleukin-4 (Todaro et al. 2007, Francipane et al. 2008, Di Stefano et al. 2010), various mechanisms of apoptosis resistance and defective apoptotic signaling (Liu et al. 2006, Chiou et al. 2008, Fulda and Pervaiz 2010, Kruyt and Schuringa 2010), protection by microenvironment and niche networks (Calabrese et al. 2007, Hambardzumyan et al. 2008, Korkaya et al. 2011, Venere et al. 2011), immune evasion (Reim et al. 2009, Schatton and Frank 2009, Schatton et al. 2010, Wu et al. 2011), low proliferative activity (Gao et al. 2010), and, ultimately, transient or long-termed quiescence, the latter also termed dormancy (Essers and Trumpp 2010, Li and Bhatia 2011).

Many of these intrinsic mechanisms as well as indirect or yet unknown mechanisms of resistance and immortality (Frank et al. 2010, Maugeri-Sacca et al. 2011, Morrison et al. 2011) allow CSCs to survive current cancer therapies and to initiate reconstitution of the original tumor, long-term tumor recurrence and metastasis (Yu et al. 2007, Eyler and Rich 2008, Li et al. 2008a, Chiu et al. 2010, Koch et al. 2010, Al-Ejeh et al. 2011, Anderson et al. 2011, Velasco-Velaquez et al. 2011).

Similar to conventional anticancer drugs, numerous novel tumor-targeted drugs were designed to target rapidly proliferating cancer cells, so that many CSCs might also be relatively insensitive to these drugs. For instance, imatinib has been developed as an inhibitor of the Bcr-Abl tyrosine kinase, which constitutes the fusion protein product of a chromosomal translocation (so called Philadelphia chromosome) that acts as a molecular switch for proliferation and differentiation of multipotent progenitor cells in chronic myeloid leukemia (CML) (Era 2002). Imatinib has been shown to eliminate proliferating, committed leukemia progenitors, but not nonproliferating CML stem cells (Graham et al. 2002, Holtz et al. 2005, Hamilton et al. 2006, Oravecz-Wilson et a. 2009), which persist after imatinib therapy (Bhatia et al. 2003), and, after initial therapeutic success, many CML patients become resistant to imatinib therapy (Deiniger et al. 2005, Li et al. 2011a).

Furthermore, trastuzumab, a humanized monoclonal antibody directed against HER2 has been developed to treat patients with HER2-overexpressing breast cancers, which represent one fourth of all breast cancer patients (Gajria and Chandarlapaty 2011, Tagliabue et al. 2011). HER2 is a member of the human epidermal growth factor receptor family of receptor tyrosine kinases that is preferentially expressed in breast and ovarian cancer and that activates signaling pathways that promote tumorigenic cellular processes such as proliferation and evasion of apoptosis (Slamon et al. 1987, Hudziak et al. 1987). However, trastuzumab mono-therapy in patients with HER2-overexpressing metastatic breast cancer shows a response rate of no more than 30% (Vogel et al. 2002), and primary or acquired resistance to trastuzumab occurs frequently in different clinical settings even when combination regimes are used (Gajria and Chandarlapaty 2011, Tagliabue et al. 2011, Wong et al. 2011). One important mechanism of trastuzumab resistance in the therapy of HER2-overexpressing breast cancer might be the failure of trastuzumab to target breast CSCs (Bedard et al. 2009, Reim et al. 2009, Oliveras-Ferraros et al. 2010), but recent evidence indicates that the HER2 pathway plays an important role in the maintenance of breast CSCs (Korkaya et al. 2008, Cicalese et al. 2009, Wang et al. 2010b). Trastuzumab has been shown to exert its antitumor activities only effectively in the presence of a normal PI3K signaling pathway and in the presence of PTEN (Nagata et al. 2004, Berns et al. 2007, Fabi et al. 2010, Kataoka 2010, Dave et al. 2011), but, as noted above, CSCs display aberrant PIK3K signaling and loss of PTEN (Hambardzumyan et al. 2008, Hill and Wu 2009, Martelli et al. 2011).

A similar scenario has been reported for the small molecule dual inhibitor of EGFR tyrosine kinase and HER2 tyrosine kinase, lapatinib, which loses its therapeutic activity in breast cancers displaying loss of PTEN (Dave et al. 2011).

Finally, sorafenib, a small molecule inhibitor of multiple tyrosine kinases involved in tumor proliferation and tumor angiogenesis, including Raf, VEGFR, PDGFR and FLT3 (Wilhelm et al. 2008) has been established in the treatment of acute myeloid leukemia (AML) because FMS-like tyrosine kinase 3 (FLT3) is overexpressed in leukemic blasts in almost all cases of AML (Carow et al. 1996, Wiernik 2010). As demonstrated recently, sorafenib is capable of reducing the number of mature AML progenitors but fails to eradicate AML stem and primitive progenitor cells due to robust protection of these cells by the bone marrow stromal microenvironment (Parmar et al. 2011), providing a further paradigm that novel tumor-targeted drugs fail to eliminate CSCs.

Thus, according to the cancer stem cell concept of carcinogenesis (Reya et al. 2001, Clarke et al. 2006, Dick 2008, Clevers 2011, Nguyen et al. 2012), CSCs represent novel and translationally relevant targets for cancer therapy, and the identification, development and therapeutic use of compounds and drugs that selectively target CSCs is a major challenge for future cancer treatment (Zhou et al. 2009, Frank et al. 2010, McDermott and Wicha 2010, Morrison et al. 2011). However, the goal for any CSC-directed therapy should be the eradication of all CSCs in a patient, and the efficacy of single agents targeting CSCs may be limited by several factors. CSCs can represent a heterogenous population that may not be homogeneously sensitive to a given anti-CSC agent (Nguyen et al. 2010, Lorico and Rappa 2011), and, under the selection pressure of agents targeting CSCs, therapy resistant CSC clones may emerge (Shackleton et al. 2009). Therefore, the eradication of all CSCs will likely require targeting of more than one intrinsic pathway operating in CSCs to reduce the probability of escape mutants (Mueller et al. 2009, Park et al. 2009, Zhou et al. 2009, Liu and Wicha 2010). Moreover, agents causing tumor regression in advanced stages of cancer likely reflect effects on the bulk tumor population, but may have minimal effect on the CSC population. In contrast, a CSC-specific therapy would show modest effect on tumor growth of the bulk tumor population in advanced stages of cancer, but may have substantial clinical benefit in early stages of cancer as well as in neoadjuvant and adjuvant clinical settings (McDermott and Wicha 2010).

Ultimately, cure of cancer will require the eradication of all malignant cells within a patient's cancer: CSCs and their progeny. Therefore it will be important and promising to combine in sophisticated clinical settings CSC targeting agents with novel tumor-targeted drugs and conventional cytotoxic drugs. Such combinations may act in concert to eradicate CSCs, more differentiated progenitors, and bulk tumor cells (Mackenzie et al. 2010, Martin-Castillo et al. 2010a, Goodwin et al. 2011, Rocha et al. 2011, Zhang et al. 2011, 2012).

Compounds that target CSCs

Various compounds and drugs that selectively target CSCs have been discovered and developed recently (Winquist et al. 2009, Zhou et al. 2009, Morrison et al. 2011), (Table 1). These agents include microbial-derived and plant-derived biomolecules (Kawasaki et al. 2008, Gupta et al. 2009,

Table 1. Examples of Compounds and Drugs that Target CSCs.

Class	Compound	Description	Targets	Clinical Status	References
Microbial-derived, ionophore antibiotic	Salinomycin	Lipophilic mono-carboxylic polyether ionophore antibiotic, isolated from *Streptomyces albus* (strain No. 80614). Ionophore with strict selectivity for alkali ions and a strong preference for K^+. Interferes with transmembrane K^+ potential and promotes the efflux of K^+ ions from mitochondria and cytoplasm. Inhibits mitochondrial oxidative phosphorylation. Inhibits p-glycoprotein and other ABC-transporters. Induces apoptosis, mitochondrial membrane depolarization and differentiation in cancer cells and CSCs. Eradicates CSCs in mice. Inhibits proximal Wnt signaling in CLL cells.	Breast CSCs, AML CSCs, gastrointestinal CSCs, colon CSCs, lung CSCs, squamous cell carcinoma CSCs, osteosarcoma CSCs, gastric CSCs, pancreatic CSCs, prostate CSCs.	Phase I/II, triple negative breast cancer.	Gupta et al. 2009, Fuchs et al. 2009, Naujokat et al. 2010, Fuchs et al. 2010, Bardsley et al. 2010, Gong et al. 2010, Riccioni et al. 2010, Wang 2011, Dong et al. 2011, Basu et al. 2011, Lu et al. 2011, Tang et al. 2011, Zhi et al. 2011, Zhang et al. 2011, 2012, Ketola et al. 2012
Plant-derived, sesquiterpene lactone	Parthenolide	Nucleophilic sesquiterpene lactone isolated from *Tanacetum parthenium* (feverfew). Inhibits activation of NF-κB by inhibition of IKK-mediated phosphorylation of IκB. Induces reactive oxygen species (ROS) exclusively in tumor cells. Inhibits STAT activation and interferes with microtubule formation. Inhibits proliferation and induces apoptosis in leukemia CSCs. Eradicates breast, prostate and myeloma CSCs.	AML CSCs, lymphoid leukemia CSCs, breast CSCs, prostate CSCs, myeloma CSCs.	Phase I, acute myeloid leukemia.	Hehner et al. 1999, Benjamin et al. 2001, Mathema et al. 2011, Guzman et al. 2005, 2007, Hassane et al. 2008, Zhou et al. 2008, Birnie et al. 2008, Kawasaki et al. 2009, Gunn et al. 2011

Table 1. contd....

Class	Compound	Description	Targets	Clinical Status	References
Monoclonal antibody	H90	Mouse monoclonal IgG1 antibody directed against CD44, a transmembrane glycoprotein and the receptor for hyaluronic acid, osteopontin fibronectin, and selectin. CD44 is expressed on different types of human CSCs. CD44 ligation by H90 induces terminal differentiation and inhibits homing, proliferation and the repopulation capacity of AML CSCs.	AML CSCs	Preclinical	Charrad et al. 2002, Gadhoum et al. 2004, Jin et al. 2006
	P245	Mouse monoclonal IgG1 antibody directed against CD44 which is expressed on breast CSCs. CD44 ligation by P245 induces tumor regression and prevention of tumor relapse in mice with human triple negative basal-like breast cancer which contains up to 40% breast CSCs.	Breast CSCs	Preclinical	Al-Hajj et al. 2003, Marangoni et al. 2009
Small molecule inhibitor	GDC-0449/ Vismodegib	Synthetic small molecule systemic hedgehog pathway inhibitor that acts as a receptor antagonist of smoothened (SMO), a serpentine receptor of the hedgehog pathway. Inhibits ABC transporters ABCG2 and P-glyco-protein. Orally available. Induces tumor regression in advanced basal cell carcinoma and medulloblastoma. Eliminates lung CSCs and hepatocellular carcinoma metastatic cells. Induces apoptosis in pancreatic CSCs.	Pancretic CSCs, lung CSCs, hepatocellular carcinoma CSCs.	Phase I-III, medullo-blastoma, basal cell carcinoma, colon, lung, ovarian, pancreatic, breast and gastric carcinoma, glioblastoma.	Robarge et al. 2009, Rudin et al. 2009, Zhang et al. 2009, Von Hoff et al. 2009, Gupta et al. 2010, Low and de Sauvage, 2010, Kelleher 2011, LoRusso et al. 2011, Philips et al. 2011, Singh et a. 2011, Tian et al. 2011

Classical drug	Metformin	Oral anti-diabetic drug of the biguanide class used for the first-line treatment of type 2 diabetes mellitus. Suppresses hepatic gluconeogenesis. Activates AMP-activated protein kinase (AMPK), inhibits mammalian target of rapamycin (mTOR) pathways, insulin like growth factors (IGFs), several protein kinases and receptor tyrosine kinases such as HER2. Up-regulates tumor-suppressive microRNAs and inhibits oncogenic microRNAs in breast and pancreatic cancer cells. Suppresses self-renewal, proliferation and epithelial-mesenchymal transition of breast CSCs. Overcomes resistance to trastuzumab and lapatinib in breast cancer. Induces tumor regression in clinical trials.	Breast CSC, prostate CSC, lung CSC, ovarian CSC, colon CSC.	Phase II, III, breast cancer, solid tumors. Combination with conventional anti-cancer drugs.	Hirsch et al. 2009, Vazquez-Martin et al. 2009, Mackenzie et al. 2010, Martin-Castillo et al. 2010a, 2010b, Vazquez-Martin et al. 2010, Oliveras-Ferraros et al. 2011, Bao et al. 2011, Belda-Iniesta et al. 2011, Hadad et al. 2011, Liu et al. 2011b, Vazquez-Martin et al. 2011a, 2011b

Ghantous et al. 2010, Li et al. 2010, Naujokat et al. 2010, Wang et al. 2010c, Li et al. 2011b), antibodies directed against CSC-specific cell surface molecules (Jin et al. 2006, Okamoto and Perez 2008, Schatton et al. 2008, Deonarain et al. 2009, Jin et al. 2009, Majeti et al. 2009, Zhou et al. 2009, Majeti 2011), small molecule inhibitors targeting key components of intrinsic signaling pathways of CSCs (Ginestier et al. 2010, Low and de Sauvage 2010, Gould and Missailidis 2011, Grinshtein et al. 2011, Martelli et al. 2011, Takebe et al. 2011a), microRNAs interfering with essential and specific functions of CSCs (Guessous et al. 2010, Li et al. 2010, Liu et al. 2011a, Liu and Tang 2011, Zimmermann and Wu 2011, Yang et al. 2012) and, surprisingly, some classical drugs, such as metformin (Hirsch et al. 2009, Vazquez-Martin et al. 2011a), disulfiram (Yip et al. 2011), acetaminophen (Takehara et al. 2011) and chloroquine (Vazquez-Martin et al. 2011b) that have been used for decades for the treatment of metabolic, inflammatory and infectious diseases.

Although these compounds and drugs have been shown to effectively target signaling pathways and/or molecules selectively operating in CSCs (Table 1), some of them are also capable of killing other types and subpopulations of cancer cells which do not display CSC properties. In particular, the biomolecules salinomycin and parthenolide as well as the biguanide metformin have been demonstrated to induce apoptosis in various types of human cancer cells (Fuchs et al. 2009, Naujokat et al. 2010, Mathema et al. 2011, Kourelis and Siegel 2011), suggesting that these compounds may contribute to the eradication of cancer more effectively than compounds targeting either CSCs or normal cancer cells. Moreover, the ionophore antibiotic salinomycin seems to have even extended capabilities of eliminating cancer, because this compound has been demonstrated to effectively target normal cancer cells (Fuchs et al. 2009, 2010, Lu et al. 2011, Kim et al. 2011a), highly multidrug and apoptosis resistant cancer cells (Fuchs et al. 2009, 2010, Gong et al. 2010), and CSCs (Gupta et al. 2009, Bardsley et al. 2010, Fuchs et al. 2010, Dong et al. 2011, Wang 2011, Zhi et al. 2011, Zhang et al. 2011, 2012).

Salinomycin as a Candidate Drug for Eliminating CSCs

From broiler to bedside: a brief history of salinomycin

During the course of a screening program for new antibiotics in 1974, Miyazaki and colleagues from the research division of Kaken Chemicals Co., Ltd., Tokyo, Japan isolated a new biologically active substance from the culture broth of *Streptomyces albus* strain No. 80614 that was termed salinomycin (Miyazaki et al. 1974). The salinomycin-producing organism was detected in and isolated from a soil sample collected at Fuji City, Shizuoka Prefecture, Japan, taxonomically classified as a member of the

Streptomyces albus genus (ROSSI-DORIA) WAKSMAN and HENRICI, and designated as the strain No. 80614 (Waksman 1961, Miyazaki et al. 1974). The production of salinomycin was carried out by tank fermentation, filtration of the culture broth of *Streptomyces albus*, purification by column chromatography on alumina/silica gel and subsequent eluation. The eluate was then concentrated *in vacuo*, dried and finally crystallized. By this isolation procedure, salinomycin was obtained in the form of colourless prism of the sodium salt (Miyazaki et al. 1974). Subsequently, it was demonstrated that salinomycin exhibit antimicrobial activity against gram-positive bacteria including *Bacillus subtilis*, *Staphylococcus aureus*, *Micrococcus flavus*, *Sarcina lutea* and *Mycobacterium spp.*, some filamentous fungi, *Plasmodium falciparum* and *Eimeria spp.*, protozoan parasites responsible for the poultry disease coccidiosis (Miyazaki et al. 1974, Danforth et al. 1977, Mehlhorn et al. 1984). The chemical structure of salinomycin was determined by X-ray crystallographic analysis, which revealed that salinomycin was a new member of the family of the monocarboxylic polyether antibiotics (Fig. 1) (Kinashi et al. 1973). Because other polyether antibiotics such as monensin and lasalocid had previously been shown to exhibit effective antimicrobial activities against *Eimeria spp.*, the anticoccidial estimation of salinomycin was carried out with chickens infected with *Eimeria tenella* oocysts. Salinomycin was effective in reducing the mortality of chickens from coccidiosis and in increasing average body weight of treated infected chickens (Miyazaki et al. 1974). Thus, a patent had been issued for the use of salinomycin to prevent coccidiosis in poultry (Miyazaki et al. 1972), and, up to today, salinomycin is used in broiler batteries and other livestock as an anticoccidial drug and is also fed to ruminants and pigs to improve nutrient absorption and feed efficiency (Danforth et al. 1977, Daugschies et al. 1998, Butaye et al. 2003, Callaway et al. 2003, Jenkins et al. 2010).

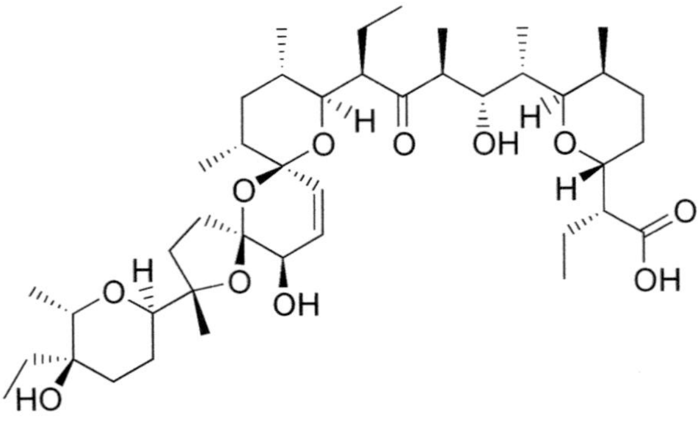

Fig. 1. Structural formula of salinomycin. The pentacyclic molecule with a unique tricyclic spiroketal ring system has a mass of 751 Da, a molecular formula of $C_{42}H_{70}O_{11}$, a melting point of 113°C and a UV absorption at 285 nm (adapted from Kinashi et al. 1973 and Miyazaki et al. 1974).

In addition, salinomycin had early been shown to act in different biological membranes, including cytoplasmic and mitochondrial membranes, as a monovalent cation ionophore with strict selectivety for alkali ions and a strong preference for K^+ (Mitani et al. 1975), thereby promoting mitochondrial and cellular K^+ efflux and inhibiting oxidative phosphorylation (Mitani et al. 1976, Pressman and Lattanzio 1978).

Similar to the monocarboxylic polyether antibiotic monensin, which exhibits complex cardiovascular effects due to its transport of Na^+ across biological membranes (Gaide et al. 1984), salinomycin had been demonstrated by Pressman and colleagues as a positive ionotropic and chronotropic agent that increased cardiac output, left ventricular systolic pressure, heart rate, mean arterial pressure, coronary artery vasodilatation and blood flow, and plasma catecholamine concentration (Fahim et al. 1986). These results had been obtained in experiments with mongrel dogs that had received a single intravenous injection of 150 $\mu g*kg^{-1}$ salinomycin (Fahim et al. 1986).

However, for several reasons, salinomycin has never been established as a drug for human diseases until now. In particular, several reports and studies published in the last three decades revealed a considerable toxicity of salinomycin in mammals such as horses, pigs, cats and alpakas after accidental oral or inhalative intake (Rollinson et al. 1987, Plumlee et al. 1995, Van der Linde-Sipman et al. 1999, Kosal and Anderson 2004, Aleman et al. 2007). A case of an accidental inhalation and swallowing of about 1 $mg*kg^{-1}$ salinomycin by a 35-year-old male human revealed severe acute and chronic salinomycin toxicity with acute nausea, photophobia, leg weakness, tachycardia and blood pressure elevation, and a chronic (day 2 to day 35) creatine kinase elevation, myoglobinuria, limb weakness, muscle pain and mild rhabdomyolysis (Story et al. 2004). Risk assessment data recently published by the Eurpean Food Safety Authority declare an acceptable daily intake (ADI) of 5 $\mu g*kg^{-1}$ salinomycin for humans, because daily intake of more than 500 $\mu g*kg^{-1}$ salinomycin by dogs leads to neurotoxic effects such as myelin loss and axonal degeneration (Dorne et al. 2011).

In view of this considerable toxicity in mammals, salinomycin has only been used for more than 30 years as a coccidostat and growth promoter in livestock, but, in terms of human drug development salinomycin rested in dormancy. Until in 2009 a seminal study by Weinberg, Lander and colleagues revealed that salinomycin selectively eliminates human breast CSCs in mice (Gupta et al. 2009). A follow-up publication demonstrated that salinomycin induces massive apoptosis in human cancer cells of different origin that display multiple mechanisms of drug and apoptosis resistance (Fuchs et al. 2009). Based on these findings, salinomycin was therapeutically applicated "first-in-man" in 2010, in the context of an experimental clinical trial with a small cohort of patients with metastatic breast, ovarian, and head and neck cancers (Steinhart et al. 2012). Intravenous application of 200 µg*kg^{-1} salinomycin every second day for three weeks resulted in partial regression of tumor metastasis (Fig. 2) (Steinhart et al. 2012) and showed only minor acute and long-term side effects, but no severe acute and long-term side effects observed with conventional chemotherapeutic drugs (Steinhart et al. 2012). Consequently, a phase I/II clinical trial with salinomycin in patients with triple negative breast cancer is envisioned to start in 2012.

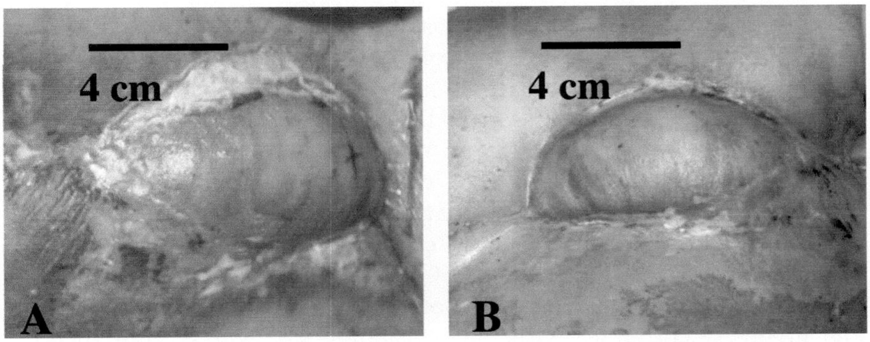

Fig. 2. Breast cancer metastasis regression by salinomycin. Orthotopic subcutaneous metastasis after mastectomy in a 40 year-old female with metastatic ductal breast carcinoma, triple negative (estrogen receptor: negative, progesterone receptor: negative, HER2: negative). (A) before, (B) after 12 intravenous applications of 200 µg*kg^{-1} salinomycin every second day (adapted from Steinhart et al. 2012).

Color image of this figure appears in the color plate section at the end of the book.

Salinomycin as a CSC killer

It came as a big surprise when Weinberg, Lander and colleagues showed in 2009 that salinomycin selectively kills human breast cancer stem-like cells (Gupta et al. 2009). In a sophisticated experimental approach, the authors used oncogenic transformed immortalized human mammary epithelial cells (termed HMLER), in which knockdown of E-cadherin by RNA interference resulted in the generation of cells undergoing epithelial-mesenchymal transition (EMT), a latent embryonic program that can endow cancer cells with migratory, invasive, self-renewal and drug resistance capabilities (Mani et al. 2008, Onder et al. 2008, Polyak and Weinberg 2009, May et al. 2011). These human breast cancer stem-like cells generated (termed HMLER-shEcad) displayed characteristic properties of CSCs, were capable of forming tumorspheres in suspension cultures, showed high and low expression of CD44 and CD24, respectively, and exhibited resistance to chemotherapeutic drugs and cytotoxic agents such as paclitaxel, doxorubicin, actinomycin D, campthotecin and staurosporine (Gupta et al. 2009). In a high-throughput screening approach, about 16,000 compounds from chemical libraries, including biological molecules and natural extracts with known bioactivity, were tested for activity and toxicity against HMLER-shEcad cells and control cells that had not undergone EMT. From a pool of 32 promising candidates, only one compound markedly and selectively reduced the viability of breast cancer stem-like HMLER-shEcad cells: salinomycin.

In subsequent experiments, it was demonstrated that salinomycin, in contrast to the anti-breast cancer drug paclitaxel, selectively reduces the proportion of $CD44^{high}/CD24^{low}$ stem-like cells in cultures of mixed populations of HMLER-shEcad cells and control cells that had not undergone EMT. In addition, pre-treatment of HMLER-shEcad cells with salinomycin resulted in inhibition of HMLER-shEcad-induced tumorsphere formation, which was not observed after pre-treatment of the cells with paclitaxel (Gupta et al. 2009). Next, comparative global gene expression profiling was employed to show that, in $CD44^{high}/CD24^{low}$ HMLER cells, salinomycin, but not paclitaxel, was capable of changing gene expression signatures characteristic of breast CSCs or mammary epithelial progenitor cells isolated from human breast cancers. For instance, expression of genes that inversely correlates with metastasis-free survival, overall survival and prognostic clinical outcome of breast cancer patients (Liu et al. 2007, Shipitsin et al. 2007), was down-regulated by salinomycin. Expression of a set of genes that promote the expansion of mammary epithelial stem cells and the formation of tumorspheres (Dontu et al. 2003) was also markedly down-regulated by salinomycin, but not by paclitaxel. In contrast, genes probably involved in mammary epithelial differentiation that encode membrane-associated and secreted proteins of the extracellular matrix were up-regulated by salinomycin (Gupta et al. 2009). Finally, as a proof of principle, it was demonstrated that salinomycin inhibits the ability of breast cancer stem-like cells to form tumors in mice. Pre-treatment of HMLER cells for 7 days with salinomycin and subsequent serial limiting dilution and injection of the cells into NOD/SCID mice resulted in a >100-fold decrease in tumor-seeding ability, relative to pretreatment of the cells with paclitaxel. Finally, salinomycin treatment of NOD/SCID mice with breast cancer tumors established by injection of human breast cancer cells resulted in a reduction of the tumor mass and metastasis, and explanted tumors showed a reduced number of breast CSCs as well as an increased epithelial differentiation (Gupta et al. 2009).

According to the primary finding that salinomycin induces massive apoptosis in human cancer cells that display different mechanisms of drug and apoptosis resistance (Fuchs et al. 2009), a subsequent study demonstrated that salinomycin is capable of overcoming ABC transporter-mediated multidrug and apoptosis resistance in human acute myeloid leukemia (AML) CSCs (Fuchs et al. 2010). One of the most important mechanism of drug resistance in leukemia stem cells and other CSCs is the expression of ABC transporters belonging to a highly conserved superfamily of transmembrane proteins capable of exporting a wide variety of macromolecules and structurally unrelated chemotherapeutic drugs from the cytosol, thereby conferring multidrug resistance, which is a major obstacle in the success of cancer chemotherapy (Dean et al. 2005, Dean 2009, Elliot et al. 2010, Moitra et al. 2011). As demonstrated in the study, expression of functional ABC transporters such as P-glycoprotein/MDR1, ABCG2/BCRP and ABCC11/MRP8 in human KG-1a AML CSCs confers resistance of the cells to various chemotherapeutic drugs, including cytosine arabinoside, doxorubicin, gemcitabine, 5-fluorouracil, topotecan, etoposide and bortezomib, but not to salinomycin, which was capable of inducing massive apoptosis in the KG-1a AML CSCs (Fuchs et al. 2010). Of note, salinomycin did not permit long-term adaptation and development of resistance of KG-1a AML CSCs to apoptosis-inducing concentrations of salinomycin, whereas the cells could be readily adapted to survive and to proliferate in the presence of initially apoptosis-inducing concentrations of doxorubicin and bortezomib (Fuchs et al. 2010).

These findings strongly suggest that salinomycin is capable of targeting breast CSCs and AML CSCs, and a series of recent studies reveal similar effects of salinomycin in other types of CSCs. In gastrointestinal stromal tumors (GIST), the most common gastrointestinal sarcomas, cells expressing CD44, CD34 and kit (activating stem cell factor receptor) have been identified as cells with self-renewal and tumorigenic capabilities (Bardsley et al. 2010). These Kit^{low} CD44+ CD34+ CSCs are resistant to inhibition of proliferation by imatinib, a tyrosine kinase inhibitor targeting oncogenic kit signaling that is commonly used in the treatment of metastatic GIST (Bardsley et al. 2010, Gronchi et al. 2010). By contrast, salinomycin nearly completely inhibited the proliferation of Kit^{low} CD44+ CD34+ CSCs without causing apoptosis, and salinomycin promoted differentiation of the cells as evidenced by the occurence of multipolarity and a fibroblast-like morphology (Bardsley et al. 2010). A combined treatment of the cells with imatinib and submaximal doses of salinomycin caused a significantly greater inhibition of proliferation than submaximal doses of salinomycin alone (Bardsley et al. 2010). Thus, the results of the study clearly show that salinomycin is capable of inhibiting proliferation and inducing terminal differentiation of GIST CSCs and also suggest that a combination of salinomycin with imatinib or other tyrosine kinase inhibitors may provide therapeutic benefit in patients with GIST.

Similar results were obtained with CD44+ CD24– ALDH1+ breast CSCs isolated from the human breast cancer cell line MCF-7. In CD44+ CD24– ALDH1+ MCF-7-derived breast CSCs, salinomycin was capable of markedly reducing the tumorsphere formation of the cells and the percentage of ALDH1+ expressing cells by nearly 50 fold (Gong et al. 2010).

Treatment of the cells with salinomycin as well as combined treatment with the cytostatic drug doxorubicin and salinomycin, but not treatment with doxorubicin alone, reduced the cloning efficiency by 10–30 fold and markedly increased apoptosis in CD44+ CD24– ALDH1+ breast CSCs (Gong et al. 2010), providing further evidence that salinomycin alone and particularly in combination with conventional anti-cancer drugs effectively targets CSCs.

Salinomycin has recently been shown to target CSCs in different types of human cancers, including gastric cancer (Zhi et al. 2011), lung adenocarcinoma (Wang 2011), osteosarcoma (Tang et al. 2011), colorectal cancer (Dong et al. 2011), squamous cell carcinoma (SCC) (Basu et al. 2011), and prostate cancer (Ketola et al. 2012), suggesting that salinomycin may be effective in CSCs of many, if not all, types of human cancers, although it is currently not known whether all cancers contain subpopulations of CSCs.

In ALDH1+ gastric CSCs, which displayed resistance to the conventional chemotherapeutic drugs 5-fluorouracil and cisplatin, salinomycin effectively inhibited tumorsphere formation, proliferation and viability of the cells (Zhi et al. 2011), similar results were obtained in ALDH1+ CSCs derived form lung adenocarcinoma cells (Wang 2011) and in osteosarcoma CSCs (Tang et al. 2011). In colorectal cancer cells and in SCC cells, salinomycin, but not oxaliplatin or cisplatin, was capable of significantly reducing the proportion of CSCs in tumorsphere assays (Dong et al. 2011, Basu et al. 2011) and in SCC xenografts in mice (Basu et al. 2011).

As noted above, cure of cancer most likely will require the eradication of different when not all cells types within a cancer, namely CSCs, more differentiated progenitors and the bulk tumor cell population that might be achieved by combining CSC targeting agents with conventional cytotoxic drugs (Zhou et al. 2009, Liu and Wicha 2010, McDermott and Wicha 2010). This hypothesis received substantial support from recent studies showing that salinomycin in combination with a conventional cytotoxic drug eradicates tumors in mice much more efficiently than the single agent (Zhang et al. 2011, 2012). In particular, salinomycin inhibited the growth of CD133+ pancreatic CSCs in colony and tumorsphere formation assays, while the cytotoxic drug gemcitabine, a nucleoside analog commonly used in the treatment of metastatic pancreatic cancer (Conroy et al. 2011), induced marked apoptosis in non-CSC 133-pancreatic cancer cells (Zhang et al. 2011). Consistently, salinomycin combined with gemcitabine eradicated tumors induced by the engraftment of human pancreatic cancer cells in mice much more efficiently than either salinomycin or gemcitabine alone (Zhang et al. 2011). Similar results were obtained in a study using CD44+ CD24– breast CSCs sorted from the human breast cancer cell line MCF-7 (Zhang et al. 2012). Salinomycin more efficiently inhibited the proliferation of CD44+

CD24– breast CSCs than of the parental MCF-7 cells, and salinomycin was also capable of inducing significant tumor regression and of reducing the number of CD44+ CD24– breast CSCs in tumors induced by inoculation of MCF-7 cells in mice (Zhang et al. 2012). Of note, salinomycin in combination with the cytotoxic drug paclitaxel almost completely eradicated the MCF-7 tumors in mice (Zhang et al. 2012).

Finally, there is growing evidence that salinomycin not only targets CSCs, but also kills more differentiated non-CSCs tumor cells, normal cancer cells, and, most importantly, cancer cells that display efficient mechanisms of resistance to cytotoxic drugs, radiation and induction of apoptosis. Salinomycin has been shown to induce massive apoptosis in acute T-cell leukemia cells (Fuchs et al. 2009) and chronic lymphocytic leukemia cells (Lu et al. 2011) isolated from leukemia patients, but failed to induce apoptosis in normal human T cells and peripheral blood lymphocytes isolated from healthy individuals (Fuchs et al. 2009, Lu et al. 2011). In different human cancer cells exhibiting resistance to cytotoxic drugs, radiation and induction of apoptosis, salinomycin has been demonstrated to induce significant apoptosis and to increase DNA damage (Fuchs et al. 2009, Kim et al. 2011b, 2011c). As in the case of breast CSCs (Gong et al. 2010), salinomycin is able to greatly enhance in normal cancer cells the cytotoxic effects of conventional cancer drugs such as doxorubicin, etoposide, paclitaxel, docetaxel and vinblastin (Kim et al. 2011b, Kim et al. 2012), envisioning a central role for salinomycin-based combination therapies in the future treatment of cancer.

Possible mechanisms of salinomycin's action against CSCs

Although the exact mechanisms underlying the elimination of CSCs by salinomycin remain poorly understood, recent work has contributed to an increased understanding of the modes of action of salinomycin in CSCs and cancer cells. It is clear that salinomycin is able to induce apoptosis in CSCs of different origin (Fuchs et al. 2010, Gong et al. 2010, Tang et al. 2011, Zhang et al. 2011), but the particular mechanisms of apoptosis induction by salinomycin in CSCs remain unclear and may differ among the cell type as demonstrated for normal cancer cells (Fuchs et al. 2009, Kim et al. 2011a). It is also evident that salinomycin is refractory to the action of ABC transporters since salinomycin is able to overcome ABC transporter-mediated multidrug and apoptosis resistance in AML CSCs (Fuchs et al. 2010), and, moreover, salinomycin has been demonstrated to be a potent inhibitor of the ABC transporter P-glycoprotein/MDR1 in different cancer cells (Riccioni et al. 2010, Kim et al. 2011c). Next, constitutive activation of the Wnt/β-catenin signaling pathway is essential for the maintenance and other specific characteristics of CSCs (Vermeulen et al. 2010, Wend et al. 2010, Takebe et al. 2011a, 2011b) and, most importantly, Wnt/β-catenin signaling confers resistance of CSCs to radiation (Chen et al. 2007, Woodward et al. 2007) and to anti-cancer drugs (Teng et al. 2010, Yeung et al. 2010). Salinomycin, however, has been shown to inhibit in chronic lymphocytic leukemia cells proximal Wnt signaling by reducing the levels of the Wnt coreceptor LRP6 and by down-regulating the expression of the Wnt target genes *LEF1*, *cyclin D1* and *fibronection*, finally leading to apoptosis (Lu et al. 2011). Most cancer cells rely more on aerobic glycolysis than on oxidative phosphorylation (the Warburg effect) (Warburg et al. 1924), but, for instance, malignant transformation of human mesenchymal stem is linked to an increase of oxidative phosphorylation (Funes et al. 2007), and glioma CSCs have been shown to mainly rely on oxidative phosphorylation (Vlashi et al. 2011), suggesting that inhibition of metabolic pathways such as oxidative phosphorylation is a promising strategy to target CSCs (Vlashi et al. 2011). In this context, salinomycin is already known to inhibit oxidative phosphorylation in mitochondria (Mitani et al. 1976) that may contribute to the efficient elimination of CSCs by salinomycin. Salinomycin is a K^+ ionophore that interferes with transmembrane K^+ potential and promotes the efflux of K^+ from mitochondria and cytoplasm (Mitani et al. 1975, 1976, Pressman and Lattanzio 1978). K^+ channels of the mitochondrial and cytoplasmic membrane are deregulated in various cancer cells (Villalonga et al. 2007, Shen et al. 2009, Haren et al. 2010), and expression of K^+ channels has been documented in CD34+/CD38– AML CSCs and in CD133+ neuroblastoma CSCs, but not in their nontumorigenic counterparts (Li et al. 2008b, Park et al. 2010). Moreover, a decrease in intracellular K^+ concentration by pharmacological induction of K^+ efflux is directly linked to induction of apoptosis and cytotoxicity in cancer cells (Bortner et al. 1997, Anderson et al. 2005), suggesting that mitochondrial and cytoplasmic K^+ efflux and K^+ depletion induced by salinomycin leads to apoptosis also in CSCs. In addition to its cytotoxic and apoptosis inducing activity in CSCs, salinomycin is apparently able to induce differentiation of CSCs and epithelial reprogramming and differentiation of cells that had undergone EMT, as recently shown for Kit^{low} CD44+ CD34+ GIST CSCs (Bardsley et al. 2010) and for cancer cells explanted from human breast cancer bearing mice treated with salinomycin (Gupta et al. 2009). This is in concert with the finding that salinomycin up-regulates the expression of genes involved in mammary epithelial differentiation that encode membrane-associated and secreted proteins of the extracellular matrix (Gupta el al. 2009). Thus, salinomycin might target and eliminate CSCs by multiple mechanisms of which only a few are currently known, including induction of apoptosis and cytotoxicity, interference with ABC transporters and K^+ channels, inhibition of oxidative phosphorylation and Wnt/β-catenin signaling pathways, and commitment of differentiation. Future research will undoubtedly uncover an increasing number of relevant mechanisms of targeting CSCs by salinomycin.

Conclusion and Future Directions

Work from the last few years highlights the possibility of selectively targeting CSCs, which are regarded as the major culprits in cancer. However, although the rather novel CSC concept of carcinogenesis is widely accepted to date, more classical mechanisms and driving forces of carcinogenesis, including genome instability, epigenetic modifications, first oncogenic hit(s), clonal evolution, replicative immortality, invasion and metastasis, immune evasion, and reprogramming of energy metabolism must be considered as a basis for defining carcinogenesis and cancer in general (Shackleton et al. 2009,Vineis et al. 2010, Yamada and Watanabe 2010, Hanahan and Weinberg 2011, Kanwal and Gupta 2011, Visvader 2011). Nevertheless, in line with the CSC concept of carcinogenesis (Reya et al. 2001, Clarke et al. 2006, Dick 2008, Clevers 2011, Nguyen et al. 2012), CSCs constitute adequately characterized cells and represent novel and translationally relevant targets for cancer therapy (Zhou et al. 2009, Frank et al. 2010, McDermott and Wicha 2010, Morrison et al. 2011). Significant advances have been made recently in the discovery, development and validation of novel compounds and drugs that target CSCs, and the clinical use of these novel agents will represent a powerful strategy for eradicating CSCs in cancer patients, thereby preventing metastasis and long-term tumor recurrence, and, hopefully, contributing to the cure of cancer.

There is growing consensus that conventional cytotoxic drugs are unable to eradicate CSCs (Zhou et al. 2009, Frank et al. 2010, Maugeri-Sacca et al. 2011), and, more disturbing, CSCs can be even selectively enriched by these drugs, as demonstrated in breast cancer patients receiving neoadjuvant systemic chemotherapy comprising conventional cytotoxic drugs (Yu et al. 2007, Li et al. 2008a). Moreover, many novel tumor-targeted drugs, including tyrosine kinase inhibitors and monoclonal antibodies raised against tumor-specific cell surface proteins, also fail to eliminate CSCs (Bhatia et al. 2003, Bedard et al. 2009, Wang et al. 2010b, Paramar et al. 2011), so that there is an urgent need for novel drugs that selectively and effectively target and eradicate CSCs, ultimately in elaborated clinical settings. In this context, one promising candidate drug is the ionophor antibiotic salinomycin, which has recently been documented to effectively eliminate CSCs in different types of human cancers, and this not only *in vitro* but also in mice bearing human cancers (Gupta et al. 2009, Basu et al. 2011, Zhang et al. 2011, 2012). To date, it is not entirely clear by which mechanisms salinomycin eliminates CSCs, but it is important to note that salinomycin, in combination with conventional cytotoxic drugs, is much more effective in eradicating tumors in mice than the single agent alone (Zhang et al. 2011, 2012). This is in accord with the postulation that efficient cancer therapy should target all cancer cell populations, including CSCs, more differentiated progenitors and bulk tumor cells that might be achieved by combining salinomycin with conventional cytotoxic drugs, novel tumor-targeted drugs and eventually radiation therapy. Importantly, salinomycin is not only able to kill CSCs, but also normal tumor cells and highly indolent tumor cells displaying resistance to cytotoxic drugs, radiation and induction of apoptosis (Fuchs et al. 2009, 2010, Gong et al. 2010, Kim et al. 2011c), hence salinomycin can be proclaimed as a triple-edged sword against cancer. More work is required to define the exact mechanisms of salinomycin's sophisticated action against CSCs, to estimate its long-term safety in humans, and finally, to exploit its probably huge therapeutic potential in combination with other drugs in all stages of human cancer.

References

Al-Ejeh, F., C.E. Smart, B.J. Morrison, G. Chenevix-Trench, J.A. Lopez, S.R. Lakhani, M.P. Brown, and K.K. Khanna. 2011. Breast cancer stem cells: treatment resistance and therapeutic opportunities. Carcinogenesis 32: 650–658.

Al-Hajj, M., M.S. Wicha, A. Benito-Hernandez, S.J. Morrison, and M.F. Clarke. 2003. Prospective identification of tumorigenic breast cancer cells. Proc Natl Acad Sci USA 100: 3983–3988.

Aleman, M., K.G. Magdesian, T.S. Peterson, and F.D. Galey. 2007. Salinomycin toxicosis in horses. J Am Vet Med Assoc 230: 1822–1826.

An, Y., and W.M. Ongkeko. 2009. ABCG2: the key to chemoresistance in cancer stem cells? Expert Opin Drug Metab Toxicol 5: 1529–1542.

Andersson, B., P. Behnam-Motlagh, R. Henriksson, and K. Grankvist. 2005. Pharmacological modulation of lung cancer cells for potassium ion depletion. Anticancer Res 25: 2609–2618.

Anderson, E.C., C. Hessman, T.G. Levin, M.M. Monroe, and M.H. Wong. 2011. The role of colorectal cancer stem cells in metastatic disease and therapeutic response. Cancers 3: 319–339.

Awad, O., J.T. Yustein, P. Shah, N. Gul, V. Katuri, A. O'Neill, Y. Kong, M.L. Brown, J.A. Toretsky, and D.M. Loeb. 2010. High ALDH activity identifies chemotherapy-resistant Ewing's sarcoma stem cells that retain sensitivity to EWS-FLI1 inhibition. PLoS One 5: e13943.

Bao, S., Q. Wu, R.E. McLendon, Y. Hao, Q. Shi, A.B. Hjelmeland, M.W. Dewhirst, D.D. Bigner, and J.N. Rich. 2006. Glioma stem cells promote radioresistance by preferential activation of the DNA damage response. Nature 444: 756–760.

Bao, B., Z. Wang, S. Ali, et al. 2011. Metformin inhibits cell proliferation, migration and invasion by attenuating CSC function mediated by deregulating miRNAs in pancreatic cells. Cancer Prev Res **Epub.**

Bardsley, M.R., V.J. Horvath, D.T. Asuzu, et al. 2010. Kit low stem cells cause resistance to Kit/platelet-derived growth factor α inhibitors in murine gastrointestinal tumors. Gastroenterology 139: 942–952.

Basu, D., K.T. Montone, L.P. Wang, et al. 2011. Detecting and targeting mesenchymal-like subpopulations within squamous cell carcinomas. Cell Cycle 10: 2008–2016.

Bedard, P.L., F. Cardoso, and M.J. Piccart-Gebhart. 2009. Stemming resistance to HER-2 targeted therapy. J Mammary Gland Biol Neoplasia 14: 55–66.

Belda-Iniesta, C., O. Pernia, and R. Simo. 2011. Metformin: a new option in cancer treatment. Clin Transl Oncol 13: 363–367.

Benjamin, H.B., B.K. Kwoka, M.K. Ndubuisia, M. Elofssona, and C.M. Crews. 2001. The anti-inflammatory natural product parthenolide from the medical herb feverfew directly binds to and inhibits IκB kinase. Chem Biol 8: 759–766.

Berns, K., H.M. Horling, B.T. Hennessy, et al. 2007. A functional genetic approach identifies the PI3K pathway as a major determinant of trastuzumab resistance in breast cancer. Cancer Cell 12: 395–402.

Bhatia R., M. Holtz, N. Niu, R. Gray, D.S. Snyder, C.L. Sawyers, D.A. Arber, M.L. Slovak, and S.J. Forman. 2003. Persistance of malignant hematopoietic progenitors in chronic myelogenous leukemia patients in complete cytogenetic remission following imatinib mesylate treatment. Blood 101: 4701–4707.

Birnie, R., S.D. Bryce, C. Roome, et al. 2008. Gene expression profiling of human prostate cancer stem cells reveals a pro-inflammatory phenotype and the importance of extracellular matrix interactions. Genome Biol 9: R83.

Bortner, C.C., F.M. Hughes, and J.A. Cidlowski. 1997. A primary role for K+ and Na+ efflux in the activation of apoptosis. J Biol Chem 272: 32436–32442.

Butaye, P., L.A. Devriese, and F. Haesebrouck. 2003. Antimicrobial growth promotors used in animal feed: effects of less well known antibiotics on gram-positive bacteria. Clin Microbiol Rev 16: 175–188.

Calabrese, C., H. Poppleton, M. Kocak, et al. 2007. A perivascular niche for brain tumor stem cells. Cancer Cell 11: 69–82.

Calcagno, A.M., C.D. Salcido, J.P. Gillet, C.P. Wu, J.M. Fostel, M.D. Mumau, M.M. Gottesman, L. Vartikovski, and S.V. Ambudkar. 2010. Prolonged drug selection of breast cancer cells and enrichment of cancer stem cell characteristics. J Natl Cancer Inst 102: 1637–1652.

Callaway, T.R., T.S. Edrington, J.L. Rychlik, K.J. Genovese, T.L. Poole, Y.S. Jung, K.M. Bischoff, R.C. Anderson, and D.J. Nesbit. 2003. Ionophores: their use as ruminant growth promotors and impact on food safety. Curr Issues Intest Microbiol 4: 43–51.

Carow, C.E., M. Levenstein, S.H. Kaufmann, J. Chen, S. Amin, P. Rockwell, L. Witte, M.J. Borowitz, C.I. Civin, and D. Small. 1996. Expression of the hematopoietic growth factor receptor FLT3 (STK-1/Flk2) in human leukemias. Blood 87: 1089–1096.

Charafe-Jauffret, E., C. Ginestier, F. Iovino, et al. 2010. Aldehyde dehydrogenase 1-positve cancer stem cells mediate metastasis and poor clinical outcome in inflammatory breast cancer. Clin Cancer Res 16: 45–55.

Charrad, R.S., Z. Gadhoum, J. Qi, A. Glachant, M. Allouche, C. Jasmin, C. Chomienne, and F. Smadja-Joffe. 2002. Effects of anti-CD44 monoclonal antibodies on differentiation and apoptosis of human myeloid leukemia cell lines. Blood 99: 290–299.

Chen, M.S., W.A. Woodward, F. Behbod, S. Peddibhotla, M.P. Alfaro, T.A. Buchholz, and J.M. Rosen. 2007. Wnt/β-catenin mediates radiation resistance of Sca1+ progenitors in an immortalized mammary gland cell line. J Cell Sci 120: 468–477.

Chen, X., S. Lingala, S. Khoobyari, J. Nolta, M.A. Zern, and J. Wu. 2011. Epithelial mesenchymal transition and hedgehog signaling activation are associated with chemoresistance of hepatoma subpopulations. J Hepatol 55: 838–845.

Chiou, S.H., C.L. Cao, Y.W. Chen, C.S. Chien, S.C. Hung, J.F. Lo, Y.J. Chen, H.H. Ku, M.T. Hsu, and T.T. Wong. 2008. Identification of CD133-positive radioresistant cells in atypical teratoid/rhabdoid tumor. PLoS One 3: e2090.

Chiu, P.P., H. Jiang, and J.E. Dick. 2010. Leukemia-initiating cells in human T-lymphoblastic leukemia exhibit glucocorticoid resistance. Blood 116: 5268–5279.

Cicalese, A., G. Bonizzi, C.E. Pasi, M. Faretta, S. Ronzoni, B. Giulini, C. Brisken, S. Minucci, P.P. Di Fiore, and P.G. Pelicci. 2009. The tumor suppressor p53 regulates polarity of self-renewing divisions in mammary stem cells. Cell 138: 1083–1095.

Clarke, M.F., J.E. Dick, P.B. Dirks, C.J. Eaves, C.H. Jamieson, D.L. Jones, J. Visvader, I.L. Weissman, and G.M. Wahl. 2006. Cancer stem cells – perspectives on current status and future directions: AACR workshop on cancer stem cells. Cancer Res 66: 9339–9344.

Clevers, H. 2011. The cancer stem cell: premises, promises and challenges. Nat Med 17: 313–319.

Conroy, T., C. Gavoille, and A. Adenis. 2011. Metastatic pancreatic cancer: old drugs, new paradigms. Curr Opin Oncol 23: 390–395.

Creighton, C.J., X. Li, M. Landis, et al. 2009. Residual breast cancers after conventional therapy display mesemchymal as well as tumor-initiating features. Proc Natl Acad Sci USA 106: 13820–13825.

Danforth, H.D., M.D. Ruff, W.M. Reid, and R.L. Miller. 1977. Anticoccidial activity of salinomycin in battery raised broiler chickens. Poult Sci 56: 926–932.

Daugschies A., U. Gässlein, and M. Rommel. 1998. Comparative efficacy of anticoccidials under the conditions of commercial broiler production and battery trials. Vet Parasitol 76: 163–171.

Dave, B., I. Migliaccio, M.C. Gutierrez, et al. 2011. Loss of phosphatase and tensin homolog or phosphoinositol-3 kinase activation and response to trastuzumab or lapatinib in human epidermal growth facor receptor 2-overexpressing locally advanced breast cancers. J Clin Oncol 29: 166–173.

Dean, M., T. Fojo, and S. Bates. 2005. Tumor stem cells and drug resistance. Nat Rev Cancer 5: 275–284.

Dean, M. 2009. ABC transporters, drug resistance, and cancer stem cells. J Mammary Gland Biol Neoplasia 14: 3–9.

Deininger, M., E. Buchdunger, and B.J. Druker. 2005. The development of imatinib as a therpeutic agent for chronic myeloid leukemia. Blood 105: 2640–2653.

Deonarain, M.P., C.A. Kousparou, and A.A. Epenetos. 2009. Antibodies targeting cancer stem cells: a new paradigm in immunotherapy? MAbs 1: 12–25.

Di Stefano, A.B., F. Iovino, Y. Lombardo, V. Eterno, T. Höger, F. Dieli, G. Stassi, and M. Todaro. 2010. Survivin is regulated by interleukin-4 in colon cancer stem cells. J Cell Physiol 225: 555–561.

Dick, J.E. 2008. Stem cell concepts renew cancer research. Blood 112: 4793–4807.

Diehn, M., et al. 2009. Association of reactive oxygen species levels and radioresistance in cancer stem cells. Nature 458: 780–783.

Dong, T.T., H.M. Zhou, L.L. Wang, B. Feng, B. Lv, and M.H. Zheng. 2011. Salinomycin selectively targets CD133+ cell subpopulations and decreases malignant traits in colorectal cancer lines. Ann Surg Oncol 18: 1797–1804.

Dontu, G., W.M. Abdallah, J.M. Foley, K.W. Jackson, M.F. Clarke, M.J. Kawamura, and M.S. Wicha. 2003. *In vitro* propagation and transcriptional profiling of human mammary stem/progenitor cells. Genes Dev 17: 1253–1270.

Dorne, J.L., M.L. Fernandez-Cruz, U. Bertelsen, D.W. Renshaw, K. Petonen, A. Anadon, A. Feil, P. Sanders, P. Wester, and J. Fink-Gremmels. 2011. Risk assessment of coccidostatics during feed cross-contamination: animal and human health aspects. Toxicol Appl Pharmacol **Epub.**

Dylla, S.J., L. Beviglia, I.K. Park, et al. 2008. Colorectal cancer stem cells are enriched in xenogeneic tumors following chemotherapy. PLoS One 3: e2428.

Elliot, A., J. Adams, and M. Al-Hajj. 2010. The ABCs of cancer stem cell resistance. IDrugs 13: 632–635.

Era, T. 2002. Bcr-Abl is a molecular switch for the decision for growth and differentiation in hematopoietic stem cells. Int J Haematol 76: 35–43.

Essers, M.A., and A. Trumpp. 2010. Targeting leukemic stem cells by breaking their dormancy. Mol Oncol 4: 443–450.

Eyler, C.E, and J.N. Rich. 2008. Survival of the fittest: cancer stem cells in therapeutic resistance and angionesis. J Clin Oncol 26: 2839–2845.

Fabi, A., G. Metro, A. Di Benedetto, C. Nistico, P. Vici, E. Melucci, B. Antoniani, L. Perracchio, I. Sperduti, and M. Milella. 2010. Clinical significance of PTEN and p-Akt co-expression in HER2-positive metastatic breast cancer patients treated with trastuzumab-based therapies. Oncology 78: 141–149.

Fahim, M., G. Del Valle, and B.C. Pressman. 1986. Comparison of the effects of the ionophore salinomycin and adrenaline on the haemodynamics and work efficiency of the dog heart. Cardiovasc Res 20: 145–152.

Francipane, M.G., M.P. Alea, Y. Lombardo, M. Todaro, J.P. Medema, and G. Stassi. 2008. Crucial role of interleukin-4 in the survival of colon cancer stem cells. Cancer Res 68: 4022–4025.

Frank, N.Y., T. Schatton, and M.H. Frank. 2010. The therapeutic promise of the cancer stem cell concept. J Clin Invest 120: 41–50.

Fuchs, D., A. Heinold, G. Opelz, V. Daniel, and C. Naujokat. 2009. Salinomycin induces apoptosis and overcomes apoptosis resistance in human cancer cells. Biochem Biophys Res Commun 390: 743–749.

Fuchs, D., V. Daniel, M. Sadeghi, G. Opelz, and C. Naujokat. 2010. Salinomycin overcomes ABC transporter-mediated multidrug and apoptosis resistance in human leukemia stem cell-like KG-1a cells. Biochem Biophys Res Commun 394: 1098–1104.

Fulda, S., and S. Pervaiz. 2010. Apoptosis signaling in cancer stem cells. Int J Biochem Cell Biol 42: 31–38.

Funes, J.M., M. Quintero, S. Henderson, et al. 2007. Transformation of human mesenchymal stem cells increases their dependency on oxidative phosphorylation for energy production. Proc Natl Acad Sci USA 104: 6223–6228.

Gadhoum, Z., J. Delaunay, E. Maquarre, L. Durand, V. Lancereaux, J. Qi, Robert-Lezenes, C. Chomienne, and F. Smadja-Joffe. 2004. The effect of anti-CD44 monoclonal antibodies on differentiation and proliferation of human acute myeloid leukemia cells. Leuk Lymphoma 45: 1501–1510.

Gaide. M.S., J.T. Lehr, S.S. Wong, F.A. Lattanzio, and A.L. Bassett. 1984. Comparative study of salinomycin and monensin on eletrophysiologic and contractile properties of canine myocardium. Eur J Pharmacol 100: 321–327.

Gajria, D., and S. Chandarlaparty. 2011. HER2-amplified breast cancer: mechanisms of trastuzumab and novel targeted therapies. Expert Rev Anticancer Ther 11: 263–275.

Gallmeier, E., P.C. Hermann, M.T. Mueller, et al. 2011. Inhibition of ataxia teleangiectasia- and Rad3-related function abrogates the in vitro and in vivo tumorigenicity of human colon cancer cells through depletion of the CD133(+) tumor-initiating cell fraction. Stem Cells 29: 418–429.

Gao, M.Q., Y.P. Choi, S. Kang, J.H. Youn, and N.H. Cho. 2010. CD24+ cells from hierarchically organized ovarian cancer are enriched in cancer stem cells. Oncogene 29: 2672–2680.

Ghantous, A., H. Gali-Muhtasib, H. Vuorela, N.A. Saliba, and N. Darwiche. 2010. What made sesquiterpene lactones reach clinical cancer trials? Drug Discov Today 15: 668–678.

Ginestier, C., M.H. Hur, E. Charafe-Jauffret, et al. 2007. ALDH1 is a marker of normal and malignant human mammary stem cells and a predictor of poor clinical outcome. Cell Stem Cell 1: 555–567.

Ginestier, C., S. Liu, M.E. Diebel, et al. 2010. CXCR1 blockade selctively targets human breast cancer stem cells in vitro and in xenografts. J Clin Invest 120: 485–497.

Goodwin, P.J., V. Stambolic, J. Lemieux, et al. 2011. Evaluation of metformin in early breast cancer: a modification of the traditional paradigm for clinical testing of anti-cancer agents. Breast Cancer Res Treat 126: 215–220.

Gong, C., H. Yao, Q. Liu, J. Chen, J. Shi, F. Su, and E. Song. 2010. Markers of tumor-initiating cells predict chemoresistance in breast cancer. PLoS One 5: e15630.

Gould, A., and S. Missailidis. 2011. Targeting the hedgehog pathway: the development of cyclopamine and the development of anti-cancer drugs targeting the hedgehog pathway. Mini Rev Med Chem 11. 200–213.

Graham, S.M., H.G. Jorgenson, E. Allan, C. Pearson, M.J. Alcorn, L. Richmond, and T.L. Holyoake. 2002. Primitive, quiescent, Philadelphia-positive stem cells from patients with chronic myeloid leukemia are insensitive to STI571 *in vitro*. Blood 99: 319–325.

Grinshtein. N., A. Datti, M. Fujitani, D. Uehling, M. Prakesh, M. Isaac, M.S. Irwin, J.L. Wrana, R. Al-Awar, and D.R. Kaplan. 2011. Small molecule kinase inhibitor screen identifies polo-like kinase 1 as a target for neuroblastoma tumor-initiating cells. Cancer Res 71: 1385–1395.

Gronchi, A., J.T. Blay, and J.C. Trent. 2010. The role of high-dose imatinib in the management of patients with gastrointestinal stromal tumor. Cancer 116: 1847–1858.

Guessous, F., Y. Zhang, A. Kofman, A. Catania, Y. Li, D. Schiff, B. Purow, and R. Abounader. 2010. microRNA-34a is tumor suppressive in brain tumors and glioma stem cells. Cell Cycle 9: 1031–1036.

Gupta, P.B., T.T. Onder, G. Jiang, K. Tao, C. Kuperwasser, R.A. Weinberg, and E.S. Lander. 2009. Identification of selective inhibitors of cancer stem cells by high-throughput screening. Cell 138: 645–659.

Gupta, S., N. Takebe, and P. Lorusso. 2010. Targeting the Hedgehog pathway in cancer. Ther Adv Med Oncol 2: 237–250.

Gunn, E.J., J.T. Williams, D.T. Huynh, M.J. Iannotti, C. Han, F.J. Barrios, S. Kendall, C.A. Glackin, D.A. Colby, and J. Kirshner. 2011. The natural products parthenolide and andrographolide exhibit anti-cancer stem cell activity in multiple myeloma. Leuk Lymphoma 52: 1085–1097.

Guzman, M.L., C.F. Swiderski, D.S. Howard, B.A. Grimes, R.M. Rossi, S.J. Szilvassy, and C.T. Jordan. 2002. Preferential induction of apoptosis for primary human leukemic stem cells. Proc Natl Acad Sci USA 99: 16220–16225.

Guzman, M.L., R.M. Rossi, L. Karnishki, X. Li, D.R. Peterson, D.S. Howard, and C.T. Jordan. 2005. The sesquiterpene lactone parthenolide induces apoptosis of human acute myelogenous leukemia stem and progenitor cells. Blood 105: 4163–4169.

Guzman, M.L., R.M. Rossi, S. Neelakantan, et al. 2007. An orally bioavailable parthenolide analog selectively eradicates acute myelogenous leukemia stem and progenitor cells. Blood 110: 4427–4435.

Hadad, S., T. Iwamoto, L. Jordan, et al. 2011. Evidence for biological effects of metformin in operable breast cancer: a pre-operative, window-of-opportunity, randomized trial. Breast Cancer Res Treat 128: 783–794.

Hambardzumyan, D., O.J. Becher, M.K. Rosenblum, P.P. Pandolfi, K. Manova-Todorava, and E.C. Holland. 2008. PI3K pathway regulates survival of cancer stem cells residing in the perivascular niche following radiation in medulloblastoma. Genes Dev 22: 436–448.

Hamilton, A., L.Elrick, S. Myssina, M. Copland, H. Jorgenson, J.V. Melo, and T. Holyoake. 2006. BCR-ABL activity and its response to drugs can be determined in CD34+ CML stem cells by CrkL phosphorylation status using flow cytometry. Leukemia 20: 1035–1039.

Hanahan, D., and R.A. Weinberg. 2011. Hallmarks of cancer: the next generation. Cell 144: 646–674.

Haren, N., H. Khorsi, M. Faouzi, A. Ahidouch, H. Sevestre, and H. Ouadid-Ahidouch. 2010. Intermediate conductance Ca2+ activated K+ channels are expressed and functional in breast adenocarcinomas: correlation with tumour grade and metastasis status. Histol Histopathol 25: 1247–1255.

Hassane, D.C., M.L. Guzman, C. Corbett, X. Li, R. Abboud, F. Young, J.L. Liesveld, M. Carroll, and C.T. Jordan. 2008. Discovery of agents that eradicate leukemia stem cells using in silico screen of public gene expression. Blood 111: 5654–5662.

Hehner, S.P., T.G. Hofmann, W. Dröge, and M.L. Schmitz. 1999. The antiinflammatory sesquiterpene lactone parthenolide inhibits NF-κB by targeting the IκB kinase complex. J Immunol 163: 5617–5623.

Hermann, P.C., S. Bhaskar, M. Cioffi, and C. Heeschen. 2010. Cancer stem cells in solid tumors. Semin. Cancer Biol 20: 77–84.

Hill, R., and H. Wu. 2009. PTEN, stem cells, and cancer stem cells. J Biol Chem 284: 11755–11759.

Hirsch, H.A., D. Illiopoulos, P.N. Tsichlis, and K. Struhl. 2009. Metformin selectively targets cancer stem cells, and acts together with chemotherapy to block tumor growth and prolong remission. Cancer Res 69: 7507–7511.

Holtz, M.S., S.J. Forman, and R. Bhatia. 2005. Nonproliferating CML CD34+ progenitors are resistant to apoptosis by a wide range of proapoptotic stimuli. Leukemia 19: 1034–1041.

Hudziak, R.M., J. Schlessinger, and A. Ullrich. 1987. Increased expression of the putative growth factor receptor p185HER2 causes transformation and tumorigenesis of NIH 3t3 cells. Proc Natl Acad Sci USA 84: 7159–7163.

Janikowa, M., and J. Skarda. 2012. Differentiation pathways in carcinogenesis and in chemo- and radioresistance. Neoplasma 59: 6–17.

Jenkins, M., S. Kloop, D. Ritter, K. Miska, and R. Fetterer. 2010. Comparison of Eimeria species distribution and salinomycin resistance in commercial broiler operations utilizing different coccidiosis control strategies. Avian Dis 54: 1002–1006.

Jin, L., K.J. Hope, Q. Zhai, F. Smadja-Joffe, and J.E. Dick. 2006. Targeting of CD44 eradicates human acute myeloid leukemic stem cells. Nat Med 12: 1167–1174.

Jin, L., E.M. Lee, H.S. Ramshaw, et al. 2009. Monoclonal antibody-mediated targeting of CD123, IL-3 receptor α chain, eliminates human acute myeloid leukemic stem cells. Cell Stem Cells 5: 31–42.

Kanwal. R., and S. Gupta. 2011. Epigenetic modifications in cancer. Clin Genet **Epub.**

Kataoka, Y., T. Mukohara, H. Shimada, N. Saijto, M. Hirai, and H. Minami. 2010. Association between gain-of-function mutations in PI3KCA and resistance to HER2-targeted agents in HER2-aplified breast cancer cell lines. Ann Oncol 21: 255–262.

Kawasaki, B.T., E.M. Hurt, T. Mistree, and W.L. Farrar. 2008. Targeting cancer stem cells with phytochemicals. Mol Interv 8: 174–184.

Kawasaki, B.T., E.M. Hurt, M. Kalathur, M.A. Duhagon, J.A. Milner, Y.S. Kim, and W.L. Farrar. 2009. Effects of sesquiterpene lactone parthenolide on prostate tumor-initiating cells: an integrated molecular profiling approach. Prostate 69: 827–837.

Kelleher, F.C. 2011. Hedgehog signaling and therapeutics in pancreatic cancer. Carcinogenesis 32: 445–451.

Ketola, K., M. Hilvo, T. Hyötyläinen, A. Vuoristo, A.L. Ruskeepää, M. Oresic, O. Kallioniemi, and K. Illjin. 2012. Salinomycin inhibits prostate cancer growth and migration via induction of oxidative stress. Br J Cancer 106: 99–106.

Kim, M.R., H.K. Choi, K.B. Cho, H.S. Kim, and K.W. Kang. 2009. Involvement of Pin1 induction in epithelial-mesemchymal transition of tamoxifen-resistant breast cancer cells. Cancer Sci 100: 1834–1841.

Kim, K.Y., S.N. Yu, S.Y. Lee, S.S. Chun, Y.L. Choi, Y.M. Park, C.S. Song, B. Chatterjee, and S.C. Ahn. 2011a. Salinomycin-induced apoptosis of human prostate cancer cells due to accumulated reactive oxygen species and mitochondrial membrane depolarization. Biochem Biophys Res Commun 413: 80–86.

Kim, J.H., M. Chae, W.K. Kim, Y.J. Kim, H.S. Kang, H.S. Kim, and S. Yoon. 2011b. Salinomycin sensitizes cancer cells to the effects of doxorubicin and etoposide treatment by increasing DNA damage and reducing p21 protein. Br J Pharmacol 162: 773–784.

Kim, W.K., J.H. Kim, K. Yoon, S. Kim, J. Ro, H.S. Kang, and S. Yoon. 2011c. Salinomycin, a p-glycoprotein inhibitor, sensitizes radiation-treated cancer cells by increasing DNA damage and inducing G2 arrest. Invest. New Drugs **Epub.**

Kim, J.H., H.I. Yoo, H.S. Kang, J. Ro, and S. Yoon. 2012. Salinomycin sensitizes antimitotic drugs-treated cancer cells by increasing apoptosis via the prevention of G2 arrest. Biochem Biophys Res Commun 418: 98–103.

Kinashi, H., N. Otake, H. Yonehara, S. Saito, and Y. Saito. 1973. The structure of salinomycin, a new member of the polyether antibiotics. Tetrahedron Lett 14: 4955–4958.

Kobune, M., R. Takimoto, K. Murase, et al. 2009. Drug resistance is dramatically restored by hedgehog inhibitors in CD34+ leukemic cells. Cancer Sci 100: 948–955.

Koch, U., M. Krause, and M. Baumann. 2010. Cancer stem cells at the crossroads of current cancer therapy failures – radiation oncology perspective. Semin Cancer Biol 20: 116–124.

Korkaya, H., A. Paulson, F. Iovino, and M.S. Wicha. 2008. HER2 regulates the mammary stem/progenitor cell population driving tumorigenesis and invasion. Oncogene 27: 6120–6130.

Korkaya, H., A. Paulson, E. Charafe-Jauffret, C. Ginestier, M. Brown, J. Dutcher, S.G. Clouthier, and M.S. Wicha. 2009. Regulation of mammary stem/progenitor cells by PTEN/Akt/β-catenin signaling. PLoS Biol 7: e1000121.

Korkaya, H., S. Liu, and M.S. Wicha. 2011. Breast cancer stem cells, cytokine networks, and the tumor microenvironment. J Clin Invest 121: 3804–3809.

Kosal, M.E., and D.E. Anderson. 2004. An unaddressed issue of agricultural terrorism: a case study on feed security. J Anim Sci. 82: 3394–3400.

Kourelis, T.V., and R.D. Siegel. 2011. Metformin and cancer: new applications for an old drug. Med Oncol **Epub.**

Kruyt, F.A., and J.J. Schuringa. 2010. Apoptosis and cancer stem cells: implications for apoptosis targeted therapy. Biochem Pharmacol 80: 423–430.

Li, L., and R. Bhatia. 2011. Stem cell quiescence. Clin Cancer Res 17: 4936–4941.

Li, X., M.T. Lewis, J. Huang, et al. 2008a. Intrinsic resistance of tumorigenic breast cancer cells to chemotherapy. J Natl Cancer Inst 100: 672–679.

Li, H., L. Liu, L. Guo, J. Zhang, W. Du, X. Li, W. Liu, X. Chen, and S. Huang. 2008b. HERG K+ channel expression in CD34+/CD38-/CD123(high) cells and primary leukemia cells and analysis of its regulation in leukemia cells. Int J Haematol 87: 387–392.

Li, Y., D. Kong, Z. Wang, and F.H. Sarkar. 2010. Regulation of microRNAs by natural agents: an emerging field in chemoprevention and chemotherapy research Pharm Res 27: 1027–1041.

Li, J., G. Xu, S. Yu, L. He, and L. Guo. 2011a. Dasatinib treatment for imatinib resistant or intolerant patients with chronic myeloid leukemia. Int J Med Res 39: 337–347.

Li, Y., M.S. Wicha, S.J. Schwartz, and D. Sun. 2011b. Implications of cancer stem cell theory for cancer chemoprevention by natural dietary compounds. J Nutr Biochem 22: 799–806.

Liu, S., and M.S. Wicha. 2010. Targeting breast cancer stem cells. J Clin Oncol 28: 4006–4012.

Liu, C., and D.G. Tang. 2011. MicroRNA regulation of cancer stem cells. Cancer Res 71: 5950–5954.

Liu, G., X. Yuan, Z. Zeng, P. Tunici, H. Ng, I.R. Abdulkadir, L. Lu, D. Irvin, K.L. Black, and J.S. Yu. 2006. Analysis of gene expression and chemoresistance of CD133+ cancer stem cells in glioblastoma. Mol Cancer 5: 1476–1498.

Liu, R., X. Wang, G.Y. Chen, et al. 2007. The prognostic role of a gene signature from tumorigenic breast-cancer cells. N Engl J Med 356: 217–226.

Liu, M., T. Sakamaki, M.C. Casimiro, et al. 2010. The canonical NF-κB pathway governs mammary tumorigenesis in transgenic mice and tumor stem cell expansion. Cancer Res 70: 10464–10473.

Liu, C., K. Kelnar, B. Liu, et al. 2011a. The microRNA-34a inhibits prostate cancer stem cells and metastasis by directly repressing CD44. Nat Med 17: 211–215.

Liu, B., Z. Fan, S.M. Edgerton, X. Yang, S.E. Lind, and A.D. Thor. 2011b. Potent anti-proliferative effects of metformin on trastuzumab-resistant breast cancer cells via inhibition of erbB2/IGF-1 receptor interactions. Cell Cycle 10: 2959–2966.

Lorico, A., and G. Rappa. 2011. Phenotypic heterogeneity of breast cancer stem cells. J Oncol 2011: 135039.

LoRusso P.M., C.M. Rudin, J.C. Reddy, et al. 2011. Phase I trial of hedgehog pathway inhibitor vismodegib (GDC-0449) in patients with refractory, locally advanced or metastatic solid tumors. Clin Cancer Res 17: 2502–2511.

Low, J.A., and F.J. de Sauvage. 2010. Clinical experience with hedgehog pathwy inhibitors. J Clin Oncol 28: 5321–5326.

Lu, D., M.Y. Choi, J. Yu, J.E. Castro, T.J. Kipps, and D.A. Carson. 2011. Salinomycin inhibits Wnt signalling and selectively induces apoptosis in chronic lymphocytic leukemia cells. Proc Natl Acad Sci USA 108: 13253–13257.

Ma, S., T.K. Lee, B.J. Zheng, K.W. Chane, and X.Y. Guan. 2008. CD133+ HCC cancer stem cells confer chemoresistance by preferential expression of the Akt/PKB survival pathway. Oncogene 27: 1749–1758.

Mackenzie, M.J., S. Ernst, C. Johnson, and E. Winquist. 2010. A phase I study of temsirolimus and metformin in advanced solid tumours. Invest New Drugs **Epub.**

Majeti, R., M.P. Chao, A.A. Alizadeh, W.W. Pang, S. Jaiswal, K.D. Gibbs, N. van Rooijen, and I.L. Weissman. 2009. CD47 is an adverse prognostic factor and therapeutic antibody target on human acute myeloid leukemia stem cells. Cell 138: 286–299.

Majeti, R. 2011. Monoclonal antibody therapy directed against human acute myeloid leukemia stem cells. Oncogene 30: 1009–1019.

Mani, S.A., W. Guo, M.J. Liao, et al. 2008. The epithelial-mesenchymal transition generates cells with properties of stem cells. Cell 133: 704–715.

Marangoni, E., N. Lecomte, L. Durant, G. de Pinieux, D. Decaudin, C. Chomienne, F. Smadja-Joffe, and M.F. Poupon. 2009. CD44 targeting reduces tumour growth and prevents post-chemotherapy relapse of human breast cancer xenografts. Br J Cancer 100: 918–922.

Martin-Castillo, B., J. Dorca, A. Vazquez-Martin, C. Oliveras-Ferraros, E. Lopez-Bonet, M. Garcia, S. Del Barco, and J.A. Menendez. 2010a. Incorporating the antidiabetic drug metformin in HER2-positive breast cancer treated with neo-adjuvant chemotherapy and trastuzumab: an ongoing clinical-translational research experience at the Catalan Institute of Oncology. Ann Oncol 21: 187–189.

Martin-Castillo, B., A. Vazquez-Martin, C. Oliveras-Ferraros, and J.A. Menendez. 2010b. Metformin and cancer. Cell Cycle 9: 1057–1064.

Martelli, A.M., C. Evagelisti, M.Y. Follo, G. Ramazzotti, M. Fini, R. Giardino, L. Manzoli, J.A. McCubrey, and L. Cocco. 2011. Targeting the phosphatidyl 3-kinase/Akt/mammalian target of rapamycin signaling network in cancer stem cells. Curr Med Chem 18: 2715–2726.

Mathema, V.B., Y.S. Koh, B.C. Thakuri, and M. Sillanpää. 2011. Parthenolide, a sesquiterpene lactone, expresses multiple anti-cancer and anti-inflammatory activities. Inflammation, **Epub.**

Maugeri-Sacca, M., P. Vigneri, and R. De Maria. 2011. Cancer stem cells and chemosensitivity. Clin Cancer Res 17: 4942–4947.

May, C.D., N. Sphyris, K.W. Evans, S.J. Werden, W. Guo, and S.A. Mani. 2011. Epithelial transition and cancer stem cells: a dangerously dynamic duo in breast cancer progression. Breast Cancer Res 13: 202.

McDermott, S.P., and M.S. Wicha. 2010. Targeting breast cancer stem cells. Mol Oncol 4: 404–419.

Mehlhorn, H., H.J. Ganster, and W. Raether. 1984. Effect of salinomycin-Na on malaria parasites (*Plasmodium falciparum and P. berghei*). Zentralbl Bakteriol Mikrobiol Hyg A 256: 305–313.

Mitani, M., T. Yamanishi, and Y. Miyazaki. 1975. Salinomycin: A new monovalent cation ionophore. Biochem Biophys Res Commun 66: 1231–1236.

Mitani, M., T. Yamanishi, Y. Mizaki, and N. Otake. 1976. Salinomycin effects on mitochondrial ion translocation and respiration. Antimicrob. Agents Chemother 9: 655–660.

Miyazaki, Y., J. Nagatsu, H. Sugawara, and M. Shibuya. 1972. Japanese patent publication No. 25392.

Miyazaki, Y., M. Shibuya, H. Sugawara, O. Kawaguchi, and C. Hirsoe. 1974. Salinomycin, a new polyether antibiotic. J Antibiot 27: 814–821.

Moitra, K., H. Lou, and M. Dean. 2011. Multidrug efflux pumps and cancer stem cells: insights into multidrug resistance and therapeutic development. Clin Pharmacol Ther 89: 491–502.

Morrison, R., S.M. Schleicher, Y. Sun, K.J. Niermann, S. Kim, D.E. Spratt, C.H. Chung, and B. Lu. 2011. Targeting the mechanisms of resistance to chemotherapy and radiotherapy with the cancer stem cell hypothesis. J Oncol 2011: 941876.

Mueller, M.T., P.C. Hermann, J. Witthauer, et al. 2009. Combined targeted treatment to eliminate tumorigenic cancer stem cells in human pancreatic cancer. Gastroenterology 137: 1102–1113.

Nagata, Y., K.H. Lan, X. Zhou, et al. 2004. PTEN activation contributes to tumor inhibition by trastuzumab, and loss of PTEN predicts trastuzumab resistance in patients. Cancer Cell 6: 117–127.

Naujokat, C., D. Fuchs, and G. Opelz. 2010. Salinomycin in cancer: A new mission for an old agent. Mol Med Report 3: 555–559.

Nguyen, N.P., F.S. Almeida, A. Chin, L.M. Nguyen, D. Cohen, U. Karlsson, and V. Vinh-Hung. 2010. Molecular biology of breast cancer stem cells: potential clinical application. Cancer Treat Rev 36: 485–491.

Nguyen, L.V., R. Vanner, P. Dirks, and C.J. Eaves. 2012. Cancer stem cells: an evolving concept. Nat Rev Cancer 12: 133–143.

Okamoto, O.K., and J.F. Perez. 2008. Targeting cancer stem cells with monoclonal antibodies: a new perspective in cancer therapy and diagnosis. Expert Rev Mol Diagn 8: 387–393.

Oliveras-Ferraros, C., A. Vazquez-Martin, B. Martin-Castillo, S. Cufi, S. Del Barco, E. Lopez-Bonet, J. Brunet, and J.A. Menendez. 2010. Dynamic emergence of the mesenchymal CD44(pos)CD24(neg/low) phenotype in HER2-gene amplified breast cancer cells with de novo resistance to trastuzumab (Herceptin). Biochem Biophys Res Commun 397: 27–33.

Oliveras-Ferraros, C., S. Cufi, A. Vazquez-Martin, V.Z. Torres-Garcia, S. Del Barco, B. Martin-Castillo, and J.A. Menendez. 2011. Micro(mi)RNA expression profile of breast cancer epithelial cells treated with the anti-diabetic drug metformin: induction of the tumor suppressor miRNA let-7a and suppression of the TGFβ-induced oncomiR-181a. Cell Cycle 10: 1144–1151.

Onder, T.T., P.B. Gupta, S.A. Mani, J. Yang, E.S. Lander, and R.A. Weinberg. 2008. Loss of E-catherin promotes metastasis via multiple downstream transcriptional pathways. Cancer Res 68: 3645–3654.

Oravecz-Wilson, K.I., S.T. Philips, O.H. Yilmaz, et al. 2009. Persistence of leukemia-initiating cells in a conditional knockin model of imatinib-responsive myeloproliferative disorder. Cancer Cell 16: 137–148.

Pajonk, F., E. Vlashi, and W.H. McBride. 2010. Radiation resistance of cancer stem cells: the 4 R´s of radiobiology revisited. Stem Cells 28: 639–648.

Pardal, R., M.F. Clarke, and S.J. Morrison. 2003. Applying the principles of stem-cell biology to cancer. Nat Rev Cancer 3: 895–902.

Park, C.Y, D. Tseng, and I.L. Weissman. 2009. Cancer stem cell-directed therapies: recent data from the laboratory and clinic. Mol Ther 17: 219–230.

Park, J.H., S.J. Park, M.K. Chung, K.H. Jung, M.R. Choi, Y. Kim, Y.G. Chai, S.J. Kim, and K.S. Park 2010. High expression of large-conductance Ca2+-activated K+ channel in the CD133+ subpopulation of SH-SY5Y neuroblastoma cells. Biochem Biophys Res Commun 396: 637–642.

Parmar, A., S. Marz, S. Rushton, C. Holzwarth, K. Lind, S. Kayser, K. Döhner, C. Peschel, R.A. Oostendorp, and K.S. Götze. 2011. Stromal niche cells protect early leukemic FLT3-ITD+ progenitor cells against first-generation FLT3 tyrosine kinase inhibitors. Cancer Res 71: 4696–4706.

Pearce, D.J., D. Taussig, C. Simpson, K. Allen, A.Z. Rohatiner, T.A. Lister, and D. Bonnet. 2005. Characterization of cells with a high aldehyde dehydrogenase activity from cord blood and acute myeloid leukemia samples. Stem Cells 23: 752–760.

Philips, G.M., I.S. Chan, M. Swiderska, et al. 2011. Hedgehog signaling antagonist promotes regression of both liver fibrosis and hepatocellular carcinoma in a murine model of primary liver cancer. PLoS One 6: e23943.

Phillips, T.M., W.H. McBride, and F. Pajonk. 2006. The response of CD24(−/low)/CD44+ breast cancer-initiating cells to radiation. J Natl Cancer Inst 98: 1777–1785.

Plumlee, K.H., B. Johnson, and F.D. Galey. 1995. Acute salinomycin toxicosis in pigs. J Vet Diagn Invest 7: 419–420.

Polyak, K., and R.A. Weinberg. 2009. Transitions between epithelial and mesenchymal states: acquisition of malignant and stem cell traits. Nat Rev Cancer 9: 265–273.

Pressman, B.C., and F.A. Lattanzio. 1978. Cardiovascular effects of ionophores. pp. 1245–1252. In: P.L Dutton, J.S. Leigh and A. Scarpa. [eds.]. Frontiers of Biological Energetics Vol 2. Academic Press, New York, USA.

Ray, A., E. Meng, E. Reed, L.A. Shevde, and R.P. Roccioni. 2011. Hedgehog signaling pathway regulates the growth of ovarian cancer spheroid forming cells. Int J Oncol 39: 797–804.

Reim, F., Y. Dombrowski, C. Ritter, et al. 2009. Immunoselection of breast and ovarian cancer cells with trastuzumab and natural killer cells: selective escape of $CD44^{high}/CD24^{low}/HER2^{low}$ breast cancer stem cells. Cancer Res 69: 8058–8066.

Reya, T., S.J. Morrison, M.F. Clarke, and I.L. Weissman. 2001. Stem cells, cancer, and cancer stem cells. Nature 414: 105–111.

Riccioni R., M.L. Dupuis, M. Bernabei, E. Petrucci, L. Pasquini, G. Mariani, M. Cianfriglia, and U. Testa. 2010. The cancer stem cell selective inhibitor salinomycin is a p-glycoprotein inhibitor. Blood Cells Mol Dis 45: 86–92.

Robarge, K.D., S.A. Brunton, G.M. Castanedo, et al. 2009. GDC-0449 – a potent inhibitor of the hedgehog pathway. Bioorg Med Chem Lett 19: 5576–5581.

Rocha, G.Z., M.M. Dias, E.R. Ropelle, F. Osorio-Costa, F.A. Rossato, A.E. Vercesi, M.J. Saad, and J.B. Cavalheira. 2011. Metformin amplifies chemotherapy-induced AMPK activation and antitumoral growth. Clin Cancer Res 17: 3993–4005.

Rollinson, J., F.G. Taylor, and J. Chesney. 1987. Salinomycin poisoning in horses. Vet Rec 121: 126–128.

Ropolo, M., A. Daga, F. Griffero, et al. 2009. Comparative analysis of DNA repair in stem and nonstem glioma cell cultures. Mol Cancer Res 7: 383–392.

Rudin, C.M., C.L. Hann, J. Laterra, et al. 2009. Treatment of medulloblastoma with hedgehog pathway inhibitor GDC-0449. N Engl J Med 361: 1173–1178.

Schatton, T., G.F. Murphy, N.Y. Frank, et al. 2008. Identification of cells initiating human melanomas. Nature 451: 345–349.

Schatton, T., and M.H. Frank. 2009. Antitumor immunity and cancer stem cells. Ann N Y Acad Sci 1176: 154–169.

Schatton, T., U. Schütte, N.Y. Frank, et al. 2010. Modulation of T-cell activation by malignant melanoma initiating cells. Cancer Res 70: 697–708.

Shackleton, M., E. Quintana, E.R. Fearon, and S.J. Morrison. 2009. Heterogeneity in cancer: cancer stem cells versus clonal evolution. Cell 138: 822–829.

Shen, Z., Q. Yang, and Q. You. 2009. Researches toward potassium channels on tumor progression. Curr Top Med Chem 9: 322–329.

Shipitsin, M., L.L. Campbell, P. Argani, et al. 2007. Molecular definition of breast tumor heterogeneity. Cancer Cell 11: 259–273.

Singh, B.N., J. Fu, R.K. Srivastava, and S. Shankar. 2011. Hedgehog signaling antagonist GDC-0449 (vismodegib) inhibits pancreatic cancer stem cells characteristics: molecular mechanisms. PLoS One 6: e27306.

Slamon, D.J., G.M. Clark, S.G. Wong, W.J. Levin, A. Ullrich, and W.L. McGuire. 1987. Huamn breast cancer: correlation of relapse and survival with amplification of the HER-2/neu oncogene. Science 244: 707–712.

Steinhart, R., L. Hojer, C. Martschoke, T. Sievers, D. Fuchs, and C. Naujokat. 2012. Breast cancer metastasis regression by salinomycin. Manuscript in preparation.

Story, P., and A. Doube. 2004. A case of human poisoning by salinomycin, an agricultural antibiotic. N Z Med J 117: 799.
Tagliabue, E., M. Campiglio, S.M. Pupa, S. Menard, and A. Balsari. 2011. Activity and resistance of trastuzumab according to different clinical settings. Cancer Treat Rev 2011. **Epub.**
Takebe, N., P.J. Harris, R.Q. Warren, and S.P. Ivy. 2011a. Targeting cancer stem cells by inhibiting Wnt, Notch, and hedgehog pathways. Nat Rev Clin Oncol 8: 97–106.
Takebe, N., R.Q. Warren, and S.P. Ivy. 2011b. Breast cancer growth and metastasis: interplay between cancer stem cells, embryonic signaling pathways and epithelial-to-mesenchymal transition. Breast Cancer Res 13: 211.
Takehara, M., T. Hoshino, T. Namba, N. Yamakawa, and T. Mizushima. 2011. Acetaminophen-induced differentiation of human breast cancer stem cells and inhibition of tumor xenograft growth in mice. Biochem Pharmacol 81: 1124–1135.
Tang, Q.L., Z.Q. Zhao, J.C. Li, et al. 2011. Salinomycin inhibits osteosarcoma by targeting its tumor stem cells. Cancer Lett 311: 113–121.
Teng, Y., X. Wang, Y. Wang, and D. Ma. 2010. Wnt/β-catenin signaling regulates cancer stem cell in lung cancer A549 cells. Biochem Biophys Res Commun 392: 373-379.
Tian, F., J. Mysliwietz, J. Ellwart, F. Gamarra, R.M. Huber, and A. Bergner. 2011. Effects of the hedgehog pathway inhibitor GDC-0449 on lung cancer cell lines are mediated by side populations. Clin Exp Med **Epub.**
Todaro, M., M.P. Alea, A.B. Di Stefano, et al. 2007. Colon cancer stem cells dictate tumor growth and resist cell death by production of interleukin-4. Cell Stem Cells 1: 389–402.
Van der Linde-Sipman, J.S., T.T. van der Ingh, J.J. van Nes, H. Verhagen, J.G. Kersten, A.C. Beynen, and R. Plekkringa. 1999. Salinomycin-induced polyneuropathy in cats: morphologic and epidemiologic data. Vet Pathol 36: 152–156.
Vasquez-Martin, A., C. Oliveras-Ferraros, S. Del Barco, B. Martin-Castillo, and J.A. Menendez. 2009. mTOR inhibitors and the anti-diabetic biguanide metformin: new insights into the molecular management of breast cancer resistance to the HER2 tyrosine kinase inhibitor lapatinib (Tykerb). Clin Transl Oncol 11: 455–459.
Vasquez-Martin, A., C. Oliveras-Ferraros, S. Cufi, S. Del Barco, B. Martin-Castillo, and J.A. Menendez. 2010. Metformin regulates breast cancer stem cell ontogeny by transcriptional regulation of the epithelial-mesenchymal transition (EMT) status. Cell Cycle 15: 3807–3814.
Vasquez-Martin, A., C. Oliveras-Ferraros, S. Del Barco, B. Martin-Castillo, and J.A. Menendez. 2011a. The anti-diabetic drug metformin suppresses self-renewal and proliferation of trastuzumab-resistant tumor-initiating breast cancer stem cells. Breast Cancer Res Treat 126: 355–364.
Vasquez-Martin, A., E. Lopez-Bonetc, S. Cufi, C. Oliveras-Ferraros, S. Del Barco, B. Martin-Castillo, and J.A. Menendez. 2011b. Repositioning chloroquine and metformin to eliminate cancer stem cell traits in pre-malignant lesions. Drug Resist Updat 14: 212–223.
Velasco-Velazquez, M.A., V.M. Popov, M.P. Lisanti, and R.G. Pestell. 2011. The role of breast cancer stem cells in metastasis and therapeutic implications. Am J Pathol 179: 2–11.
Venere, M., H.A. Fine, P.B. Dirks, and J.N. Rich. 2011. Cancer stem cells in gliomas: identifying and understanding the apex cell in cancer´s hierarchy. Glia 59: 1148–1154.
Vermeulen, L., E. De Sousa, F. Melo, et al. 2010. Wnt activity defines colon cancer stem cells and is regulated by the microenvironment. Nat Cell Biol 12: 468–476.
Viale, A., F. De Franco, A. Orleth, et al. 2009. Cell-cycle restriction limits DNA damage and maintains self-renewal of leukaemia stem cells. Nature 457: 51–56.
Villalonga, N., J.C. Ferreres, J.M. Argiles, E. Condom, and A. Felipe. 2007. Potassium channels are a new target field in anticancer drug design. Recent Pat Anticancer Discov 2: 212–223.
Vineis, P., A. Schatzkin, and J.D. Potter. 2010. Models of carcinogenesis: an overview. Carcinogenesis 31: 1703–1709.
Visvader, J.E. 2011. Cells of origin in cancer. Nature 469: 314–322.

Vlashi, E., C. Lagadec, L. Vergnes, et al. 2011. Metabolic state of glioma stem cells and nontumorigenic cells. Proc Natl Acad Sci USA 108: 16062–16067.

Von Hoff, D.D., P.M. LoRusso, C.M. Rudin, et al. 2009. Inhibition of the hedgehog pathway in advanced basal-cell carcinoma. N Engl J Med 361: 1164–1172.

Vogel, C.L., M.A. Cobleight, D. Tripathy, J.C. Gutheil, L.N. Harris, L. Fehrenbacher, D.J. Slamon, M. Murphy, W.F. Novotny and M. Burchmore. 2002. Efficacy and safety of trastuzumab as a single agent in first-line treatment of HER2-overexpressing metastatic breast cancer. J Clin Oncol 20: 719–726.

Waksman, S.A. 1961. The actinomycetes. Vol. 2. Classification, identification and descriptions of genera and species. The Williams and Wilkins Co., Baltimore, USA.

Wang, Z., D. Kong, S. Banerjee, A. Ahmad, A.S. Azmi, S. Ali, J.L. Abbruzzese, G.E. Gallick, and F.H. Sarker. 2009. Acquisition of epithelial-mesemchymal transition phenotype of gemcitabine-resistant pancreatic cancer cells is linked with activation of the Notch signaling pathway. Cancer Res 69: 2400–2407.

Wang, J., T.P. Wakerman, J.D. Lathia, A.B. Hjelmeland, X. F. Wang, R.R. White, J.N. Rich, and B.A. Sullenger. 2010a. Notch promotes radioresistance of glioma stem cells. Stem Cells 28: 17–28.

Wang, K.H., A.P. Kao, C.C. Chang, J.N. Lee, M.F. Hou, C.Y. Long, H.S. Chen, and E.M. Tsai. 2010b. Increasing CD44+/CD24(-) tumor stem cells, and upregulation of COX-2 and HDAC6, as major functions of HER2 in breast tumorigenesis. Mol Cancer 9: 288.

Wang, Z., R. Ravula, M. Cao, M. Chow, and Y. Huang. 2010c. Transporter-mediated multidrug resistance and its modulation by Chinese medicines and other herbal products. Curr Drug Discov Technol 7: 54–66.

Wang, Y. 2011. Effects of salinomycin on cancer stem cell in human lung adenocarcinoma A549 cells. Med Chem 7: 106–111.

Warburg, O., K. Posener, and E. Negelein. 1924. Über den Stoffwechsel von Tumoren. Biochem Z 152: 319–344.

Wend P., J.D. Holland, U. Ziebold, and W. Birchmeier. 2010. Wnt signaling in stem and cancer stem cells. Semin. Cell Dev Biol 21: 855–863.

Wiernik, P.H. 2010. FLT3 inhibitors for the treatment of acute myeloid leukemias. Clin Adv Hematol Oncol 8: 429–436.

Wilhelm, S.M., L. Adnane, P. Newell, A. Villanueva, J.M. Llovet, and M. Lynch. 2008. Preclinical overview of sorafenib, a multikinase inhibitor that targets both Raf and VEGF and PDGF receptor tyrosine kinase signaling. 2008. Mol Cancer Ther 7: 3129–3140.

Winquist, R.J., D.M. Boucher, M. Wood, and B.F. Furey. 2009. Targeting cancer stem cells for more effective therapies: taking out cancer´s locomotive engine. Biochem Pharmacol 78: 326–334.

Wong, H., R. Leung, A. Kwong, J. Chiu, R. Liang, C. Swanton, and T. Yau. 2011. Integrating molecular mechanisms and clinical evidence in the management of trastuzumab resistant or refractory HER-2+ metastatic breast cancer. Oncologist 16: 1535–1546.

Woodward, W.A., M.S. Chen, F. Behbod, M.P. Alfaro, T.A. Buchholt, and J.M. Rosen. 2007. WNT/β-catenin mediates radiation resistance of mouse mammary progenitor cells. Proc Natl Acad Sci USA 104: 618–623.

Wu, A., J. Wei, L.Y. Kong, Y. Wang, W. Priebe, W. Qiao, R. Sawaya, and A.B. Heimberger. 2011. Glioma cancer stem cells induce immunosuppressive macrophges/microglia. Neuro Oncol 12: 1113–1125.

Yamada,Y., and A. Watanabe. 2010. Epigenetic codes in stem cells and cancer stem cells. Adv Genet 70: 177–199.

Yang, Y.P., Y. Chien, G.Y. Chiou, et al. 2012. Inhibition of cancer stem cell-like properties and reduced chemoradioresistance of glioblastoma using microRNA 145 with cationic polyurethane-short branch PEI. Biomaterials 33: 1462–1476.

Yeung, J., M.T. Esposito, A. Gandillet, B.B. Zeisig, E. Griessinger, D. Bonnet, and C.W. So. 2010. β-catenin mediates the establishment and drug resistance of MLL leukemic stem cells. Cancer Cell 18: 606–618.

Yip, N.C., I.S. Fombon, P. Liu, S. Brown, V. Kannappan, A.L. Armesilla, B. Xu, J. Cassidy, J.L. Darling, and W. Wang. 2011. Disulfiram modulated ROS-MAPK and NFκB pathways and targeted breast cancer cells with cancer stem cell-like properties. Br J Cancer 104: 1564–1574.

Yu, F., H. Yao, P. Zhu, et al. 2007. let-7 regulates self renewal and tumorigenicity of breast cancer cells. Cell 131: 1109–1123.

Zhang, Y., J. Laterra, and M.G. Pomper. 2009. Hedgehog pathway inhibitor HhAntag691 is a potent inhibitor of ABCG2/BCRP and ABCB1/Pgp. Neoplasia 11: 96–101.

Zhang, G.N., Y. Liang, L.J. Zhou, S.P. Chen, G. Chen, T.P. Zhang, T. Kang, Y.X. Zeng, and Y.P. Zhao. 2011. Combination of salinomycin and gemcitabine eliminates pancreatic cancer cells. Cancer Lett 313: 137–144.

Zhang, Y., H. Zhang, X. Wang, J. Wang, X. Zhang, and Q. Zhang. 2012. The eradication of breast cancer and cancer stem cells using octreotide modified paclitaxel active targeting micelles and salinomycin passive targeting micelles. Biomaterials 33: 679–691.

Zhao, C., A. Chen, C.H. Jamieson, et al. 2009. Hedgehog signaling is essential for maintenance of cancer stem cells in myeloid leukemia. Nature 458: 776–779.

Zhi, Q.M., X.H. Chen, J. Ji, J.N. Zhang, J.F. Li, Q. Cai, B.Y. Liu, Q.L. Gu, Z.G. Zhu, and Y.Y. Yu. 2011. Salinomycin can effectively kill ALDH(high) stem-like cells on gastric cancer. Biomed Pharmacother 65: 509–515.

Zhou, J., H. Zhang, P. Gu, J. Bai, J.B. Margolick, and Y. Zhang. 2008. NF-κB pathway inhibitors preferentially inhibit breast cancer stem-like cells. Breast Cancer Res Treat 111: 419–427.

Zhou, B.B., H. Zhang, M. Damelin, K.G. Geles, J.C. Grindley, and P.B. Dirks. 2009. Tumour-initiating cells: challenges and opportunities for anticancer drug discovery. Nat Rev Drug Discov 8: 806–823.

Zhou, Y., Y. Zhou, T. Shingu, L. Feng, Z. Chen, M. Ogasawara, M.J. Keating, S. Kondo, and P. Huang. 2011. Metabolic alterations in highly tumorigenic glioblastoma cells: preference for hypoxia and high dependency on glycolysis. J Biol Chem 286: 32843–32853.

Zimmermann, A.L., and S. Wu. 2011. MicroRNAs, cancer and cancer stem cells. Cancer Lett 300: 10–19.

Chapter 10

Targeting Breast Cancer Stem Cells

Kelli E. Valdez,[a,*] Hanan Elsarraj,[b] Shane Stecklein[c] and Fariba Behbod[d]

Introduction

The existence of breast cancer stem cells (BCSCs) has profound implications for cancer biology and therapy because it is likely that eradication of BCSCs is the critical determinant in achieving a cure (Diehn and Clarke 2006). Because they share the properties of their normal stem cell counterparts, i.e., long term self-renewal potential and quiescence, CSCs may be resistance to current drug interventions and allow tumor recurrences (Reya et al. 2001, Clarke and Fuller 2006, Dalerba et al. 2007). Current research is taking advantage of these intrinsic properties to develop therapies targeting BCSCs. In the following sections, we focus on two promising directions of BCSC research. The first is the attempt to sensitize BCSCs to current therapies by understanding the unique ability of these cells to escape chemotherapy induced cell death. The second section discusses the potential targeting of pathways that participate in self-renewal and cell fate as a part of BCSC therapy. The hope is that innovative therapies, based on a better understanding of CSCs, will lead to enhanced and long-term cure rates in breast cancer.

University of Kansas Medical Center, 3901 Rainbow Blvd, MS 3003, Kansas City, KS 66160.
[a]E-mail: kvaldez@kumc.edu
[b]E-mail: helsarraj@kumc.edu
[c]E-mail: sstecklein@kumc.edu
[d]E-mail: fbehbod@kumc.edu
*Corresponding author

Effects of Current Therapies on BCSC

Radiotherapy

Treatment with ionizing radiation (IR) is an integral component of breast cancer therapy for patients with both early stage and advanced disease. Nearly all patients who undergo breast conserving surgery (i.e., lumpectomy) for invasive and *in situ* disease will receive radiation, as will many patients undergoing mastectomy (Reya et al. 2001, Carlson and McCormick 2005). In both cohorts, the addition of IR has been shown to dramatically reduce local recurrence, which has been independently associated with improved overall survival (Clarke et al. 2005, Gebski et al. 2006). In patients with advanced and metastatic disease, radiation is indispensable in slowing progression and providing palliation (van der Linden et al. 2005, Rades et al. 2006, Wadasadawala et al. 2007).

Despite the proven efficacy of radiotherapy in reducing risk of local recurrence and improving overall survival, approximately 10% of women who receive adjuvant radiation will develop recurrent disease within 10 years of treatment (Debeb et al. 2009). Though radiologic and clinical data may indicate total absence of residual disease, recurrence in the presence of sterilizing doses of IR implies the existence of one or more radiation-resistant populations within human breast cancers. Emerging evidence suggests that such recurrent lesions are incited by a rare population of breast cancer stem cells that, by virtue of their relative quiescence and enhanced developmental plasticity, can recover from radiotherapy and proliferate.

The lethal effects of IR are due to production of unrepairable DNA lesions involving DNA double strand breaks (DSBs). Though irradiation can directly ionize DNA, formation of reactive oxygen species (ROS) from water in the immediate vicinity of DNA is responsible for most of the sustained damage. Detoxification of ROS before modification of biomolecules and repair of DSBs are two major mechanisms by which normal and cancer cells overcome sublethal doses of IR. It has previously been reported that the $CD44^{+}/CD24^{-/low}$ population of breast cancer stem cells exhibit lower ROS levels before and after irradiation, implying that such CSCs possess higher intrinsic levels of free radical scavengers (Phillips et al. 2006). Moreover, it has been noted that radiation produced fewer γH2AX foci (the hallmark of DSB induction) in breast CSCs and that such foci resolved faster in CSCs than in non-CSC populations (Diehn and Clarke 2006, Phillips et al. 2006, Woodward et al. 2007). Non-homologous end joining (NHEJ) and homologous recombination (HR) are the two major DSB repair pathways in higher eukaryotes. NHEJ, a relatively error-prone process that directly ligates strands following a break, is responsible for the bulk of DSB repair following irradiation, while the error-free HR pathway, involving BRCA1,

BRCA2 and RAD51, plays a modest role in repair of IR-induced lesions. Interestingly, it has been noted from *in vitro* studies of human breast cancer cell lines that HR, and not NHEJ, appears to be hyperactive in $CD44^{+}/CD24^{-/low}$ cells and may be responsible for the intrinsic radiation resistance of BCSCs (Yin and Glass 2011). It is interesting to speculate that normal stem cells, and by extension, malignant stem cells, may favor HR to NHEJ as the former is far superior in preventing DSB-induced mutation. Though the upstream signaling pathways that regulate intrinsic radiation resistance in stem cells remain largely uncharacterized, Woodward and colleagues have implicated the Wnt/β-catenin pathway (Woodward et al. 2007). Interestingly, a previous study noted that heightened Wnt signaling triggered a DNA damage response in human breast epithelial cells, ultimately leading to oncogenic conversion (Ayyanan et al. 2006). Study of additional interactions between developmental signaling nodes in stem cells and the DNA damage response are certain to elaborate our understanding of the pathogenesis of breast cancer, and may also illuminate novel interventions to target BCSCs using radiation.

The efficacy of radiation in different subtypes of breast cancer lends further support to the dogma that BCSCs are resistant to radiotherapy. The tremendous molecular diversity of human breast cancers was first illuminated by Perou and colleagues. These studies demonstrated that specific molecular taxonomies of breast cancer are associated with aggressive biological features and poor prognosis (Perou et al. 2000, Sorlie et al. 2001). Specifically, the basal-like breast cancers, which are highly enriched in triple-negative breast cancers (TNBCs) (i.e., estrogen receptor α (ERα)-negative, progesterone receptor (PR)-negative and ERBB2/HER2-negative), and HER2-positive breast cancers are the most resistant to conventional therapy and are associated with the highest risk of local recurrence even after optimal treatment (Woodward et al. 2003, Nguyen et al. 2008). Another study has even noted that survival benefit from adjuvant radiotherapy is confined to breast cancer patients with ER-positive disease (Kyndi et al. 2008). Interestingly, TNBCs are highly enriched in the $CD44^{+}/CD24^{-/low}$ population of cells which have been shown to be enriched in breast CSCs when compared to luminally-differentiated (e.g., ERα-positive) breast cancers.

Sensitizing BCSCs to radiation

Identification of signaling pathways leading to the radiotherapy resistance of breast cancer stem cells, specifically regulation of ROS and DNA damage, is particularly valuable because it may be a way to target BCSCs. It has been demonstrated that Notch signaling is activated in irradiated BCSCs, and that inhibition of Notch1 signaling may be a way to radiosensitize these cells (Debeb et al. 2009). Additionally, gain-of-function of Siah1 and Siah1L, two ubiquitin-proteasome pathway targets, sensitized breast cancer cells to the cytotoxic effects of irradiation, while knockdown of Siah1 decreased radiosensitivity in breast cancer cells. Siah1 increased DNA double-strand breaks induced by irradiation in these breast cancer cells, and finally resulted in an increased radiosensitivity (He et al. 2010). Targeting ALDH activity using either DEAB or ATRA in $ALDH^{hi}CD44^{+}$ cells isolated from two human breast cancer cell lines results in significant sensitization to therapy and radiation, in many cases to the same level of sensitization as non-resistant $ALDH^{low}CD44^{-}$ cells. This provides functional evidence that ALDH activity influences breast cancer cell resistance to chemotherapy as well as to radiation (Croker and Allan 2012).

Hyperthermia as an adjunct to radiation therapy has recently been a focus of interest in cancer research. Interestingly, Atkinson et al. used two different preclinical breast cancer models to study radiation resistance with and without thermal enhancement. Thermal enhancement with gold-coated nanoshells as a radiation sensitizer significantly decreased the radiation resistance of CSC-enriched subpopulations. Nanoshell-induced heating appears to prevent breast tumor cells from repairing DNA double-strand breaks induced by ionizing radiation, resulting in an increase in their radiation sensitivity (Atkinson et al. 2010).

Chemotherapy

Breast cancer patients with no evidence of metastatic disease (i.e., stages I-III) are treated with curative intent by surgical excision and adjuvant radiotherapy. Though such early stage disease is defined by the absence of overt metastatic involvement, the existence of micrometastatic deposits is likely in many of these patients. To avoid recurrence from and progression of micrometastases, adjuvant systemic treatments including chemotherapy, endocrine therapies and biologic agents are utilized in an attempt to eradicate disease that was not surgically excised or biologically sterilized with radiotherapy. In patients with more advanced, but still potentially curable disease (e.g., stage IIIB and operable stage IIIC), such systemic therapies may be used in the neoadjuvant setting to reduce tumor burden followed by surgery and radiotherapy.

Though radiotherapy dramatically drastically reduces local recurrence rates and has been shown to improve survival in large meta-analyses, at present, approximately 40% of breast cancer patients will experience recurrence; of these, 60–70% involve the development of distant metastases that are considered incurable (Gerber et al. 2010). Despite receiving optimal surgery and adjuvant radiotherapy, the mortality rate at 15 years remains between 26–31% for patients with lymph node-negative disease at initial surgery, and 48–55% for patients with node-positive disease (Clarke et al. 2005). A number of potential explanations for such recurrence have been posited, including under-treatment, failure to remove or radiate local deposits of tumor cells, and intrinsic resistance to radiation administered in both the adjuvant and neoadjuvant settings.

As is the case for radiotherapy, several studies have noted enrichment of BCSCs following treatment of human breast cancer cell line xenografts and genetic animal models of breast cancer with cytotoxic chemotherapy. Epirubicin treatment of SKBR3 $HER2^{+}$ breast cancer xenografts led to an enrichment of BCSCs, and similarly BCSCs were noted to promote resistance to cisplatin in a *Brca1/p53* model of mouse mammary tumorigenesis (Yu et al. 2007, Shafee et al. 2008). In human studies, a report by Jenny Chang's group has demonstrated that breast cancer cells isolated 12 weeks after treatment with neoadjuvant chemotherapy are enriched for the $CD44^{hi}$/$CD24^{-/low}$ population and exhibit increased mammosphere forming potential compared to paired specimens acquired before cytotoxic therapy (Li et al. 2008). In an extension of these studies, the same group reported that specimens acquired after endocrine or cytotoxic therapy are enriched for the mammary stem cell/claudin-low gene expression signature, including mesenchymal-associated genes (Creighton et al. 2009).

Sensitizing BCSCs to chemotherapy

One proposed mechanism by which CSCs to escape the effects of cytotoxic chemotherapeutic drugs appears to be the high levels of ABC transporter proteins compared to their normal counterparts. ABC transporters function as efflux pumps to protect cells from drug damage. This property results in exclusion of the fluorescent dye Hoechst 33342 and Rhodamine 123, and can be detected by flow-cytometry (Goodell et al. 1996, Hirschmann-Jax et al. 2004, Challen and Little 2006). The ability to modulate ABC transporters could potentially help overcome drug resistance in BCSCs. However, previous clinical trials using compounds that block ABC transporters non-specifically have failed due to serious side effects. Recently, the highly selective potassium ionophore salinomycin was identified by high-throughput screening to selectively eradicate BCSCs (Gupta et al. 2009). While the mechanism of salinomycin's action remains unclear, it appears it may induce terminal differentiation accompanied by cell-cycle arrest rather than trigger cytotoxicity. Importantly, salinomycin's actions were demonstrated using an *in vivo* model of breast cancer, in which it not only appear to induce terminal differentiation in the primary tumor, but also prevented metastasis. Of note, salinomycin has recently been shown to reduce experimental metastases in other types of cancer (Dong et al. 2011, Kim et al. 2011, Lu et al. 2011).

Endocrine therapies

The mammary gland is a unique organ in that the majority of its development occurs at puberty, followed by functional differentiation and drastic remodeling during menstruation, pregnancy, lactation and involution. The estrogen and progesterone steroid hormones secreted by the gonads and paracrine growth factors play a critical role in regulating the initial development of the mammary gland, as well as many aspects of its cellular and functional plasticity (Coleman et al. 1988, Brisken et al. 1998, Brisken et al. 1999, Mallepell et al. 2006). Experimental knockout of the estrogen receptor (ER) in mice causes complete failure of mammary gland development beyond the primitive ductal structures seen in early gestation, confirming the intimate dependence of this structure on sex hormones (Korach et al. 1996). In the mature human resting mammary gland, 10–20% of luminal epithelial cells will co-express the ER and progesterone receptor (PR); interestingly, these cells do not actively divide, implying that expression of such receptors is confined to a developmentally mature population of cells with restricted replicative potential (Clarke et al. 1997, Russo et al. 1999). Unlike the normal ductal and lobular epithelium, actively dividing ER^+ cells are commonly noted in both pre-malignant proliferations of the mammary epithelium (e.g., atypical ductal hyperplasia) as well as in *in situ* and invasive breast cancers. The presence of ER in a breast neoplasm, as determined by immunohistochemistry on biopsy and/or surgical specimens, is highly predictive of response rates to anti-estrogen directed therapies including selective estrogen receptor modulators (SERMs, e.g., tamoxifen) and aromatase inhibitors (e.g., anastrazole). Despite the impressive preventative and therapeutic potential of these agents, acquired resistance to endocrine agents remains a formidable challenge in curative treatment of ER^+ breast cancer (Gelber et al. 1996, Howell et al. 2005). Emerging evidence suggests that functional inactivation of estrogen signaling and/or estrogen deprivation in ER^+ breast cancers may select for cells with stem-like features that do not rely on sex hormone signaling for survival or proliferation.

Expression of the epidermal growth factor receptor 1 (EGFR) and HER2 are inversely correlated with expression of ER; EGFR is highly expressed in TNBCs, and HER2 is amplified and overexpressed in $HER2^+$ breast cancers (Dowsett et al. 2006, Giltnane et al. 2007). *In vitro* studies of MCF7 ER^+ breast cancer cells demonstrated that tamoxifen-resistant MCF7 cells overexpressed EGFR and HER2 mRNA and protein 5–10 fold compared to parental MCF7 cells (Knowlden et al. 2003). Similarly, acquired resistance to other anti-estrogen agents including fulvestrant and aromatase inhibitors can be conferred by the HER2 pathway (Massarweh et al. 2006). Additionally, the activity of several signaling networks that can be activated through the EGFR/HER2 pathway, including MAPK, PI3K, and Src, can post-translationally modify the AF1 domain of ER and allow ligand-independent activation of ER, circumventing the activity of anti-estrogen therapies (Campbell et al. 2001, Hiscox et al. 2006, Sarwar et al. 2006, Pancholi et al. 2008). There is emerging evidence that the HER2 protein plays an important role in regulating CSC function, as expression of HER2 has been positively correlated with both ALDH1 expression (a CSC marker) and that HER2-positivity could enrich for cells with sphere forming capacity, *in vitro* clonogenicity and tumorigenicity (Ginestier et al. 2007, Korkaya et al. 2008). *In vitro* models of acquired estrogen resistance have demonstrated that treatment with the dual EGFR/HER2 inhibitor lapatinib caused reactivation of ER activity and restoration of anti-estrogen sensitivity (Leary et al. 2010). In a clinical study of patients with ER^+ disease who had experienced recurrence within six months of discontinuing tamoxifen, the addition of lapatinib to letrozole (an aromatase inhibitor) significantly improved progression free survival, despite the fact that these patients did not exhibit HER2-positivity (Johnston et al. 2009). Lapatinib was also noted to reduce the $CD44^{hi}/CD24^{-/low}$ population and mammosphere forming capacity in $HER2^+$ primary breast cancers (Li et al. 2008). Thus, despite gross tumor regression caused by abrogation of estrogen signaling or estrogen deprivation, such approaches may induce or select for cells with hyperactive EGFR and/or HER2 signaling nodes that are responsible for recurrence. Dual targeting of estrogen signaling and HER2, even in tumors which do not exhibit HER2-positivity, may prove to be efficacious regimens for treating and preventing endocrine resistance in ER^+ tumors.

Targeting Molecules that Participate in Self-renewal and Cell Fate

The intrinsic resistance of CSCs to traditional therapies may explain why marked regression of gross tumor volume during therapy does not always correspond to improved survival. This highlights the necessity of developing agents that specifically target the CSC population. Since the defining characteristic of both normal and cancer-associated stem cells is their ability to undergo self-renewal, targeting morphogenetic pathways that are dysregulated in the CSCs, and in effect permit aberrant self-renewal, may provide novel strategies to combat cancer therapeutic resistance and tumor recurrence. Below we review the signaling pathways established to be critical for stem cell self-renewal, and the current status of drug development targeting theses pathways.

Notch

Aberrant Notch signaling has also been implicated in acquired resistance to endocrine therapies. The Notch pathway has been documented to play a critical role in specification of cellular lineage within the normal human mammary gland. Notch signaling is an evolutionarily conserved pathway that is essential for embryonic development, organogenesis, and tissue homeostasis. Recently, there has been considerable interest in elucidating how developmental signaling pathways, including Notch, may aberrantly contribute to tumorigenesis. Mounting evidence suggests that Notch deregulation may engender critical tumor hallmarks, including oncogene expression, angiogenesis, stem cell maintenance, deregulated cell cycle progression, and anti-apoptotic mechanisms (Lee et al. 2008). In mammals, there are four Notch receptors (Notch1-4), three Delta-like ligands (Dll1, Dll3, and Dll4), and two ligands of the Jagged family (Jag1 and Jag2; Lobry et al. 2011). A possible link between Notch signaling and breast cancer was first noted when the INT3 protein in *MMTV LTR-INT3* induced mouse mammary tumors was found to be *Notch4* (Gallahan and Callahan 1997). Later on, it was found that overexpression of Notch1^{intra} or Notch3^{intra} induces mammary tumors at high frequency in transgenic mice (Hu et al. 2006). These observations have stimulated intense interest in exploring Notch signaling as a therapeutic target of breast cancer.

Notch signaling is triggered upon ligand–receptor interaction which induces two sequential proteolytic cleavages of the membrane anchored notch receptor; the first in the extracellular domain mediated by metalloproteases of the ADAM (A Disintegrin And Metalloproteinase) family, and the second within the transmembrane domain mediated by a γ-secretase presenilin complex. The second cleavage allows the release and translocation of the intracellular domain of Notch (NIC) into the nucleus where it associates with a transcription complex containing RBPJk. This interaction converts the multiprotein co-repressor complex into a co-activator complex resulting in increased expression of Notch target genes (Izrailit and Reedijk 2011).

The best-characterized Notch targets are the Hes (Hes1–7) and Hey subfamilies (Hey1, Hey2, HeyL, HesL/HelT, Dec1/BHLHB2, Dec2/BHLHB3) (Guo et al. 2011). Both Hes and Hey proteins are basic helix-loop-helix (bHLH) type of trancriptional repressors and act as Notch effectors by negatively regulating expression of downstream target genes such as tissue-specific transcription factors (Iso et al. 2003). Their role in breast cancer is still unclear. Hes1 is a tumor suppressor in epithelial cells and acts as a mediator of 17β-estradiol (E2) proliferative effects on breast cancer cells. In contrast, Hes6 is a potential oncogene that inhibits Hes1, increases the proliferation of breast cancer cells and is regulated by estrogen (Hartman et al. 2009).

In luminal-type breast cancers, Notch signaling appears to be suppressed, while it is often hyperactive in basal-like breast cancers. This finding is congruent with *in vitro* studies in which pharmacologic inhibition of the Notch pathway with γ-secretase inhibitors or Notch neutralizing antibodies reduced mammosphere formation in primary human DCIS, and cold reduce CSC activity and overall tumor formation *in vivo*. Similarly, in enriched CSCs from breast cancer cell lines and invasive cancer specimens, Notch1 and Notch4 activity were increased 4- to 8-fold. An interesting and clinically relevant link has been identified between Notch signaling and HER2, as Notch has been shown to positively regulate HER2 and the ability of trastuzumab to reduce mammosphere formation can be abrogated by antagonism of the Notch pathway (Magnifico et al. 2009). Similarly, *in vitro* studies have demonstrated that Notch1 activity is significantly increased in breast cancer cell lines treated with trastuzumab or lapatinib, and that inhibition of the Notch pathway resensitizes cells to trastuzumab (Osipo et al. 2008). Consistent with the inverse relationship of ER expression and Notch activity in human breast cancers, it has been shown that estrogen signaling negatively regulates Notch signaling. *In vitro*, treatment with tamoxifen or fulvestrant can rescue estrogen-induced downregulation of Notch1 and Notch4, and in mouse xenotransplantation assays, the addition of a γ-secretase inhibitor to tamoxifen appears to be superior to tamoxifen monotherapy (Rizzo et al. 2008).

Notch in mammary stem cells and BCSCs

Notch signaling has been proposed to maintain the stemness of BCSCs. Recent studies have investigated the role of Notch signaling in normal mammary stem and progenitor cells by either knockdown of the canonical Notch effector gene *Cbf-1* or constitutive activation of the pathway. Downregulation of Notch signaling in MaSCs led to increased repopulating activity *in vivo* and aberrant ductal morphogenesis, providing evidence that this pathway normally plays a role in restricting expansion of the MaSC pool. In contrast, Notch activity was found to be substantially higher in luminal epithelial cells of the mammary gland, with prominent expression of the active form of Notch1 and its target genes in luminal progenitor cells *in vivo*. Moreover, constitutive signaling was found to specifically target luminal progenitor cells for expansion and self-renewal, eventually leading to hyperplasia and tumorigenesis (Bouras et al. 2008).

Recent studies examined the differential activity of Notch1 and Notch4 receptors particularly in the BCSC population sorted from breast cancer cell lines and primary breast cancer samples. Notch4 signaling found to have a greater effect on CSC activity than Notch1, and it was proposed that Notch4 regulates exit of CSCs into the proliferating progenitor population, whereas Notch1 activity regulates progenitor proliferation and luminal differentiation (Harrison, Farnie et al. 2010). Furthermore, Notch inhibition by γ-secretase inhibitors (GSI) and shRNA reduces BCSC number and activity *in vitro* and *in vivo*. GSI treatment inhibited Notch1 and not Notch4 activity in breast cancer cell lines, and reduced CSC number *in vitro* by <50%. These findings suggest that specifically targeting the Notch4 receptor in BCSCs for treatment of breast cancer will be superior to inhibiting γ-secretase or targeting the Notch1 receptor. The Notch pathway has also been linked to radiation resistance in breast cancer cell lines; BCSCs, isolated by nonadherent culture, have increased resistance to radiation compared with the more differentiated nonstem cells and Notch signaling is increased in these cells (Harrison et al. 2010).

Notch in the therapeutic point of view

Pharmacologic inhibition of either estrogen biosynthesis or signaling would be expected to alleviate repression of Notch signaling, which in turn, can activate expression of HER2 and lead to survival and acquisition of CSC-like properties. Future clinical trials employing combination agents which interfere with sex hormone signaling, Notch and HER2 simultaneously may provide yet unseen efficacy in the curative treatment of ER^+ breast cancers.

The Notch pathway has become an attractive drug target for breast cancer treatment, and the use of Notch inhibitors is predicted to be effective in reducing tumor growth, perhaps in conjunction with other treatments. In $ER\alpha^+$ breast cancer cells, estradiol inhibited Notch activity and Notch-1^{IC} nuclear levels and affected Notch-1 cellular distribution, the same study showed that inhibition of γ-secretase has anti-neoplastic effects on breast cancer cell lines (Rizzo et al. 2008). Furthermore, high-level JAG1 expression may identify tumors susceptible to treatments that target the Notch pathway.

In summary, potential approaches may include: γ-secretase inhibitors, which exploit the dependence of NOTCH receptor on enzyme processing for its activity; Notch ligand/Fc fusion proteins that inhibit binding and signaling through the NOTCH receptor, and inhibitory siRNA structures that target mRNA encoding NOTCH receptors or their ligands (Dickson et al. 2007).

Wnt

The Wnt signaling pathway was first discovered by Nusse and Varmus in 1982 when they identified proviral insertion of the mouse mammary tumor virus (MMTV) into the promoter of Int-1 (Integration-1) in murine mammary adenocarcinoma. Int-1 is orthologous to the Drosophila segment polarity gene Wingless (Wg), and "Wnt" is a fusion of the terms Int and Wg (Izrailit and Reedijk 2011). Wnt signaling pathways play important roles in multiple aspects of both MaSC and CSC biology. Wnts promote stem cell amplification and branching by paracrine induction of β-catenin signaling in basal cells. In contrast, β-catenin signaling in luminal progenitors is essential for alveologenesis and is probably induced by Wnt-independent mechanisms (Incassati et al. 2010).

The canonical Wnt signaling pathway

The canonical signaling network includes 19 Wnt ligands, 10 Frizzled (Fzd) receptors and two co-receptors, low density lipoprotein receptor-related protein (Lrp) 5 and 6 (Izrailit and Reedijk 2011). When Wnt proteins are unable to bind to their receptors at the cell surface, β-catenin is constitutively sequestered in a "destruction complex" that consists of the adenomatous polyposis coli (APC) tumor suppressor, Axin2, glycogen synthase kinase-3β (GSK3β), and casein kinase 1 (CK1). However, when Wnt proteins are secreted properly from cells they form a ternary complex with Fzd and Lrp 5/6, which results in the activation of dishoveled (Dvl) followed by the inhibition of GSK3β and the stabilization of cytosolic β-catenin. The β-catenin then translocates into the nucleus where it interacts with T-cell factor/lymphoid enhancing factor (TCF/LEF) to induce the expression of downstream target genes that regulate cell cycle, proliferation, and differentiation (King et al. 2011).

The non-canonical Wnt signaling pathway

To activate the non-canonical Wnt pathway, the Fzd receptors respond to Wnt proteins in the presence of the Wnt co-receptor Ror2, a tyrosine kinase receptor. The non-canonical Wnt pathway has two intracellular signaling cascades that consist of the Wnt/Ca^{2+} pathway and the Wnt/PCP pathway. The Wnt/Ca^{2+}pathway is mediated by heterotrimeric G-proteins that are activated by the Wnt/Fzd/Dvl complex. This increases intracellular Ca^{2+} levels that subsequently activate Ca2+-dependent protein kinases. The Wnt/PCP pathway is stimulated by Wnt5a or Wnt11, and diverges following the activation of Dvl to transduce signals that activate Rho and Rac, which in turn activate Rho-associated kinase (ROCK) and c-Jun N-terminal kinase

(JNK), respectively. The ensuing cytoskeletal rearrangement executes cell polarity and epithelial-to-mesenchymal transition; the latter contributes to tumor metastasis (King et al. 2011).

Wnt signaling in mammary stem cells renewal and differentiation

MaSC renewal requires the presence of Lrp5/6 in conjunction with Frizzled receptors. Lrp5-/- mammary glands exhibit fewer TEBs and diminished side branching, presumably as a result of dampened Wnt/ß-catenin activity within MaSCs or progenitors residing in the cap cell layer of the terminal end buds (TEB). Moreover, ablation of Lrp5 significantly reduces the percentage of cells within the basal compartment of the mammary gland, suggesting that Wnt/β-catenin dependent signaling might also direct cell fate decisions in addition to conferring self-renewal properties (Roarty and Rosen 2010).

The downstream targets of the canonical Wnt signaling pathway comprise several genes important for cancer progression, such as *c-myc* and *cyclin D1 (He et al. 1998, Shtutman et al. 1999)*. It been shown that increased β-catenin/Tcf-4 activity leads to overexpression of c-*MYC*, which then promotes neoplastic growth in colorectal epithelial cells (He et al. 1998). Elevated transcription from the cyclin D1 gene was shown to be induced by the β-catenin/LEF1 complex (Shtutman et al. 1999). The majority of Wnt target genes appear to be cell type specific. It is not clear whether "universal" Wnt/Tcf target genes exist. The best current candidate in vertebrates is Axin2/conductin (Clevers 2006). Recent studies have demonstrated that Axin2/conductin, in addition to being a scaffold protein negatively regulating β-catenin activation, is a transcriptional target gene of the Wnt signaling (Hadjihannas and Behrens 2006). MMP-7 is another gene controlled by Wnt signaling. It is a matrix metalloproteinase expressed in the malignant epithelium of the esophagus, stomach, colorectum, kidney, and breast (Beeghly-Fadiel et al. 2009). β-catenin interacts with the MMP-7 promotor, and upregulates its expression (Crawford et al. 1999). Other relevant β-catenin targets include WISP1, Slug, Twist, Sox17, and Limb Heart Bud (Xu et al. 2000, Xie et al. 2001, DiMeo et al. 2009, Fu et al. 2010, Rieger et al. 2010).

In many malignancies, mutations in Wnt pathway mediators like APC, β-catenin, and Axin, are frequently observed and play a critical role in their pathogenesis; however, these overt Wnt pathway mutations are rarely observed in breast cancer (Roarty and Rosen 2010). Instead, constitutive expression of Wnt signaling enhances the self-renewal of mammary progenitor cells and continuous stimulation of this pathway leads to formation of breast tumors (Lindvall et al. 2007, Jones and Kemp 2008). A study of a mouse model of breast cancer indicated that breast cancers that arise from stem-progenitor cells undergo transformation through dyregulation of Wnt signaling, whereas epithelial cell derived breast tumors are triggered by oncogenic activation of HER2 (Li et al. 2005). Furthermore, blockade of Wnt signaling has an impact on the *in vitro* proliferation and motility of the cancer cells. The *in vivo* effects of Wnt pathway blockade were even more dramatic since it had observed a strong decrease in the mammary tumor-forming potential and an impairment of lung metastases (Matsuda et al. 2009).

Recent studies revealed that Wnt pathway genes are overexpressed in TNBC, as well as Fzd 7, Lrp 6 and Tcf 7. Inhibition of either Fzd 7 or Lrp 6 will block Wnt signaling and suppress Wnt signal-mediated cell proliferation in TNBC cell lines (Yang et al. 2011). On the therapeutic point of view, studies demonstrate that progenitor cells in the mammary gland are more resistant to clinically relevant doses of radiation than are non-progenitors, which constitute the bulk of the mammary gland, and that overexpression of the Wnt/β-catenin pathway can enhance the radioresistance of progenitor cells (Chen et al. 2007, Woodward et al. 2007). These findings suggest that the Wnt/β-catenin signaling pathway may be an attractive target for directed anti-stem cell therapeutics.

Hedgehog signaling

The Hedgehog (Hh) signaling pathway is another highly conserved developmental pathway which acts by controlling cell fate, pattering, proliferation, survival and differentiation. In the adult, Hh signaling plays a critical role in the maintenance of somatic stem cells. The secreted glycoproteins Sonic, Indian, and Desert are the three mammalian Hh homologs, and have distinct patterns of tissue expression and likely distinct biological function. Pathway activation is initiated upon ligand binding to the Hh receptor Patched (Ptch1 and Ptch2). In the absence of Hh ligand, Ptch represses signal transduction by inhibiting the trans-membrane protein, Smoothened (Smo). Binding of Hh ligand to Ptch relieves Smo inhibition, resulting in Smo activation of the transcription factors Gli-1, -2, and -3 and alterations in target gene transcription (Ingham and McMahon 2001).

As recently reviewed, there is increasing evidence that inhibition of Hh signaling in breast tumors may interfere with the maintenance of a putative cancer stem cell compartment (Kasper et al. 2009). Human BCSCs, as identified by the $CD44^{+}CD24^{-/low}lin^{-}$ phenotype, show increased gene expression of PTCH1, GLI1, and GLI2 compared to remaining tumor cells isolated from primary breast cancers, and have been shown to be dependent upon Hh signaling for proliferation (Liu et al. 2006). Independently conducted analyses of array-based comparative genomic hybridizations have demonstrated frequent loss of Ptch1 and amplification of Gli1 chromosomal regions in primary breast cancer samples and breast cancer cell lines (Naylor et al. 2005, Nessling 2005). Additionally, in association with Ptch1 promoter methylation, Ptch1 protein expression is lost in 58% of invasive carcinomas (Wolf 2007). This indirect evidence that aberrant Hh signaling promotes breast cancer is supported by mouse models in which partial knockdown of Ptch1 results in severe histological defects in mammary ductal structure, including hyperplasia and dysplasia similar to human breast lesions (Lewis et al. 1999). Furthermore, overexpression of Gli1 in mouse mammary epithelial cells induces hyperplastic lesions, defective terminal end bud development and tumor formation (Fiaschi et al. 2009). Constitutive activation of Smo leads to defects in mammary gland morphology and altered stem/progenitor cell number (Moraes et al. 2007).

Of the discussed signaling pathways, the Hh may show the most promise as a CSC targeting therapy. Indeed, numerous Hh antagonists are currently in development and clinical testing. Inhibition of Hh signaling has been shown to increase the response of cancer cell lines to classical chemotherapies (Sims-Mourtada et al. 2007). Much of this work is focusing on Smo as a therapeutic target for the treatment of cancers that are dependent on Hh ligand activation or induced by mutations at the receptor level. The first identified Hh inhibitor, the naturally occurring plant alkaloid cyclopamine, directly binds and inactivates Smo (Chen et al. 2002). However, while cyclopamine has been shown to inhibit growth in cell lines and xenografts from several tumor types, including breast, clinical trials have not been possible because the drug would have to be administered at systemically toxic concentrations for effective tumor inhibition (Kolterud 2007, Sanchez 2001, Zhang 2009). Cyclopamine derivatives, along with several other synthetic Smo antagonists are currently being evaluated in Phase I and II trials (Williams et al. 2003, Tremblay et al. 2008, Barginear et al. 2009, Rudin et al. 2009, Scales and de Sauvage 2009, Von Hoff et al. 2009, Merchant and Matsui 2010). Additionally, several small molecule compounds targeting the Hh pathway are under development. Combining cyclopamine and Smoothened-targeting shRNA in a human glioma xenograft model shows that Hh signaling can regulate the self-renewal of the CSC population, and also reduced proliferation and prevents recurrence and metastasis (Stecca et al. 2007). To address cancers in which Hh activation occurs downstream of Smo, several compound screens have been performed to identify inhibitors that target Gli transcription factors (Lauth et al. 2007). Additionally, small molecules are being developed targeting the Gli-mediated transcription of Hh target genes (Hyman et al. 2009).

Bmi-1

Compared to Hh, the importance of Bmi-1, a transcriptional repressor of the polycomb group of transcription factors and a key regulator of self-renewal in both normal and malignant stem cells, in breast cancer tumorigenesis has been more controversial (Alkema et al. 1993, Liu et al. 2006). Bmi-1 regulates expression of genes controlling cell proliferation, survival, and differentiation; and is involved in the stable maintenance of repression of gene transcription. Two critical tumor suppressor pathways, p16 Ink4a and p19 ARF, with known roles in stem cell maintenance and cancer are repressed by Bmi-1 (Bracken et al. 2007, Ginestier et al. 2007). As reviewed by Park et al. (Park et al. 2004), Bmi-1 is required for the self-renewal of adult murine hematopoietic stem cells and neuronal stem cells (Molofsky et al. 2003, Park et al. 2003). It is postulated to act by preventing premature senescence of somatic stem cells, possibly by repressing expression of genes involved in senescence or by regulating telomerase expression. Additionally, Bmi-1 expression is associated with epigenetic silencing of important tumor suppressors (Sparmann and van Lohuizen 2006, Simon and Kingston 2009).

Recently, Liu et al. (Liu et al. 2006) observed increased expression of Bmi-1 in human BCSCs, as identified by the $CD44^{+}CD24^{-/low}lin^{-}$ phenotype, leading to speculation of Bmi-1's involvement in self-renewal of breast cancer stem cells. They established by mammosphere and transplantation assays that Bmi-1 overexpression promotes mammary stem cell self-renewal and proliferation. Additionally, Hh signaling increases expression of Bmi-1 in isolated mammary epithelial stem cells and CSCs. Given the range of tissue expression and phenotypic changes in *Bmi-1* deficient mice (van der Lugt et al. 1994), Bmi-1 likely has a role in maintaining self-renewal of other types of adult stem cells. Mammary gland development is impaired in *Bmi-1* knockout mice (Pietersen et al. 2008). These mice develop a rudimentary mammary tree during embryogenesis that later fails to invade the fat pad at puberty. The wild-type phenotype can be restored, however, either by pregnancy or Ink4a/Arf deletion, indicating the presence of mammary stem cells in *Bmi-1* deficient mice. Although as demonstrated by limiting dilution studies, these mammary stem cells appear to have reduced activity.

Bmi-1 expression has been linked to mammary carcinogenesis in several studies (Dimri et al. 2002, Datta et al. 2007). However, some find Bmi-1 expression to be associated with a favorable prognosis (Kim et al. 2004, Choi et al. 2009), while others have reported the opposite (Glinsky 2005, Silva 2007). Recently, Nalwoga (2010) found expression of Bmi-1 was related to features of good prognosis, such as low histological grade, low mitotic count, and ER positivity (Nalwoga et al. 2010). In addition, Glinsky et al. (2005) found that expression of a BMI-1-driven 11 gene signature was associated with risk of metastases in breast carcinoma.

Since it is overexpressed in many cancer types and plays an essential role in stem cell self-renewal and survival, Bmi-1 stands out as a promising target to control cancer stem cell function. However, as a non-enzymatic protein, targeting of Bmi-1 by traditional drug discovery methods is considered to be a challenge. Alternative therapeutic strategies that directly or indirectly impact the expression of Bmi-1 protein are being developed. The use of small interfering RNA (siRNA) to selectively reduce Bmi-1 levels *in vivo* and *in vitro* has proven effective in inhibiting tumor cell proliferation using fibrosarcoma cells and a HT1080 xenograft model, respectively (Cao 2011). Histone deacetylase inhibitors (HDACi) downregulate expression of Bmi-1 in both immortal non-transformed breast epithelial cells and breast cancer cells, and lead to derepression of polycomb group targets (Bommi et al. 2010). Recently, the drug artemisinin was also shown to inhibit Bmi-1 expression at both the protein and transcript levels (Wu et al. 2011). However, both HDACi and artemisinin are likely to be highly pleiotropic and could cause many unwanted side effects, making them unsuitable for clinical development. As no small molecules with acceptable specificities have been reported, discovery of novel compounds that selectively inhibit Bmi-1 expression is needed.

Steroid hormone regulation

Recently, several groups have published studies examining the role the ovarian steroid hormones have in indirectly influencing mammary stem cell numbers and cancer progression. Asselin-Labat et al. found that ovariectomy markedly diminished stem cell number and outgrowth potential *in vivo*, whereas stem cell activity increased in mice treated with estrogen plus progesterone (Asselin-Labat et al. 2010). Similarly, Joshi et al. observed a stem-cell enrichment in the mouse mammary gland at dioestrus (when progesterone is elevated), or with exogenous progesterone, both occur in the presence of estrogen (Joshi et al. 2010). Both of these studies examined the influence of steroid hormones on normal mammary stem cells and provided mRNA evidence that these paracrine effects may be mediated by the RANKL/RANK and the amphiregulin signaling pathways. In a progestin-driven mammary cancer model, Schramek et al. demonstrated that genetic inactivation of RANK in mammary gland epithelial cells prevents progestin-induced epithelial proliferation, impairs expansion of the BCSC-enriched population, and sensitizes these cells to DNA-damage-induced cell death (Schramek et al. 2010). Furthermore, deletion of RANK from the mammary epithelium decreased the incidence and delayed the onset of progestin-driven mammary cancer. Using breast cancer cell lines and freshly isolated breast cancer cells Fillmore et al. found that estrogen or FGF9 expanded the pool of BCSCs (Fillmore et al. 2010). Each of these studies independently reached the similar conclusion that hormone-deprived mammary cells, whether normal or malignant, have impaired self-renewal capacity. While most focused on the RANKL/RANK system, both amphiregulin and FGF9 were also implicated as paracrine signaling pathways contributing to this observation. The common model appears to involve estrogen and progesterone acting on the ER, PR-expressing luminal cells, which in turn release molecules involved in one or all of these potential pathways. These molecules in turn activate BCSC self-renewal.

These studies suggest that RANKL inhibition may have therapeutic applications in the clinical setting of breast cancer. Osteoprotegerin (OPG), a decoy receptor for RANKL which blocks it's binding to RANK, may have clinical use for breast cancer treatment (Kostenuik 2005). In fact, lower RANK and high OPG mRNA levels correlate with longer overall survival and disease-free survival in breast cancer patients (Santini et al. 2011). Additionally, denosumab, a human anti-RANKL antibody has already been approved by the US FDA and European Commission for osteoporosis treatment. Given the results in xenograft tumors and cell lines, denosumab may also attenuate tumor initiation in women.

Cytokine signaling

In addition to signaling from the epithelial component, BCSCs are also regulated by signaling from neighboring fibroblasts, immune and endothelial cells. Several cytokines secreted by these cells have been implicated in regulating BCSCs, including IL-6 and IL-8 (Wels et al. 2008). The IL-8 receptor, CXCR1, is expressed higher in ALDH-expressing cells from numerous breast cancer cells lines and exogenous IL-8 increases the CSC population (Charafe-Jauffret et al. 2009). This suggested that inhibition of IL-8/CXCR1 signaling could target BCSCs. Ginestier and colleagues also showed that <2% of breast cancer cells expressed CXCR1, which overlapped with the ALDH-positive fraction (Ginestier et al. 2010). Interestingly, repertaxin, a small molecule inhibitor of CXCR1, is able to decrease viability of the whole population even though only a small fraction expressed CXCR1.

The chemokine CXCL12 is expressed by stromal fibroblasts and endothelial cells and, through the activation of its receptor CXCR4, has a prominent role in metastisis (Dewan et al. 2006). CXCR4 is overexpressed in many human breast cancers, and its expression level correlates with tumor aggressiveness and lung metastasis in mouse models. Additionally, treatment of breast cancer cells with CXCR4-neutralizing antibodies or down-regulation of the receptor by siRNA reduces lung metastasis (Muller et al. 2001, Smith et al. 2004). CXCR4 expression has also been associated with the detection of disseminated tumor cells in bone marrow in breast cancer patients (Andre et al. 2006). While it is not clear whether these cells are tumorigenic, the majority are $CD44^+/CD24^-$, suggesting that BCSCs are sensitive to chemokine-induced homing.

CXCR1 signals through AKT thereby inhibiting FOXO3A localization, and FAS ligand expression. Repertaxin treatment inhibits AKT signaling resulting in nuclear FOXO3A and FAS ligand expression (Jonsson et al. 2005). Chemotherapy is known to kill cells through FAS ligand, but also induces IL-8 and thereby protects CSC from FAS ligand. This suggested that repertaxin might block this effect and target the CSC population. Repertaxin reduced the CSC population *in vitro* and in tumor xenografts (Ginestier et al. 2010). Used alone, repertaxin has little effect on tumor growth, but significantly reduces tumor volume in combination with docetaxel. In addition, repertaxin reduces metastatic burden and secondary tumor formation. Taken together, these results suggest that repertaxin sensitizes CSC to the bystander effect via FAS ligand and that CXCR1 blockade may represent a new approach to targeting and eliminating BCSCs.

MicroRNAs and BCSCs

MicroRNAs (miRNA) are small, noncoding RNA molecules that target specific messenger RNA (mRNA) for translational repression. Emerging evidence suggests that miRNAs also play essential roles in stem cell self-renewal and differentiation by negatively regulating the expression of key stem cell regulating genes (Liu et al. 2012). Recent studies showed that miR-200c, a member of the miR-200 family, is down-regulated in human BCSCs, normal human and murine mammary stem/progenitor cells, and embryonal carcinoma cells (Shimono et al. 2009). It also been shown that miR-200c modulates the expression of BMI1, an important regulator of stem cell self-renewal. Furthermore, the miR-200 family of miRNAs were found to directly target the mRNA of the E-cadherin transcriptional repressors ZEB1/ZEB2 (Park et al. 2008). The ectopic expression of miR-200 results in up-regulation of E-cadherin in cancer cell lines and its inhibition has induced EMT (Park et al. 2008, Shimono et al. 2009).

Another miRNA, Let-7 is markedly reduced in BCSCs, increased in differentiated cells, and its overexpression inhibits BCSC self-renewal and their ability to form mammospheres (Yu et al. 2007). Targets of let-7 include HMGA2 which promotes EMT, Ras which promotes self renewal, and IL6 which activates Stat3 to promote cell motility, growth and self-renewal (Pal et al. 2010).

Similar to let-7, miR-30 is reduced in BCSCs, and enforced constitutive expression of miR-30 in BCSCs inhibits their self-renewal capacity by reducing ubiquitin-conjugating enzyme 9 (Ubc9), and induces apoptosis through silencing integrin beta3 (ITGB3), which are two known targets for miR-30 (Yu et al. 2010).

The association of miRNA dysregulation with breast cancer stemness, and the functional analysis of specific miRNAs in BCSCs suggest that they may provide a promising new approach for cancer therapy. Anti-miRNA 2-*O*-methyl or locked nucleic acid oligonucleotides used to inactivate oncomirs in breast tumors may inhibit tumor growth. Conversely, the induction of tumor suppressor miRNA expression using viral or liposomal delivery of tissue-specific tumor suppressors to affected tissue may result in the prevention of progression or even shrinking of breast tumors (Lowery et al. 2008).

An additional stem cell/miRNA therapeutic possibility is the use of mesenchymal stem cells as potential tumor-targeted vectors for miRNA therapeutics, because of their ability to home in on the tumor site and evade the host immune response (Dwyer et al. 2007).

Furthermore, nanoparticles are a developing field for miRNA targeting therapy. The transferrin receptor (TfR) levels are elevated in various types of cancer cells. Systemic administration of a "nanoscale nonviral liposome-based complex", attached to an anti-transferrin receptor antibody fragment has been used to target tumor cells *in vivo* (Pirollo et al. 2006). Interestingly, a modified hybrid with anti-HER-2 siRNA was tested *in vivo* on mouse xenograft models of human breast, pancreatic, lung and colonic cancer cell lines. It was found that the modified hybrid can sensitize human tumor cells to chemotherapeutics with dramatic tumor growth inhibition (Pirollo et al. 2006, Pirollo et al. 2007).

Concluding Remarks

Understanding and characterizing the mechanisms of CSC development, self-renewal, survival, and differentiation within its microenvironment is leading to the development of new therapies for many human cancers, including breast cancer. If BCSCs are key cells in the induction of chemoresistance, as hypothesized, current chemotherapy may not be the best approach to fully eradicate existing breast tumors. The elucidation of the mechanisms responsible for therapy resistance by BCSCs is in progress, and is promising for use in combination with existing therapies to prevent breast cancer recurrence. Additionally, the identification of tumor initiating cells and their aberrant signaling pathways controlling their fate is leading to the development of individualized therapeutic strategies for prevention of breast cancer progression and recurrence. Both of these approaches provide potential targets for the modulation of BCSC fate. These novel approaches targeting BCSCs, and their surrounding microenvironment, may become the basis for generation of effective and clinically applicable therapies that prevent breast cancer relapse and metastasis and enhance long-term patient survival.

References

Alkema, M.J., J. Wiegant, et al. 1993. Characterization and chromosomal localization of the human proto-oncogene BMI-1. Hum Mol Genet 2(10): 1597–1603.

Andre, F., N. Cabioglu, et al. 2006. Expression of chemokine receptors predicts the site of metastatic relapse in patients with axillary node positive primary breast cancer. Ann Oncol 17(6): 945–951.

Asselin-Labat, M.L., F. Vaillant, et al. 2010. Control of mammary stem cell function by steroid hormone signalling. Nature 465(7299): 798–802.

Atkinson, R.L., M. Zhang, et al. 2010. Thermal enhancement with optically activated gold nanoshells sensitizes breast cancer stem cells to radiation therapy. Sci Transl Med 2(55): 55ra79.

Ayyanan, A., G. Civenni, et al. 2006. Increased Wnt signaling triggers oncogenic conversion of human breast epithelial cells by a Notch-dependent mechanism. Proc Natl Acad Sci USA 103(10): 3799–3804.

Barginear, M.F., M. Leung, et al. 2009. The hedgehog pathway as a therapeutic target for treatment of breast cancer. Breast Cancer Res Treat 116(2): 239–246.

Beeghly-Fadiel, A., W. Lu, et al. 2009. Matrix metalloproteinase-2 polymorphisms and breast cancer susceptibility. Cancer Epidemiol Biomarkers Prev 18(6): 1770–1776.

Bommi, P.V., M. Dimri, et al. 2010. The polycomb group protein BMI1 is a transcriptional target of HDAC inhibitors. Cell Cycle 9(13): 2663–2673.

Bouras, T., B. Pal, et al. 2008. Notch signaling regulates mammary stem cell function and luminal cell-fate commitment. Cell Stem Cell 3(4): 429–441.

Bracken, A.P., D. Kleine-Kohlbrecher, et al. 2007. The Polycomb group proteins bind throughout the INK4A-ARF locus and are disassociated in senescent cells. Genes Dev 21(5): 525–530.

Brisken, C., S. Kaur, et al. 1999. Prolactin controls mammary gland development via direct and indirect mechanisms. Dev Biol 210(1): 96–106.

Brisken, C., S. Park, et al. 1998. A paracrine role for the epithelial progesterone receptor in mammary gland development. Proc Natl Acad Sci USA 95(9): 5076–5081.

Campbell, R.A., P. Bhat-Nakshatri, et al. 2001. Phosphatidylinositol 3-kinase/AKT-mediated activation of estrogen receptor alpha: a new model for anti-estrogen resistance. J Biol Chem 276(13): 9817–9824.

Carlson, R.W., and B. McCormick. 2005. Update: NCCN breast cancer Clinical Practice Guidelines. J Natl Compr Canc Netw 3 Suppl 1: S7–11.

Challen, G.A., and M.H. Little. 2006. A side order of stem cells: the SP phenotype. Stem Cells 24(1): 3–12.

Charafe-Jauffret, E., C. Ginestier, et al. 2009. Breast cancer cell lines contain functional cancer stem cells with metastatic capacity and a distinct molecular signature. Cancer Res 69(4): 1302–1313.

Chen, J.K., J. Taipale, et al. 2002. Inhibition of Hedgehog signaling by direct binding of cyclopamine to Smoothened. Genes Dev 16(21): 2743–2748.

Chen, M.S., W.A. Woodward, et al. 2007. Wnt/beta-catenin mediates radiation resistance of Sca1+ progenitors in an immortalized mammary gland cell line. J Cell Sci 120(Pt 3): 468–477.

Choi, Y.J., Y.L. Choi, et al. 2009. Expression of Bmi-1 protein in tumor tissues is associated with favorable prognosis in breast cancer patients. Breast Cancer Res Treat 113(1): 83–93.

Clarke, M., R. Collins, et al. 2005. Effects of radiotherapy and of differences in the extent of surgery for early breast cancer on local recurrence and 15-year survival: an overview of the randomised trials. Lancet 366(9503): 2087–2106.

Clarke, M.F., and M. Fuller. 2006. Stem cells and cancer: two faces of eve. Cell 124(6): 1111–1115.

Clarke, R.B., A. Howell, et al. 1997. Dissociation between steroid receptor expression and cell proliferation in the human breast. Cancer Res 57(22): 4987–4991.

Clevers, H. 2006. Wnt/beta-catenin signaling in development and disease. Cell 127(3): 469–480.

Coleman, S., G.B. Silberstein, et al. 1988. Ductal morphogenesis in the mouse mammary gland: evidence supporting a role for epidermal growth factor. Dev Biol 127(2): 304–315.

Crawford, H.C., B.M. Fingleton, et al. 1999. The metalloproteinase matrilysin is a target of beta-catenin transactivation in intestinal tumors. Oncogene 18(18): 2883–2891.

Creighton, C.J., X. Li, et al. 2009. Residual breast cancers after conventional therapy display mesenchymal as well as tumor-initiating features. Proc Natl Acad Sci USA 106(33): 13820–13825.

Croker, A.K., and A.L. Allan. 2012. Inhibition of aldehyde dehydrogenase (ALDH) activity reduces chemotherapy and radiation resistance of stem-like ALDH(hi)CD44 (+) human breast cancer cells. Breast Cancer Res Treat 133(1): 75–87.

Dalerba, P., R. W. Cho, et al. 2007. Cancer stem cells: models and concepts. Annu Rev Med 58: 267–284.

Datta, S., M.J. Hoenerhoff, et al. 2007. Bmi-1 cooperates with H-Ras to transform human mammary epithelial cells via dysregulation of multiple growth-regulatory pathways. Cancer Res 67(21): 10286–10295.

Debeb, B.G., W. Xu, et al. 2009. Radiation resistance of breast cancer stem cells: understanding the clinical framework. J Mammary Gland Biol Neoplasia 14(1): 11–17.

Dewan, M.Z., S. Ahmed, et al. 2006. Stromal cell-derived factor-1 and CXCR4 receptor interaction in tumor growth and metastasis of breast cancer. Biomed Pharmacother 60(6): 273–276.

Dickson, B.C., A.M. Mulligan, et al. 2007. High-level JAG1 mRNA and protein predict poor outcome in breast cancer. Mod Pathol 20(6): 685–693.

Diehn, M., and M.F. Clarke. 2006. Cancer stem cells and radiotherapy: new insights into tumor radioresistance. J Natl Cancer Inst 98(24): 1755–1757.

DiMeo, T.A., K. Anderson, et al. 2009. A novel lung metastasis signature links Wnt signaling with cancer cell self-renewal and epithelial-mesenchymal transition in basal-like breast cancer. Cancer Res 69(13): 5364–5373.

Dimri, G.P., J.L. Martinez, et al. 2002. The Bmi-1 oncogene induces telomerase activity and immortalizes human mammary epithelial cells. Cancer Res 62(16): 4736–4745.

Dong, T.T., H.M. Zhou, et al. 2011. Salinomycin selectively targets 'CD133+' cell subpopulations and decreases malignant traits in colorectal cancer lines. Ann Surg Oncol 18(6): 1797–1804.

Dowsett, M., J. Houghton, et al. 2006. Benefit from adjuvant tamoxifen therapy in primary breast cancer patients according oestrogen receptor, progesterone receptor, EGF receptor and HER2 status. Ann Oncol 17(5): 818–826.

Dwyer, R.M., S.M. Potter-Beirne, et al. 2007. Monocyte chemotactic protein-1 secreted by primary breast tumors stimulates migration of mesenchymal stem cells. Clin Cancer Res 13(17): 5020–5027.

Fiaschi, M., B. Rozell, et al. 2009. Development of mammary tumors by conditional expression of GLI1. Cancer Res 69(11): 4810–4817.

Fillmore, C.M., P.B. Gupta, et al. 2010. Estrogen expands breast cancer stem-like cells through paracrine FGF/Tbx3 signaling. Proc Natl Acad Sci USA 107(50): 21737–21742.

Fu, D.Y., Z.M. Wang, et al. 2010. Sox17, the canonical Wnt antagonist, is epigenetically inactivated by promoter methylation in human breast cancer. Breast Cancer Res Treat 119(3): 601–612.

Gallahan, D., and R. Callahan. 1997. The mouse mammary tumor associated gene INT3 is a unique member of the NOTCH gene family (NOTCH4). Oncogene 14(16): 1883–1890.

Gebski, V., M. Lagleva, et al. 2006. Survival effects of postmastectomy adjuvant radiation therapy using biologically equivalent doses: a clinical perspective. J Natl Cancer Inst 98(1): 26–38.

Gelber, R.D., B.F. Cole, et al. 1996. Adjuvant chemotherapy plus tamoxifen compared with tamoxifen alone for postmenopausal breast cancer: meta-analysis of quality-adjusted survival. Lancet 347(9008): 1066–1071.

Gerber, B., M. Freund, et al. 2010. Recurrent breast cancer: treatment strategies for maintaining and prolonging good quality of life. Dtsch Arztebl Int 107(6): 85–91.

Giltnane, J.M., L. Ryden, et al. 2007. Quantitative measurement of epidermal growth factor receptor is a negative predictive factor for tamoxifen response in hormone receptor positive premenopausal breast cancer. J Clin Oncol 25(21): 3007–3014.

Ginestier, C., M.H. Hur, et al. 2007. ALDH1 is a marker of normal and malignant human mammary stem cells and a predictor of poor clinical outcome. Cell Stem Cell 1(5): 555–567.

Ginestier, C., S. Liu, et al. 2010. CXCR1 blockade selectively targets human breast cancer stem cells *in vitro* and in xenografts. J Clin Invest 120(2): 485–497.

Goodell, M.A., K. Brose, et al. 1996. Isolation and functional properties of murine hematopoietic stem cells that are replicating *in vivo*. J Exp Med 183(4): 1797–1806.

Guo, S., M. Liu, et al. 2011. Role of Notch and its oncogenic signaling crosstalk in breast cancer. Biochim Biophys Acta 1815(2): 197–213.

Gupta, P.B., T.T. Onder, et al. 2009. Identification of selective inhibitors of cancer stem cells by high-throughput screening. Cell 138(4): 645–659.

Hadjihannas, M.V., and J. Behrens. 2006. CIN By WNT: growth pathways, mitotic control and chromosomal instability in cancer. Cell Cycle 5(18): 2077–2081.

Harrison, H., G. Farnie, et al. 2010. Regulation of breast cancer stem cell activity by signaling through the Notch4 receptor. Cancer Res 70(2): 709–718.

Hartman, J., E.W. Lam, et al. 2009. Hes-6, an inhibitor of Hes-1, is regulated by 17beta-estradiol and promotes breast cancer cell proliferation. Breast Cancer Res 11(6): R79.

He, H.T., E. Fokas, et al. 2010. Siah1 proteins enhance radiosensitivity of human breast cancer cells. BMC Cancer 10: 403.

He, T.C., A.B. Sparks, et al. 1998. Identification of c-MYC as a target of the APC pathway. Science 281(5382): 1509–1512.

Hirschmann-Jax, C., A.E. Foster, et al. 2004. A distinct side population of cells with high drug efflux capacity in human tumor cells. Proc Natl Acad Sci USA 101(39): 14228–14233.

Hiscox, S., L. Morgan, et al. 2006. Elevated Src activity promotes cellular invasion and motility in tamoxifen resistant breast cancer cells. Breast Cancer Res Treat 97(3): 263–274.

Howell, A., J. Cuzick, et al. 2005. Results of the ATAC (Arimidex, Tamoxifen, Alone or in Combination) trial after completion of 5 years' adjuvant treatment for breast cancer. Lancet 365(9453): 60–62.

Hu, C., A. Dievart, et al. 2006. Overexpression of activated murine Notch1 and Notch3 in transgenic mice blocks mammary gland development and induces mammary tumors. Am J Pathol 168(3): 973–990.

Hyman, J.M., A.J. Firestone, et al. 2009. Small-molecule inhibitors reveal multiple strategies for Hedgehog pathway blockade. Proc Natl Acad Sci USA 106(33): 14132–14137.

Incassati, A., A. Chandramouli, et al. 2010. Key signaling nodes in mammary gland development and cancer: beta-catenin. Breast Cancer Res 12(6): 213.

Ingham, P.W., and A.P. McMahon. 2001. Hedgehog signaling in animal development: paradigms and principles. Genes Dev 15(23): 3059–3087.

Iso, T., L. Kedes, et al. 2003. HES and HERP families: multiple effectors of the Notch signaling pathway. J Cell Physiol 194(3): 237–255.
Izrailit, J., and M. Reedijk. 2011. Developmental pathways in breast cancer and breast tumor-initiating cells: Therapeutic implications. Cancer Lett.
Johnston, S., J. Pippen, Jr., et al. 2009. Lapatinib combined with letrozole versus letrozole and placebo as first-line therapy for postmenopausal hormone receptor-positive metastatic breast cancer. J Clin Oncol 27(33): 5538–5546.
Jones, K.A., and C.R. Kemp. 2008. Wnt-induced proteolytic targeting. Genes Dev 22(22): 3077–3081.
Jonsson, H., P. Allen, et al. 2005. Inflammatory arthritis requires Foxo3a to prevent Fas ligand-induced neutrophil apoptosis. Nat Med 11(6): 666–671.
Joshi, G., S. Pradhan, et al. 2010. Role of the oestrogen receptor (ESR1 PvuII and ESR1 325 C->G) and progesterone receptor (PROGINS) polymorphisms in genetic susceptibility to migraine in a North Indian population. Cephalalgia 30(3): 311–320.
Kasper, M., V. Jaks, et al. 2009. Hedgehog signalling in breast cancer. Carcinogenesis 30(6): 903–911.
Kim, J.H., S.Y. Yoon, et al. 2004. Overexpression of Bmi-1 oncoprotein correlates with axillary lymph node metastases in invasive ductal breast cancer. Breast 13(5): 383–388.
Kim, K.Y., S.N. Yu, et al. 2011. Salinomycin-induced apoptosis of human prostate cancer cells due to accumulated reactive oxygen species and mitochondrial membrane depolarization. Biochem Biophys Res Commun 413(1): 80–86.
King, T.D., W. Zhang, et al. 2011. Frizzled7 as an emerging target for cancer therapy. Cell Signal.
Knowlden, J.M., I.R. Hutcheson, et al. 2003. Elevated levels of epidermal growth factor receptor/c-erbB2 heterodimers mediate an autocrine growth regulatory pathway in tamoxifen-resistant MCF-7 cells. Endocrinology 144(3): 1032–1044.
Korach, K.S., J.F. Couse, et al. 1996. Estrogen receptor gene disruption: molecular characterization and experimental and clinical phenotypes. Recent Prog Horm Res 51: 159–186; discussion 186–158.
Korkaya, H., A. Paulson, et al. 2008. HER2 regulates the mammary stem/progenitor cell population driving tumorigenesis and invasion. Oncogene 27(47): 6120–6130.
Kostenuik, P.J. 2005. Osteoprotegerin and RANKL regulate bone resorption, density, geometry and strength. Curr Opin Pharmacol 5(6): 618–625.
Kyndi, M., F.B. Sorensen, et al. 2008. Estrogen receptor, progesterone receptor, HER-2, and response to postmastectomy radiotherapy in high-risk breast cancer: the Danish Breast Cancer Cooperative Group. J Clin Oncol 26(9): 1419–1426.
Lauth, M., A. Bergstrom, et al. 2007. Inhibition of GLI-mediated transcription and tumor cell growth by small-molecule antagonists. Proc Natl Acad Sci USA 104(20): 8455–8460.
Leary, A.F., S. Drury, et al. 2010. Lapatinib restores hormone sensitivity with differential effects on estrogen receptor signaling in cell models of human epidermal growth factor receptor 2-negative breast cancer with acquired endocrine resistance. Clin Cancer Res 16(5): 1486–1497.
Lee, C.W., K. Simin, et al. 2008. A functional Notch-survivin gene signature in basal breast cancer. Breast Cancer Res 10(6): R97.
Lewis, M.T., S. Ross, et al. 1999. Defects in mouse mammary gland development caused by conditional haploinsufficiency of Patched-1. Development 126(22): 5181–5193.
Li, N., Y. Zhang, et al. 2005. Beta1 integrins regulate mammary gland proliferation and maintain the integrity of mammary alveoli. EMBO J 24(11): 1942–1953.
Li, X., M.T. Lewis, et al. 2008. Intrinsic resistance of tumorigenic breast cancer cells to chemotherapy. J Natl Cancer Inst 100(9): 672–679.
Lindvall, C., W. Bu, et al. 2007. Wnt signaling, stem cells, and the cellular origin of breast cancer. Stem Cell Rev 3(2): 157–168.
Liu, S., S.G. Clouthier, et al. 2012. Role of microRNAs in the regulation of breast cancer stem cells. J Mammary Gland Biol Neoplasia 17(1): 15–21.

Liu, S., G. Dontu, et al. 2006. Hedgehog signaling and Bmi-1 regulate self-renewal of normal and malignant human mammary stem cells. Cancer Res 66(12): 6063–6071.

Lobry, C., P. Oh, et al. 2011. Oncogenic and tumor suppressor functions of Notch in cancer: it's NOTCH what you think. J Exp Med 208(10): 1931–1935.

Lowery, A.J., N. Miller, et al. 2008. MicroRNAs as prognostic indicators and therapeutic targets: potential effect on breast cancer management. Clin Cancer Res 14(2): 360–365.

Lu, D., M.Y. Choi, et al. 2011. Salinomycin inhibits Wnt signaling and selectively induces apoptosis in chronic lymphocytic leukemia cells. Proc Natl Acad Sci USA 108(32): 13253–13257.

Magnifico, A., L. Albano, et al. 2009. Tumor-initiating cells of HER2-positive carcinoma cell lines express the highest oncoprotein levels and are sensitive to trastuzumab. Clin Cancer Res 15(6): 2010–2021.

Mallepell, S., A. Krust, et al. 2006. Paracrine signaling through the epithelial estrogen receptor alpha is required for proliferation and morphogenesis in the mammary gland. Proc Natl Acad Sci USA 103(7): 2196–2201.

Massarweh, S., C.K. Osborne, et al. 2006. Mechanisms of tumor regression and resistance to estrogen deprivation and fulvestrant in a model of estrogen receptor-positive, HER-2/neu-positive breast cancer. Cancer Res 66(16): 8266–8273.

Matsuda, Y., T. Schlange, et al. 2009. WNT signaling enhances breast cancer cell motility and blockade of the WNT pathway by sFRP1 suppresses MDA-MB-231 xenograft growth. Breast Cancer Res 11(3): R32.

Merchant, A.A., and W. Matsui. 2010. Targeting Hedgehog--a cancer stem cell pathway. Clin Cancer Res 16(12): 3130–3140.

Molofsky, A.V., R. Pardal, et al. 2003. Bmi-1 dependence distinguishes neural stem cell self-renewal from progenitor proliferation. Nature 425(6961): 962–967.

Moraes, R.C., X. Zhang, et al. 2007. Constitutive activation of smoothened (SMO) in mammary glands of transgenic mice leads to increased proliferation, altered differentiation and ductal dysplasia. Development 134(6): 1231–1242.

Muller, A., B. Homey, et al. 2001. Involvement of chemokine receptors in breast cancer metastasis. Nature 410(6824): 50–56.

Nalwoga, H., J.B. Arnes, et al. 2010. Expression of aldehyde dehydrogenase 1 (ALDH1) is associated with basal-like markers and features of aggressive tumours in African breast cancer. Br J Cancer 102(2): 369–375.

Naylor, M.J., S.R. Oakes, et al. 2005. Transcriptional changes underlying the secretory activation phase of mammary gland development. Mol Endocrinol 19(7): 1868–1883.

Nguyen, P.L., A.G. Taghian, et al. 2008. Breast cancer subtype approximated by estrogen receptor, progesterone receptor, and HER-2 is associated with local and distant recurrence after breast-conserving therapy. J Clin Oncol 26(14): 2373–2378.

Osipo, C., P. Patel, et al. 2008. ErbB-2 inhibition activates Notch-1 and sensitizes breast cancer cells to a gamma-secretase inhibitor. Oncogene 27(37): 5019–5032.

Pal, A., K.E. Valdez, et al. 2010. Targeting the perpetrator: breast cancer stem cell therapeutics. Curr Drug Targets 11(9): 1147–1156.

Pancholi, S., A.E. Lykkesfeldt, et al. 2008. ERBB2 influences the subcellular localization of the estrogen receptor in tamoxifen-resistant MCF-7 cells leading to the activation of AKT and RPS6KA2. Endocr Relat Cancer 15(4): 985–1002.

Park, I.K., S.J. Morrison, et al. 2004. Bmi1, stem cells, and senescence regulation. J Clin Invest 113(2): 175–179.

Park, I.K., D. Qian, et al. 2003. Bmi-1 is required for maintenance of adult self-renewing haematopoietic stem cells. Nature 423(6937): 302–305.

Park, S.M., A.B. Gaur, et al. 2008. The miR-200 family determines the epithelial phenotype of cancer cells by targeting the E-cadherin repressors ZEB1 and ZEB2. Genes Dev 22(7): 894–907.

Perou, C.M., T. Sorlie, et al. 2000. Molecular portraits of human breast tumours. Nature 406(6797): 747–752.

Phillips, T.M., W.H. McBride, et al. 2006. The response of CD24(–/low)/CD44+ breast cancer-initiating cells to radiation. J Natl Cancer Inst 98(24): 1777–1785.

Pietersen, A.M., B. Evers, et al. 2008. Bmi1 regulates stem cells and proliferation and differentiation of committed cells in mammary epithelium. Curr Biol 18(14): 1094–1099.

Pirollo, K.F., A. Rait, et al. 2007. Materializing the potential of small interfering RNA via a tumor-targeting nanodelivery system. Cancer Res 67(7): 2938–2943.

Pirollo, K.F., G. Zon, et al. 2006. Tumor-targeting nanoimmunoliposome complex for short interfering RNA delivery. Hum Gene Ther 17(1): 117–124.

Rades, D., T. Veninga, et al. 2006. Prognostic factors predicting functional outcomes, recurrence-free survival, and overall survival after radiotherapy for metastatic spinal cord compression in breast cancer patients. Int J Radiat Oncol Biol Phys 64(1): 182–188.

Reya, T., S.J. Morrison, et al. 2001. Stem cells, cancer, and cancer stem cells. Nature 414(6859): 105–111.

Rieger, M.E., A.H. Sims, et al. 2010. The embryonic transcription cofactor LBH is a direct target of the Wnt signaling pathway in epithelial development and in aggressive basal subtype breast cancers. Mol Cell Biol 30(17): 4267–4279.

Rizzo, P., H. Miao, et al. 2008. Cross-talk between notch and the estrogen receptor in breast cancer suggests novel therapeutic approaches. Cancer Res 68(13): 5226–5235.

Roarty, K., and J.M. Rosen. 2010. Wnt and mammary stem cells: hormones cannot fly wingless. Curr Opin Pharmacol 10(6): 643–649.

Rudin, C.M., C.L. Hann, et al. 2009. Treatment of medulloblastoma with hedgehog pathway inhibitor GDC-0449. N Engl J Med 361(12): 1173–1178.

Russo, J., X. Ao, et al. 1999. Pattern of distribution of cells positive for estrogen receptor alpha and progesterone receptor in relation to proliferating cells in the mammary gland. Breast Cancer Res Treat 53(3): 217–227.

Santini, D., G. Schiavon, et al. 2011. Receptor activator of NF-kB (RANK) expression in primary tumors associates with bone metastasis occurrence in breast cancer patients. PLoS One 6(4): e19234.

Sarwar, N., J.S. Kim, et al. 2006. Phosphorylation of ERalpha at serine 118 in primary breast cancer and in tamoxifen-resistant tumours is indicative of a complex role for ERalpha phosphorylation in breast cancer progression. Endocr Relat Cancer 13(3): 851–861.

Scales, S.J., and F.J. de Sauvage. 2009. Mechanisms of Hedgehog pathway activation in cancer and implications for therapy. Trends Pharmacol Sci 30(6): 303–312.

Schramek, D., A. Leibbrandt, et al. 2010. Osteoclast differentiation factor RANKL controls development of progestin-driven mammary cancer. Nature 468(7320): 98–102.

Shafee, N., C.R. Smith, et al. 2008. Cancer stem cells contribute to cisplatin resistance in Brca1/p53-mediated mouse mammary tumors. Cancer Res 68(9): 3243–3250.

Shimono, Y., M. Zabala, et al. 2009. Downregulation of miRNA-200c links breast cancer stem cells with normal stem cells. Cell 138(3): 592–603.

Shtutman, M., J. Zhurinsky, et al. 1999. The cyclin D1 gene is a target of the beta-catenin/LEF-1 pathway. Proc Natl Acad Sci USA 96(10): 5522–5527.

Simon, J.A., and R.E. Kingston. 2009. Mechanisms of polycomb gene silencing: knowns and unknowns. Nat Rev Mol Cell Biol 10(10): 697–708.

Sims-Mourtada, J., J.G. Izzo, et al. 2007. Sonic Hedgehog promotes multiple drug resistance by regulation of drug transport. Oncogene 26(38): 5674–5679.

Smith, M.C., K.E. Luker, et al. 2004. CXCR4 regulates growth of both primary and metastatic breast cancer. Cancer Res 64(23): 8604–8612.

Sorlie, T., C.M. Perou, et al. 2001. Gene expression patterns of breast carcinomas distinguish tumor subclasses with clinical implications. Proc Natl Acad Sci USA 98(19): 10869–10874.

Sparmann, A., and M. van Lohuizen. 2006. Polycomb silencers control cell fate, development and cancer. Nat Rev Cancer 6(11): 846–856.

Stecca, B., C. Mas, et al. 2007. Melanomas require HEDGEHOG-GLI signaling regulated by interactions between GLI1 and the RAS-MEK/AKT pathways. Proc Natl Acad Sci USA 104(14): 5895–5900.

Tremblay, M.R., M. Nevalainen, et al. 2008. Semisynthetic cyclopamine analogues as potent and orally bioavailable hedgehog pathway antagonists. J Med Chem 51(21): 6646–6649.

van der Linden, Y.M., S.P. Dijkstra, et al. 2005. Prediction of survival in patients with metastases in the spinal column: results based on a randomized trial of radiotherapy. Cancer 103(2): 320–328.

van der Lugt, N.M., J. Domen, et al. 1994. Posterior transformation, neurological abnormalities, and severe hematopoietic defects in mice with a targeted deletion of the bmi-1 proto-oncogene. Genes Dev 8(7): 757–769.

Von Hoff, D.D., P.M. LoRusso, et al. 2009. Inhibition of the hedgehog pathway in advanced basal-cell carcinoma. N Engl J Med 361(12): 1164–1172.

Wadasadawala, T., S. Gupta, et al. 2007. Brain metastases from breast cancer: management approach. J Cancer Res Ther 3(3): 157–165.

Wels, J., R.N. Kaplan, et al. 2008. Migratory neighbors and distant invaders: tumor-associated niche cells. Genes Dev 22(5): 559–574.

Williams, J.A., O.M. Guicherit, et al. 2003. Identification of a small molecule inhibitor of the hedgehog signaling pathway: effects on basal cell carcinoma-like lesions. Proc Natl Acad Sci USA 100(8): 4616–4621.

Woodward, W.A., M.S. Chen, et al. 2007. WNT/beta-catenin mediates radiation resistance of mouse mammary progenitor cells. Proc Natl Acad Sci USA 104(2): 618–623.

Woodward, W.A., E.A. Strom, et al. 2003. Locoregional recurrence after doxorubicin-based chemotherapy and postmastectomy: Implications for breast cancer patients with early-stage disease and predictors for recurrence after postmastectomy radiation. Int J Radiat Oncol Biol Phys 57(2): 336–344.

Wu, J., D. Hu, et al. 2011. Down-regulation of BMI-1 cooperates with artemisinin on growth inhibition of nasopharyngeal carcinoma cells. J Cell Biochem 112(7): 1938–1948.

Xie, D., K. Nakachi, et al. 2001. Elevated levels of connective tissue growth factor, WISP-1, and CYR61 in primary breast cancers associated with more advanced features. Cancer Res 61(24): 8917–8923.

Xu, L., R.B. Corcoran, et al. 2000. WISP-1 is a Wnt-1- and beta-catenin-responsive oncogene. Genes Dev 14(5): 585–595.

Yang, L., X. Wu, et al. 2011. FZD7 has a critical role in cell proliferation in triple negative breast cancer. Oncogene 30(43): 4437–4446.

Yin, H., and J. Glass. 2011. The phenotypic radiation resistance of CD44+/CD24(-or low) breast cancer cells is mediated through the enhanced activation of ATM signaling. PLoS One 6(9): e24080.

Yu, F., H. Deng, et al. 2010. Mir-30 reduction maintains self-renewal and inhibits apoptosis in breast tumor-initiating cells. Oncogene 29(29): 4194–4204.

Yu, F., H. Yao, et al. 2007. let-7 regulates self renewal and tumorigenicity of breast cancer cells. Cell 131(6): 1109–1123.

Chapter 11

Targeting Leukemic Stem Cells by Breaking Their Dormancy

Stephan Wurzer[1,a,2] and Marieke A.G. Essers[1,b,2,*]

Hematopoietic Stem Cells in Mouse and Human

Hematopoietic stem cells in the mouse

The life-long maintenance of adult self-renewing tissues, including the hematopoietic system, the skin and the intestinal epithelium is dependent on adult stem cells. These cells have the intrinsic capability to self-renew, therefore producing cells of the same kind, as well as to differentiate, giving rise to more committed progenitor and differentiated cells to maintain their host tissue. Many terminally differentiated cells of the hematopoietic system, such as granulocytes, have rapid turnover rates and therefore need to be constantly replaced during homeostasis. Additionally, during hematopoietic stress conditions such as bleeding or toxic insult, the demand for production of new blood cells is further increased. In order to maintain the integrity of the system also under such conditions, hematopoietic stem and progenitor cells have to be able to sense such injuries and rapidly produce new blood cells.

The murine hematopoietic stem cell (HSC) is to date one of the best characterized stem cells and serves as a model for other tissue stem cells.

[1]Deutsches Krebsforschungszentrum (DKFZ), Im Neuenheimer Feld 280, 69120 Heidelberg, Germany.
[a]E-mail: s.wurzer@dkfz.de
[b]E-mail: m.essers@dkfz.de
[2]HI-STEM—Heidelberg Institute for Stem Cell Technologies and Experimental Medicine, Im Neuenheimer Feld 280, 69120 Heidelberg, Germany.
*Corresponding author

During homeostasis in adult mice, the majority of HSCs reside in the bone marrow microenvironment, where they are protected from pathogens and toxic stress (Fig. 1).

In the previous decades, a combination of functional assays together with flow cytometric analysis has allowed the isolation of very pure fractions of hematopoietic stem cells (Weissman and Shizuru 2008). The most important assay to determine HSC function is the capability of a HSC to reconstitute the bone marrow of lethally irradiated hosts. In this assay, recipient mice are lethally irradiated with a dose that kills the hematopoietic cells of the recipient mouse, including the HSCs, and is afterwards transplanted with cells from a donor mouse. Transplanted HSCs are able to rescue the host and to give long time engraftment, showing the capability of these cells to self-renew. In contrast, transplantation of progenitor cells leads only to

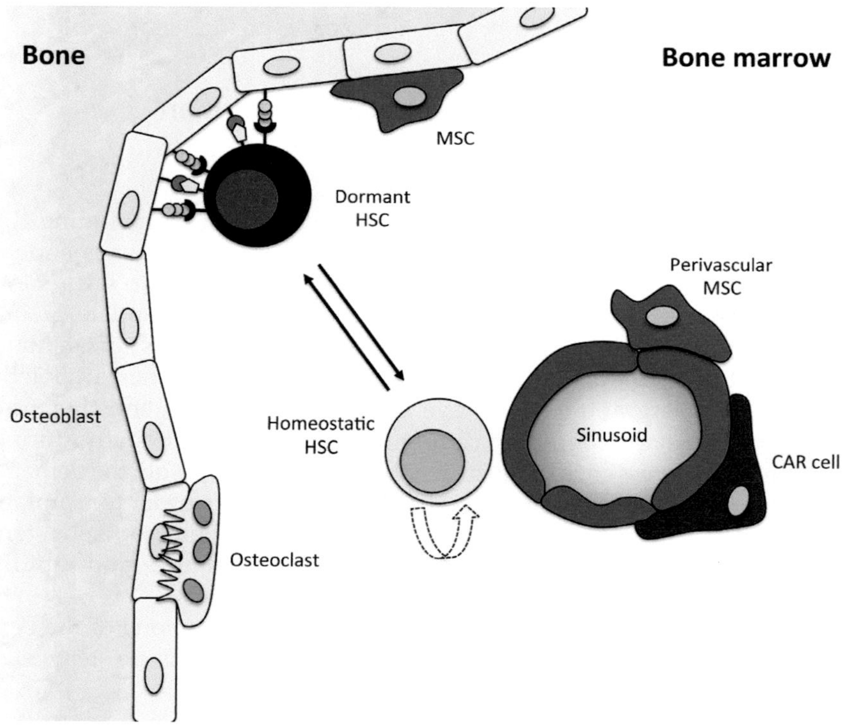

Fig. 1. A model for dormant and homeostatic hematopoietic stem cells in the bone marrow. Hematopoietic stem cells reside in the bone marrow in close proximity to the osteoblasts. Whereas dormant HSCs are in an osteoblastic niche composed of mainly osteoblasts and MSCs, homeostatic HSCs are within perivascular niches closer to sinusoids, CAR cells and perivascular MSCs.

Color image of this figure appears in the color plate section at the end of the book.

short term reconstitution of the system. Furthermore HSCs can produce all of the different lineages of the hematopoietic system, showing their multipotency.

By combining transplantation assays with fluorescence activated cell sorting (FACS) of potential HSCs in lethally irradiated mice, HSCs were initially found to reside in the lineage^{neg} Sca-1^{+} c-Kit^{+} (LSK) population of cells in the bone marrow (Uchida et al. 1994). These cells do not express any surface markers of fully differentiated cells (e.g., CD3, 4 or 8 expressed by lymphocytes or CD11b or Gr-1 expressed by monocytes) but do express c-Kit as well as Sca-1 on their surface. This definition of HSCs was further refined by the addition of newly identified markers, enhancing the purity of long-term reconstituting stem cells. More than 20% of LSK CD34 negative cells are able to give long-term multi-lineage reconstitution in lethally irradiated mice (Osawa et al. 1996). Moreover almost 50% of the signal-lymphocyte-activating molecule (SLAM) CD150 positive and CD48 negative fraction of LSKCD34^{-} cells (LSKCD150^{+}CD48^{-}CD34^{-}) are long-term reconstituting stem cells, as shown in limiting dilution and single cell transplantation assays (Kiel et al. 2005, Wilson et al. 2009). The frequency of this population in the bone marrow is less than one of 10^5 cells of the bone marrow, illustrating the low abundance of stem cells and the steep hierarchy within the hematopoietic system (Wilson et al. 2009).

Dormancy of mouse hematopoietic stem cells

Whereas HSCs were thought to divide slowly, their detailed cycling behavior in time has not been known until recently. To investigate the frequency of divisions of quiescent HSCs over time, label retaining assays were performed, similar to those experiments performed to identify stem cells from the intestine or the skin (Fuchs 2009). In these assays stem cells are labeled either chemically or genetically. In the following "chase" period, the label is effectively halved with every cell division, thus diluting the label in proliferating cells after four to five divisions, with only those cells that do not actively divide keeping the label. These cells are hence called label-retaining cells (Fig. 2).

To perform these LRC experiments for HSCs mice were chemically labeled using a BrdU pulse for 2 weeks, followed by a BrdU free chase of up to 300 days (Wilson et al. 2008). Interestingly only a small proportion of the LSKCD150^{+}CD48^{-}CD34^{-} HSCs retained the label over long periods of time. Mathematical modeling of the decay of the label revealed that within this stem cell compartment there are two cell populations with different cycling behaviors. Although all HSCs are quiescent, the majority of HSCs, which comprises about 85% of the cells, divides more frequently, while another population comprising about 15% of HSCs divides only around

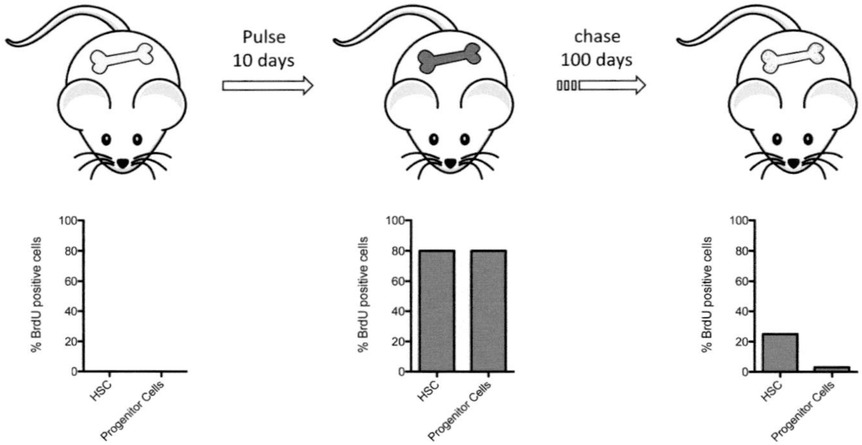

Fig. 2. Long term label retaining (LRC) assay for HSCs. HSCs are labeled *in vivo* with the thymidine analogue bromodeoxyuridine (BrdU) to label all HSCs and progenitor cells. After 10 days BrdU incorporation is stopped by removing the BrdU source, followed by a subsequent chase period of on average 100 days (70–300 days). During the chase period dividing cells gradually loose the BrdU label with every cell division. After more than four divisions the BrdU label is diluted out and no longer detectable by FACS. Thus HSCs that have divided 4 times or less in the chase period of 100 days will still have a detectable BrdU label and are thus called label retaining cells (LRCs).

once every 150 days under homeostatic conditions (Wilson et al. 2008). So while the majority of HSCs—the homeostatic HSCs—are quiescent but still slowly cycling to maintain the hematopoietic system, a small fraction—the dormant HSCs—are almost permanently in the G_0 phase of the cell cycle, dividing less than once in 5 months, or only about 5 times in the lifetime of a mouse (Fig. 1) (Foudi et al. 2009, Wilson et al. 2008). Similar results were obtained using a genetic mouse model, where the expression of histone H2B coupled to GFP (H2B-GFP) is chemically inducible (Foudi et al. 2009, Wilson et al. 2008). Interestingly, transplantation experiments using the H2B-GFP mouse model showed that the label retaining dormant subpopulation of LSKCD150$^+$CD48$^-$CD34$^-$ HSCs contained the cells with the highest long-term multi-lineage stem cell activity (Foudi et al. 2009, Wilson et al. 2008). Thus the most dormant HSCs are also the most potent ones, linking dormancy to higher self-renewal capacity. Moreover, their metabolic activity is highly reduced and their replication machinery inactive (Trumpp et al. 2010).

The bone marrow stem cell niche

The exact localization of dormant and homeostatic HSC within the bone marrow is still a highly controversial topic. HSCs are preferentially found in the trabecular cavities of long bones and putative HSCs have been identified both at the endosteum, the region located in immediate proximity to bone lining osteoblasts, as well as close to sinusoidal endothelium (Ehninger and Trumpp 2011). While some sinusoids are found close to the endosteum, the majority are located at a greater distance to the endosteal surface, suggesting the existence of two different HSC niches in the bone marrow; an endosteal niche and a perivascular niche (Fig. 1). Stainings on bone sections suggest that HSCs reside in the bone marrow as individual stem cells. They are connected to the surrounding stromal cells, especially osteoblasts, by a network of adhesion molecules including N-cadherin and integrins. Another important component of the niche besides direct cell–cell contact with stromal cells is the production of cytokines and chemokines such as CXCL12, stem cell factor (SCF), TPO or Ang-1 by the stromal cells of the niche. Together, these interactions with the niche environment are crucial for the maintenance of HSCs as well as keeping them quiescent and in a metabolically inactive state (Wilson et al. 2009). In addition to endosteal osteoblasts the presence of other cells, including nestin-expressing mesenchymal stem cells (MSCs) as well as CXCL12 abundant reticular (CAR) cells, is known to be crucial for HSC maintenance. Nestin^{+} MSCs are mainly found around blood vessels, and in lower abundance close to the endosteum. CAR cells, which are mesenchymal progenitor cells, are more abundant and are mainly located in the bone marrow itself. Depletion of either of these cells impairs HSC maintenance. Due to the many similarities of these two populations it is possible that they are, at least partly, overlapping cell types (Ehninger and Trumpp 2011).

In conclusion it is possible that two highly similar HSC niches exist in the bone marrow. The endosteal niche, harboring dormant HSC close to osteoblasts and supporting dormancy by secreted factors as well as direct cell to cell contact, and the perivascular niche, harboring homeostatic HSCs close to sinusoidal endothelial cells and CAR cells, promoting HSC self-renewal (Fig. 1) (Ehninger and Trumpp 2011).

Stress to the hematopoietic system including bleeding, infections, inflammation, or chemotherapeutic substances lead to loss of hematopoietic cells, especially of highly cycling populations like progenitor cells. As a feedback response to such an insult, dormant HSCs enter an active cell cycle within hours after the injury, leading to the production of actively dividing progenitor cells, which in turn replace the cells lost during the insult to the system. Dormant HSCs seem to have only a minor contribution to hematopoiesis during homeostasis, where they preserve their self-renewal capacity by avoiding cell divisions, but serve as an important reserve pool for emergency situations. Additionally their dormancy and very low metabolic profile offers them protection from various toxic substances.

Treatment of mice with the chemotherapeutic agent 5-Fluoro-Uracil (5-FU) efficiently kills proliferating cells, including highly cycling progenitor cells, but does not harm the dormant HSC compartment. This myeloid depletion leads to rapid entry of dormant HSCs into the cell cycle and recovery of the hematopoietic system. Human HSCs seem to be regulated in a similar way, as 5-FU chemotherapy also leads to rapid myelo-ablation, but fast recovery after cessation of the treatment (Essers and Trumpp 2010). Recently related studies in the skin and the intestine also showed that these compartments harbor actively cycling as well as dormant stem cell populations, suggesting that dormancy and activation of stem cells may also occur outside of the hematopoietic system (Fuchs 2009, Tian et al. 2011).

Human hematopoietic stem cells

Human HSCs reside mainly in the marrow of long bones and the hipbone, similar to murine HSCs. The cell surface phenotype of human HSCs has been investigated primarily using *in vitro* colony formation assays as well as *in vivo* xeno-transplantation assays (Weissman and Shizuru 2008). The use of immune-compromised mice in the transplantation of human cells into a murine host is crucial in order to avoid rejection of transplanted cells due to an immune response of the host immune system towards the graft. Frequently used mouse strains include NOD/SCID mice, which lack T-cells and B-cells, and more recently the highly immune compromised NOD/SCID common gamma (NSG) mice. NSG mice lack functional natural killer cells in addition to T-cells and B-cells, as well as multiple signaling pathways that are important in innate immunity, allowing the development of functional human hemato-lymphopoiesis in these mice (Ishikawa et al. 2005). Transplantations of human bone marrow cells into sub-lethally irradiated NOD/SCID mice showed that the lineage negative, CD34 positive and CD38 negative population is highly enriched for hematopoietic stem cells with long term reconstitution capability (Bhatia et al. 1997). The frequency of stem cells in the $lin^{neg}CD34^{+}CD38^{-}$ population of human bone marrow is still much lower in comparison to the purity achievable for murine HSCs using the markers described above. Therefore, the search for additional markers to further purify human HSCs is ongoing. Together with CD34, the cell surface protein Thy1 (CD90) has been shown to enrich for human stem cells *in vitro* as well as in transplantation experiments (Weissman and Shizuru 2008). Moreover, alpha-6-integrin (CD49f) has been recently shown to further enrich the purity of $lin^{neg}CD34^{+}CD38^{-}$ HSC irrespective of their expression of Thy1 (Notta et al. 2011).

Although murine HSCs are in general quiescent, they contain a sub-population, which is even more dormant during homeostasis. Whether this is also the case for human HSCs is difficult to address, as the label retention assays carried out in mice cannot be used in humans. Hence xeno-transplanted mice, which resemble the situation of human HSCs in the bone marrow as close as possible, were used to investigate the cycling behavior and dormancy of human HSCs. In these xenografts human HSCs are more actively cycling then mouse HSCs, with almost no cells reaching a state of long-term quiescence [unpublished data]. However the lack of dormancy in human HSCs could well be inherent to the xenograft-assay, in which human HSCs are put into a murine bone marrow niche. As interaction with and secretion of proteins from stromal cells is crucial for the maintenance of dormancy, it cannot be ruled out that some crucial factors for human HSC quiescence are missing in this setting. Further experiments in mice in which the bone marrow niche is humanized, for example by using human mesenchymal stromal cells, will shed further light on this question.

Leukemic Stem Cells in Humans

The concept of a cancer stem cell

Research in recent years has suggested that not only are normal self-renewing tissues build up in a hierarchical way, containing a population of stem cells responsible for the life-long maintenance of the tissue, but that also different types of cancer show a small population of cells with stem cell properties, the so-called cancer stem cells. The first evidence for the concept of a cancer stem cell came from acute myeloid leukemia (AML) (Wang and Dick 2005). Transplantation experiments with different sub-populations of AML cells showed that only a fraction of the AML cells was capable of transplanting the disease in immuno-compromised mice. These cells were called leukemia initiating cells. In contrast, transplantation of cells from the bulk of the leukemia containing the more differentiated leukemic progeny did not lead to development of AML. Since these initial observations in AML, the concept of a cancer stem cell population has been tested and proven in several other forms of leukemia, for example chronic myeloid leukemia (CML), as well as many other types of solid cancers like breast cancer, colon cancer, prostate cancer and brain tumors (Cho and Clarke 2008, Reya et al. 2001, Trumpp and Wiestler 2008). These observations suggest that many tumors show a hierarchical organization with a cancer stem cell population (CSCs) at the top of the hierarchy. CSCs are the only cells in the tumor with tumor propagation potential. While these cells are genetically identically to the bulk of the tumor, they have gained additional stem cell like characteristics, which not only gives them the potential to propagate, but also making them highly resistant to radiotherapy and anti-proliferative chemotherapy. Thus, even though these therapies dramatically reduce the size of a tumor by eliminating the bulk of the tumor cells, CSCs are often left behind because of their resistance to these therapies (Fig. 3).

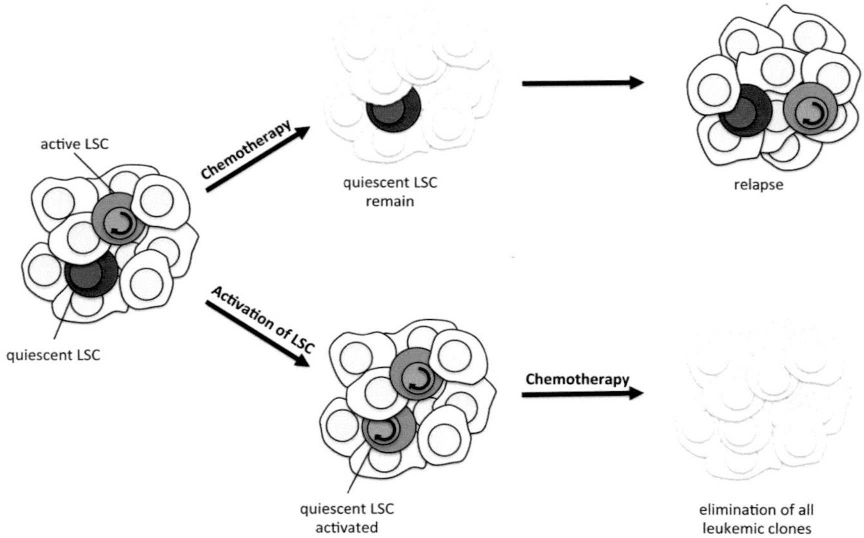

Fig. 3. A combined treatment strategy to eliminate leukemic stem cells (LSCs). LSCs are highly resistant to anti-proliferative chemotherapy. This is thought to be the main reason for relapse in leukemia. Similar to HSCs, one of the reasons for their resistance might be the state of dormancy these LSCs are in. Therefore by specifically targeting the dormancy of LSCs in the first step of the treatment, getting these cells into an active stage of the cell cycle, one might be able to eliminate the cycling LSCs in a second step by the classical anti-proliferative chemotherapy.

Therefore CSCs are thought to be the main cause for relapse and metastasis formation following an initial successful therapy.

Even though the existence of CSCs is now well accepted in the cancer field, the identification and characterization of CSCs is still in the early stages of development. Depending on tumor type and mutations present in the tumor, there is a great variation in the expression of markers expressed specifically on CSCs. Thus, for many cancers, no reliable markers have yet been identified to identify the CSC. It also not yet sure whether the expression of stem cell markers in a tumor is stable over time and during progression of the disease, making the search for reliable CSC markers more difficult. In addition, for most CSCs there is no reliable *in vivo* experimental set-up available to test the stem cell function of a CSC. Furthermore, not all types of cancer show a hierarchical organization. It has recently been suggested that in tumors progressing from a more benign to a more malignant and aggressive stage, the hierarchy within the different cell types of the tumor is flattening, with more differentiated tumor cells starting to show tumor initiating cell properties as well (Kelly et al. 2007, Passegue et al. 2009, Quintana et al. 2008). However, due to the highly similar functional properties of normal tissue stem cells and cancer stem cells it will be of great interest to see whether these similar properties are the reason for both populations of stem cells to be resistant to current therapies.

Resistance of stem cells to therapy

For both, tissue stem cells as well as cancer stem cells, a number of reasons have been suggested why these cells are resistant to classical cancer therapies like irradiation and chemotherapy. Most of these mechanisms are linked to general characteristics of stem cells. For example, stem cells express high levels of various ABC transporter pumps that export small molecules out of the cell. These small molecules also include many of the drugs used in chemotherapy, indicating that these drugs might not reach critical concentrations within the stem cells to fulfill their mode of action against these cells. Moreover, stem cells as well as CSCs are believed to reside in a hypoxic niche environment. These low oxygen levels prevent efficient induction of DNA damage upon radiotherapy and may also lower the exposure of stem cells to cancer drugs. Furthermore, it is believed that stem cells have different, more effective DNA damage response pathways in place to deal with DNA damage induced by injuries or therapy attacks, compared to their more differentiated progeny.

Finally, another important characteristic of normal stem cells that is suggested to play an important role in the resistance of stem cells against chemotherapy is quiescence. Most stem cells are believed to reside in their niche environment in a state of long-term quiescence, called dormancy. This state is thought to encourage stem cell resistance to anti-proliferative drugs, usually designed to attack highly proliferating cells. Furthermore, this state of quiescence is accompanied by a low metabolic activity in the cells and thus a reduced need for oxygen in the stem cells. Therefore the inhibition of signaling pathways or molecules involved in these processes by cancer drugs may not be effective, as these stem cells do not depend on them for their survival (Essers and Trumpp 2010).

In summary, these characteristics help stem cells to survive over extensive periods of time in the niche, thus remaining in the most desirable state for tissue maintenance over the whole lifetime of the animal. However, at the same time they also give cancer stem cells the same properties to survive over longer periods of time even under the stress conditions of different cytotoxic therapies. Figuring out more about the therapy resistance mechanisms of cancer stem cells may help to develop new treatment strategies to tackle the whole tumor, including the CSCs, thus preventing disease relapse and metastatic outgrowth (Fig. 3).

Leukemic stem cells and their resistance mechanisms

Several hematological malignancies have been shown to contain a small sub-population of cells with stem cell properties, often with a similar phenotype as the normal HSCs. These leukemic stem cells (LSCs) have been shown to be highly resistant to current therapies and thus often form a major hurdle for the eradication of the leukemia. Acute myeloid leukemia (AML) was the first leukemia for which the existence of such LSCs was shown. Transplantation of $CD34^{+}CD38^{-}$ enriched bone marrow cells from newly diagnosed patients was able to transplant the AML disease in immunocompromised NOD-SCID mice (Fig. 4) (Bonnet and Dick 1997).

Until the recent development and use of more sensitive transplantation models such as the NSG mice, it was accepted that the $CD34^{+}CD38^{-}$ cells were the only sub-fraction of cells in the bone marrow of AML patients containing LSCs. Several groups have now shown that in contrast to normal hematopoietic cells, some AML patient derived $CD34^{+}CD38^{+}$ cells contain stem cell activity, demonstrating the as yet unelucidated heterogeneity of marker expression on LSCs (Taussig et al. 2008). To gain more insight into similarities and differences between normal and malignant stem cells, it will be crucial to assess the tumor initiating capability of the investigated cell populations in additional to the phenotypic analysis. It is of note that most markers used to identify LSCs in AML are similar to the ones used to

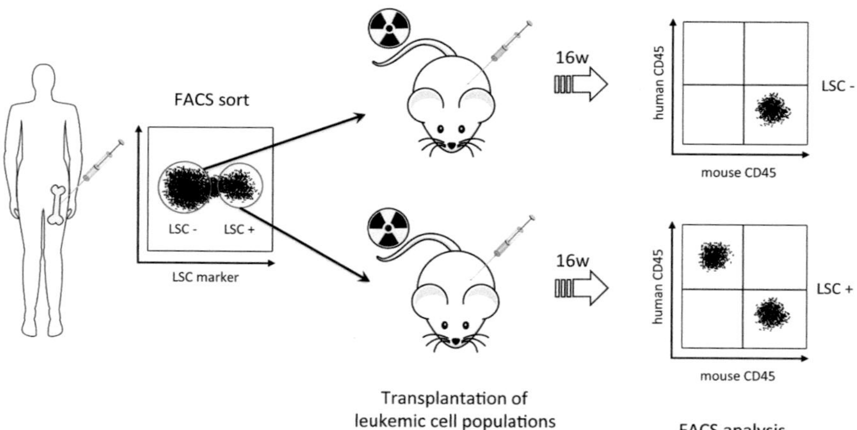

Fig. 4. Analysis of leukemic stem cells in xenograft mouse models. For the generation of leukemic xenografts, $CD34^{+}$ leukemic cells are isolated from bone marrow punctures from leukemia patients, sorted based on specific LSC markers and transplanted into sub-lethally irradiated (250rad), highly immune-deficient NOD/SCID/γ_c^{null} mice. Efficient long-term engraftment can then be analyzed by FACS in peripheral blood or bone marrow at 16 weeks post transplantation.

isolate normal HSCs. An assumption is made that upon isolation of LSCs from newly diagnosed AML patients, most cells present in the BM with a stem cell phenotype will be derived from the leukemic clone, although there may be great inter-patient variation in the amount of LSCs present within this cell fraction. Even though it is not yet clear whether this is true in the patient, experiments in xenografts have shown that AML LSCs are mainly quiescent and this may therefore be one of the reasons for their resistance to chemotherapy.

In the case of chronic myeloid leukemia (CML), treatment outcome in patients has indicated the presence of leukemic stem cells. CML is characterized by a single translocation leading to the formation of the BCR-ABL fusion kinase (Daley et al. 1990). The constitutively active BCR-ABL kinase is the driving force behind this leukemia and is thus an ideal target for drug development (Ren 2005). This led to the development of the tyrosine kinase inhibitor Imatinib (Gleevec) in the late nineties (Druker et al. 2001, Druker et al. 1996). Imatinib very efficiently targets BCR-Abl positive CML cells, leading to a rapid hematologic and cytogenetic response (O'Brien et al. 2003). Nonetheless, after years of treatment with Imatinib, residual leukemic cells remain, which can lead to a relapse of the disease on cessation of treatment (Goldman 2009). The reason for the resistance of CML stem cells is highly debated and several mechanisms have been suggested. CML stem cells have been suggested to be dependent on other autocrine or paracrine signaling through cytokine receptors for their survival, and thus perturbing these pathways might be a way to break their resistance. Alternatively, CML stem cells might express much higher levels of BCR-ABL, making it impossible for Imatinib to block all of these fusion proteins in a cell, though evidence for this explanation remains very thin. A very likely explanation for the resistance of CML stem cells is, as in the case of normal HSCs, dormancy. CML stem cells are thought to be slowly cycling and *in vitro*, treatment with Imatinib drives the cells further into quiescence. Therefore, breaking their dormancy and/or interfering with the Imatinib induced quiescence might be a way to sensitize these cells to chemotherapy. Mathematical models of the effect of drug treatment on CML clones suggested that one explanation for the resistance of CML stem cells might be a selective effect of Imatinib solely on proliferating leukemic cells (Roeder et al. 2006). Furthermore, these models suggest that in CML, sustained treatment might eventually lead to eradication of all leukemic clones, if no Imatinib resistance-mutations are acquired. This would take a substantial, continuous treatment over time, until all leukemic stem cells have entered proliferation as part of normal cycling behavior and thus becoming susceptible to Imatinib. Interestingly, these models further support the hypothesis that cell-cycle-activation of leukemic stem cells in combination with Imatinib treatment might be a way of eliminating the CML stem cells (Roeder et al. 2006).

Attacking Resistance of Leukemic Stem Cells by Breaking Their Dormancy

Dormancy, interactions with the stem cell niche and a low metabolic profile might be important reasons why normal HSCs and LSCs in particular are highly resistant to chemotherapy. Since these two populations of stem cells are quite similar, finding out more about the mechanism of dormancy in normal HSCs may thus help to better understand how to break the dormancy in LSCs. In recent years several methods of activation of HSCs have been shown, suggesting new ways of targeting the leukemic stem cells by combining these activation treatments with known chemotherapies, thus creating a combined therapy approach to eliminate leukemic stem cells. Several of these potential options will be discussed here.

Different ways of breaking dormancy of stem cells

G-CSF

Granulocyte-colony-stimulating-factor (G-CSF) was one of the first agents described to affect quiescent mouse HSCs. Treatment of mice with G-CSF leads to a rapid entry of quiescent HSCs into the cell cycle, followed by mobilization of the cells out of the bone marrow into the blood stream and to the spleen (Morrison et al. 1997). The exact mechanism of G-CSF induced activation of HSCs is not fully understood. However mobilization of the HSCs following the activation is induced by the release of proteolytic enzymes such as elastase, cathepsin G or matrix metalloproteases (MMPs) by neutrophils. These released enzymes cleave several interactions between HSCs and the osteoblasts in the stem cell niche, such as the binding of SDF1 with CXCR4, the interaction of VCAM-1 and VLA-4 or the dimerization of N-Cadherins on both side of the niche. Furthermore, treatment with G-CSF leads to depletion of trophic endosteal macrophages as well as endosteal osteoblasts, which is accompanied by a marked reduction of HSC trophic cytokines. Together, this leads to mobilization of the HSCs (Morrison et al. 1997). Interestingly, once activated and mobilized by G-CSF, mouse HSCs become susceptible to chemotherapeutic treatments such as the anti-proliferative drug 5-FU or cytarabine and can then be efficiently eliminated.

Several studies have been carried out to test the combination of G-CSF induced activation and mobilization and chemotherapy to eradicate LSCs, both in CML as well as AML models. In CML, stem cells in which resistance to the tyrosine kinase inhibitor Imatinib has been correlated with low cell cycle activity, pre-treatment of these cells with G-CSF *in vitro* lead to a cell cycle activation and rendered the cells more susceptible to elimination by Imatinib (Jorgensen et al. 2006). Thus, these experiments suggested that by combining G-CSF and Imatinib, CML stem cells could be eradicated. Unfortunately, the first clinical trial of combined treatment of CML patients with G-CSF and Imatinib did not show efficient elimination of leukemic clones (Drummond et al. 2009). A plausible explanation for the failure of this trial could be timing. Exact timing of the activation by G-CSF and start of the treatment with Imatinib is likely very crucial in this combined treatment approach and thus needs to be optimized, for example using xenograft models, before a final conclusion can be drawn.

Interestingly, in xenograft experiments with AML patient samples, treatment of the mice with G-CSF *in vivo* resulted in the activation of quiescent AML stem cells located in their endosteal bone marrow niches. Furthermore, in combination with the chemotherapeutic agent cytarabine, AML stem cells could be eliminated (Saito et al. 2010). Even though the elimination of the leukemic stem cell pool was not complete, the combined treatment did lead to an increased survival of the mice over time. Of particular interest in this setting was the observation that normal HSCs were less affected by the combination treatment of G-CSF with cytarabine, leaving a therapeutic window, which could be exploited to specifically eliminate LSCs. However, in these experiments the LSC were of human origin whereas the host hematopoietic system was murine. It remains to be determined whether human HSCs are equally more resistant to this combination treatment in comparison to leukemic stem cells.

In conclusion, several data have now shown the activation effect of G-CSF on HSCs as well as LSCs and initial studies in mice have suggested a possible therapeutic window for combined treatment of G-CSF and chemotherapeutic agents for the eradication of LSCs. Nevertheless, efficiency of G-CSF as well as timing of the combined treatment needs to be improved in order to eradicate the whole LSC pool and thus cure the leukemia.

Arsenic trioxide (As_2O_3)

The tumor suppressor PML (promyelocytic leukemia) is localized to PML bodies in cells and plays a role in controlling apoptosis, cellular proliferation and senescence (Ito Bernardi 2009). Inhibition of PML by arsenic trioxide (As_2O_3), which leads to proteosomal degradation of PML, has been a successful treatment in patients with promyelocytic leukemia, a form of leukemia driven by the PML-retinoic acid receptor (RARα) fusion protein. PML is highly expressed in mouse HSCs and loss of PML in mice induces increased cycling of HSCs accompanied by an increase in mTOR activity (Ito et al. 2008). Similar effects can be achieved by inhibiting PML: pharmacological inhibition of PML by Arsenic trioxide (As_2O_3) leading to increased proliferation of human HSCs *in vitro* (Ito et al. 2008).

In a mouse model for CML, the role of PML in leukemic stem cells was tested. Interestingly, loss of PML resulted in a strong increase in cycling CML stem cells. Furthermore, As_2O_3 treatment in combination with the chemotherapeutic drug Ara-C increased the susceptibility of leukemic stem cells to the treatment with Ara-C, leading to their eradication. These effects of combined treatment of As_2O_3 and Ara-C were not only shown in the mouse but also on primary human LSC *in vitro* (Ito et al. 2008), indicating the possibility of a therapeutic window for a combination therapy with induced activation of LSCs by a block of PML. Whether human leukemias, not driven by oncogenic PML, will indeed be sensitive to this strategy will need to be determined in more detail.

Interferons

Interferons are important players in the activation of the immune system in the regulation of resistance to viral infections (Stark et al. 1998). In addition, *in vitro* treatment of many cell lines and primary hematopoietic cells leads to an inhibition of proliferation of these cells. This observation has made Interferons not only attractive as therapy of viral infections, but also an interesting drug for cancer treatments.

Interestingly, it was recently shown that, in contrast to previously published data based on observations made *in vitro*, both IFN-α as well as IFN-γ have the opposite effect on dormant HSCs *in vivo*. Within hours after administration of either one of the Interferons, quiescent HSCs enter an active cell cycle (Fig. 5) (Essers et al. 2009). Even the most dormant HSC population in the bone marrow, the so-called label-retaining HSCs (LRCs) (Wilson et al. 2008), is efficiently activated in response to IFN-α (Essers et al. 2009). Moreover, HSCs from mice deficient for IFN-γ signaling have a decreased baseline proliferation rate (Baldridge et al. 2010).

Activation of HSCs upon IFN-α treatment depends on IFN-α-Receptor-STAT1 signaling and leads to transcription of a set of IFNα target genes (Fig. 5). Furthermore, it is accompanied by and depending on the up-regulation of the cell surface protein Sca-1, as HSCs in mice lacking Sca-1 no longer respond to IFN-α, thus identifying Sca-1 as a key player in the IFNα induced activation of HSCs *in vivo* (Fig. 5). Unlike following G-CSF treatment, IFN activated HSCs remain in the bone marrow and do not enter

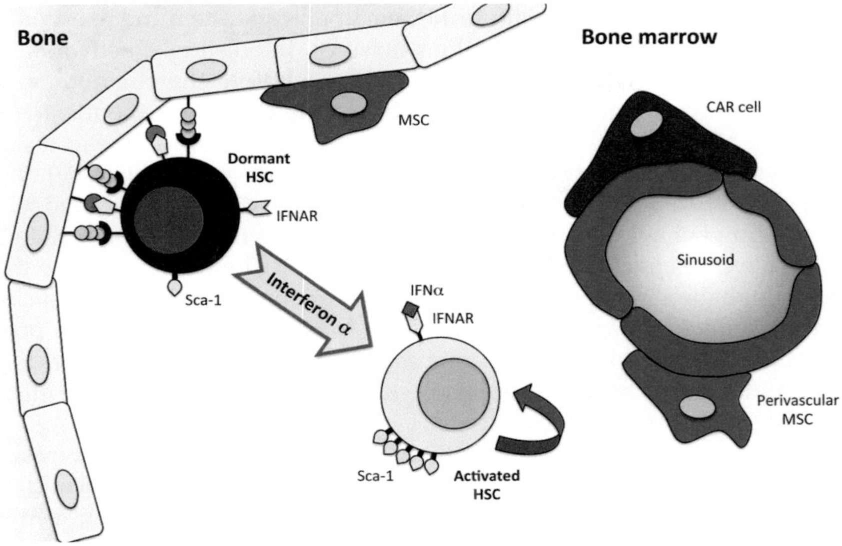

Fig. 5. Interferon-induced activation of dormant HSCs. Treatment of mice with the cytokine IFN-α induces a rapid activation of the dormant HSCs in the BM. The IFN-α treatment leads to increased signaling via the IFNAReceptor-STAT1 signaling pathway, followed by the induction of a set of IFN-α target genes. The effect of IFN-α on HSCs is accompanied by increased expression of the cell surface marker Sca-1, which is required for the IFN-α mediated proliferation of the HSCs.

Color image of this figure appears in the color plate section at the end of the book.

circulation. However it does remain likely that IFNα alters the interaction between HSCs and bone marrow niche cells. This might explain the striking differences between *in vitro* and *in vivo* effects of IFNα on HSCs.

Chronic treatment of mice with IFN-α does not impair the function of their HSCs, since these HSCs still show reconstitution in transplantation experiments. However in a competitive setting these chronically activated HSCs are outcompeted by HSCs that have not been previously activated (Essers et al. 2009). A possible explanation for this is that chronically activated HSCs are outcompeted for dormant niche space and are thus lost. Similar observations have been made with HSCs lacking the Interferon Response Factor-2 (IRF2), a transcriptional repressor of IFN-α signaling (Sato et al. 2009).

Since IFN-α can drive HSCs out of dormancy and dormancy might be one of the main reasons for the chemo-resistance of HSCs a combination therapy of IFN-α with the anti-proliferative chemotherapeutic drug 5-Fluoro-Uracil (5-FU) was tested on normal HSCs in mice. When mice are administered with 5-FU, proliferating cells of the hematopoietic system are efficiently eliminated, but the dormant HSC compartment stays unharmed and can subsequently reconstitute the hematopoietic system and re-establish hematopoiesis, and the mouse thus survives. Interestingly, activation of dormant HSCs, e.g., by priming with IFN-α before the chemotherapy treatment now renders them susceptible to the chemotherapeutic agent. This combined treatment rapidly leads to death of the mice due to severe pan-cytopenia and depletion of the HSC pool (Essers et al. 2009). The susceptibility of HSCs to chemotherapy in this assay highly correlated with the cycling behavior of these cells, making them vulnerable only as long as they were actively cycling (Essers et al. 2009). These data indicate that there might be a therapeutic window for combining proliferation-inducing cytokines like IFN-α with chemotherapy to eliminate the stem cell pool. Therefore, if dormancy in leukemic stem cells is one of the main reasons for chemotherapy resistance, then a combination therapy with IFN-α might benefit the therapy outcome as well.

The concept of treatment with IFN-α in combination with tyrosine kinase inhibitors became of interest in the field of CML. First clinical trials testing a combination of IFN-α with imatinib showed promising results, including higher major molecular response rates, which are indicative of a lower level of residual disease (Simonsson et al. 2011). However whether these can be attributed to the activation of residual leukemic stem cells or other complimentary functions of the drugs is not yet clear. In these trials the drugs were given concurrently and the optimal timing for the combination of the drugs is not yet determined. This is especially crucial as CML cells were recently shown to be desensitized to IFN-α due to down-regulation of the interferon receptor by a BCR-ABL dependent mechanism (Bhattacharya et al. 2011). While immune-modulatory functions of IFN-α on healthy immune cells might not be affected, the effects on leukemic stem cells themselves might thus be highly dependent on the exact timing of treatment cycles. For leukemic stem cells inhibition of the BCR-ABL kinase might have to precede the activation by IFN-α, followed again by treatment with Imatinib to eliminate leukemic stem cell clones. Further trials in animal models might give valuable insights to optimize the timing of this therapy, which might greatly enhance therapeutic outcome with regard to the elimination of residual leukemic clones.

Thoughts for the Future

HSCs and LSCs are two different stem cell populations with great similarities. A combination of their stem cell features such as dormancy, niche interactions and altered metabolism might be important mechanisms of their resistance to chemotherapy and irradiation. Thus breaking their dormancy or influencing their interactions with the surrounding niche suggest new ways of overcoming the resistance of stem cells against the current therapy strategies. Initial observations made mainly in mouse models for HSCs and different leukemias look promising. However several issues still need to be tackled. For example, one major concern in combination therapies is the toxicity on normal HSCs. The different mechanisms of activation described above activate both HSCs as well as LSC, although not always to the same extent. Still, strategies are required that limit the damage on normal HSCs, while they maximize the effect on the leukemic clones. One such strategy is using a therapeutic window, where toxicity is already reached for LSCs while HSCs are either not activated equally efficient, or they are more resistant to the treatment. Alternatively the activation can be combined with drugs, which specifically target leukemic cells, for example Imatinib in the treatment of CML. This approach has the advantage that the toxicity on normal HSCs should be minimal, while residual leukemic stem cells could be efficiently targeted. Another crucial aspect for combination therapies is the timing and dosing of the individual drugs, as well as the sequence of their administration. Leukemic stem cells might only be susceptible to treatments in a narrow time window, as shown for the elimination of normal HSCs by a combination of IFNα and 5-FU treatment. The optimal timing therefore needs to be accurately determined for every drug combination, and could potentially depend even on the individual patient.

Nevertheless, sensitizing leukemic stem cells for elimination by chemotherapy by tackling their dormancy and niche interactions is a promising new route towards the elimination of the complete leukemic clone. Future experiments in both mouse models and clinical trials will further fine-tune these ideas and give rise to optimal strategies for combined treatments, thus leading to a cure of the leukemia.

Acknowledgement

The authors would like to thank Dr Aine Prendergast and Ines Brückmann for critically reading the manuscript. This research was supported by the German Ministry for Education and Research, BMBF-grant on Medical Systems Biology "HaematoSys" (BMBF-FKZ 0315452), and the Dietmar Hopp Foundation.

References

Baldridge, M.T., K.Y. King, N.C. Boles, D.C. Weksberg, and M.A. Goodell. 2010. Quiescent haematopoietic stem cells are activated by IFN-gamma in response to chronic infection. Nature 465: 793–797.

Bhatia, M., J.C. Wang, U. Kapp, D. Bonnet, and J.E. Dick. 1997. Purification of primitive human hematopoietic cells capable of repopulating immune-deficient mice. Proceedings of the National Academy of Sciences of the United States of America 94: 5320–5325.

Bhattacharya, S., H. Zheng, C. Tzimas, M. Carroll, D.P. Baker, and S.Y. Fuchs. 2011. Bcr-abl signals to desensitize chronic myeloid leukemia cells to IFNalpha via accelerating the degradation of its receptor. Blood 118: 4179–4187.

Bonnet, D., and J.E. Dick. 1997. Human acute myeloid leukemia is organized as a hierarchy that originates from a primitive hematopoietic cell. Nature medicine 3: 730–737.

Cho, R.W., and M.F. Clarke, 2008. Recent advances in cancer stem cells. Curr Opin Genet Dev 18: 48–53.

Daley, G.Q., R.A. Van Etten, and D. Baltimore. 1990. Induction of chronic myelogenous leukemia in mice by the P210bcr/abl gene of the Philadelphia chromosome. Science 247: 824–830.

Druker, B.J., C.L. Sawyers, H. Kantarjian, D.J. Resta, S.F. Reese, J.M. Ford, R. Capdeville, and M. Talpaz. 2001. Activity of a specific inhibitor of the BCR-ABL tyrosine kinase in the blast crisis of chronic myeloid leukemia and acute lymphoblastic leukemia with the Philadelphia chromosome. The New England journal of medicine 344: 1038–1042.

Druker, B.J., S. Tamura, E. Buchdunger, S. Ohno, G.M. Segal, S. Fanning, J. Zimmermann, and N.B. Lydon. 1996. Effects of a selective inhibitor of the Abl tyrosine kinase on the growth of Bcr-Abl positive cells. Nature medicine 2: 561–566.

Drummond, M.W., N. Heaney, J. Kaeda, F.E. Nicolini, R.E. Clark, G. Wilson, P. Shepherd, J. Tighe, L. McLintock, T. Hughes, et al. 2009. A pilot study of continuous imatinib vs pulsed imatinib with or without G-CSF in CML patients who have achieved a complete cytogenetic response. Leukemia : official journal of the Leukemia Society of America, Leukemia Research Fund, UK 23: 1199–1201.

Ehninger, A., and A. Trumpp. 2011. The bone marrow stem cell niche grows up: mesenchymal stem cells and macrophages move in. J Exp Med 208: 421–428.

Essers, M.A., S. Offner, W.E. Blanco-Bose, Z. Waibler, U. Kalinke, M.A. Duchosal, and A. Trumpp. 2009. IFNalpha activates dormant haematopoietic stem cells *in vivo*. Nature 458: 904–908.

Essers, M.A., and A. Trumpp. 2010. Targeting leukemic stem cells by breaking their dormancy. Mol Oncol 4: 443–450.

Foudi, A., K. Hochedlinger, D. Van Buren, J.W. Schindler, R. Jaenisch, V. Carey, and H. Hock. 2009. Analysis of histone 2B-GFP retention reveals slowly cycling hematopoietic stem cells. Nat Biotechnol 27: 84–90.

Fuchs, E. 2009. The tortoise and the hair: slow-cycling cells in the stem cell race. Cell 137: 811–819.

Goldman, J.M. 2009. Treatment strategies for CML. Best Pract Res Clin Haematol 22: 303–313.

Ishikawa, F., M. Yasukawa, B. Lyons, S. Yoshida, T. Miyamoto, G. Yoshimoto, T. Watanabe, K. Akashi, L.D. Shultz, and M. Harada. 2005. Development of functional human blood and immune systems in NOD/SCID/IL2 receptor {gamma} chain(null) mice. Blood 106: 1565–1573.

Ito, K., R. Bernardi, A. Morotti, S. Matsuoka, G. Saglio, Y. Ikeda, J. Rosenblatt, D.E. Avigan, J. Teruya-Feldstein, and P.P. Pandolfi. 2008. PML targeting eradicates quiescent leukaemia-initiating cells. Nature 453: 1072–1078.

Jorgensen, H.G., M. Copland, E.K. Allan, X. Jiang, A. Eaves, C. Eaves, and T.L. Holyoake 2006. Intermittent exposure of primitive quiescent chronic myeloid leukemia cells to granulocyte-colony stimulating factor in vitro promotes their elimination by imatinib mesylate. Clinical cancer research : an official journal of the American Association for Cancer Research 12: 626–633.

Kelly, P.N., A. Dakic, J.M. Adams, S.L. Nutt, and A. Strasser. 2007. Tumor growth need not be driven by rare cancer stem cells. Science 317: 337.

Kiel, M.J., O.H. Yilmaz, T. Iwashita, C. Terhorst, and S.J. Morrison. 2005. SLAM family receptors distinguish hematopoietic stem and progenitor cells and reveal endothelial niches for stem cells. Cell 121: 1109–1121.

Morrison, S.J., D.E. Wright, and I.L. Weissman. 1997. Cyclophosphamide/granulocyte colony-stimulating factor induces hematopoietic stem cells to proliferate prior to mobilization. Proceedings of the National Academy of Sciences of the United States of America 94: 1908–1913.

Notta, F., S. Doulatov, E. Laurenti, A. Poeppl, I. Jurisica, and J.E. Dick. 2011. Isolation of single human hematopoietic stem cells capable of long-term multilineage engraftment. Science 333: 218–221.

O'Brien, S.G., F. Guilhot, R.A. Larson, I. Gathmann, M. Baccarani, F. Cervantes, J.J. Cornelissen, T. Fischer, A. Hochhaus, T. Hughes, et al. 2003. Imatinib compared with interferon and low-dose cytarabine for newly diagnosed chronic-phase chronic myeloid leukemia. The New England journal of medicine 348: 994–1004.

Osawa, M., K. Hanada, H. Hamada, and H. Nakauchi. 1996. Long-term lymphohematopoietic reconstitution by a single CD34-low/negative hematopoietic stem cell. Science 273: 242–245.

Passegue, E., S. Rafii, and M. Herlyn. 2009. Cancer stem cells are everywhere. Nature medicine 15: 23.

Quintana, E., M. Shackleton, M.S. Sabel, D.R. Fullen, T.M. Johnson, and S.J. Morrison. 2008. Efficient tumour formation by single human melanoma cells. Nature 456: 593–598.

Ren, R. 2005. Mechanisms of BCR-ABL in the pathogenesis of chronic myelogenous leukaemia. Nature reviews Cancer 5: 172–183.

Reya, T., S.J. Morrison, M.F. Clarke, and I.L. Weissman. 2001. Stem cells, cancer, and cancer stem cells. Nature 414: 105–111.

Roeder, I., M. Horn, I. Glauche, A. Hochhaus, M.C. Mueller, and M. Loeffler. 2006. Dynamic modeling of imatinib-treated chronic myeloid leukemia: functional insights and clinical implications. Nature medicine 12: 1181–1184.

Saito, Y., N. Uchida, S. Tanaka, N. Suzuki, M. Tomizawa-Murasawa, A. Sone, Y. Najima, S. Takagi, Y. Aoki, A. Wake, et al. 2010. Induction of cell cycle entry eliminates human leukemia stem cells in a mouse model of AML. Nat Biotechnol 28: 275–280.

Sato, T., N. Onai, H. Yoshihara, F. Arai, T. Suda, and T. Ohteki. 2009. Interferon regulatory factor-2 protects quiescent hematopoietic stem cells from type I interferon-dependent exhaustion. Nature medicine 15: 696–700.

Simonsson, B., T. Gedde-Dahl, B. Markevarn, K. Remes, J. Stentoft, A. Almqvist, M. Bjoreman, M. Flogegard, P. Koskenvesa, A. Lindblom, et al. 2011. Combination of pegylated IFN-alpha2b with imatinib increases molecular response rates in patients with low- or intermediate-risk chronic myeloid leukemia. Blood 118: 3228–3235.

Stark, G.R., I.M. Kerr, B.R. Williams, R.H. Silverman, and R.D. Schreiber. 1998. How cells respond to interferons. Annual review of biochemistry 67: 227–264.

Taussig, D.C., F. Miraki-Moud, F. Anjos-Afonso, D.J. Pearce, K. Allen, C. Ridler, D. Lillington, H. Oakervee, J. Cavenagh, S.G. Agrawal, et al. 2008. Anti-CD38 antibody-mediated clearance of human repopulating cells masks the heterogeneity of leukemia-initiating cells. Blood 112: 568–575.

Tian, H., B. Biehs, S. Warming, K.G. Leong, L. Rangell, O.D. Klein, and F.J. de Sauvage. 2011. A reserve stem cell population in small intestine renders Lgr5-positive cells dispensable. Nature 478: 255–259.

Trumpp, A., M. Essers, and A. Wilson. 2010. Awakening dormant haematopoietic stem cells. Nature reviews Immunology 10: 201–209.

Trumpp, A., and O.D. Wiestler. 2008. Mechanisms of Disease: cancer stem cells—targeting the evil twin. Nature clinical practice Oncology 5: 337–347.

Uchida, N., H.L. Aguila, W.H. Fleming, L. Jerabek, and I.L. Weissman. 1994. Rapid and sustained hematopoietic recovery in lethally irradiated mice transplanted with purified Thy-1.1lo Lin-Sca-1+ hematopoietic stem cells. Blood 83: 3758–3779.

Wang, J.C., and J.E. Dick. 2005. Cancer stem cells: lessons from leukemia. Trends in cell biology 15: 494–501.

Weissman, I.L., and J.A. Shizuru. 2008. The origins of the identification and isolation of hematopoietic stem cells, and their capability to induce donor-specific transplantation tolerance and treat autoimmune diseases. Blood 112: 3543–3553.

Wilson, A., E. Laurenti, G. Oser, R.C. van der Wath, W. Blanco-Bose, M. Jaworski, S. Offner, C.F. Dunant, L. Eshkind, E. Bockamp, et al. 2008. Hematopoietic stem cells reversibly switch from dormancy to self-renewal during homeostasis and repair. Cell 135: 1118–1129.

Wilson, A., E. Laurenti, and A. Trumpp. 2009. Balancing dormant and self-renewing hematopoietic stem cells. Curr Opin Genet Dev 19: 461–468.

Index

H

I

K

L

M

N

O

P

R

S

T

V

W

X

Color Plate Section

Chapter 1

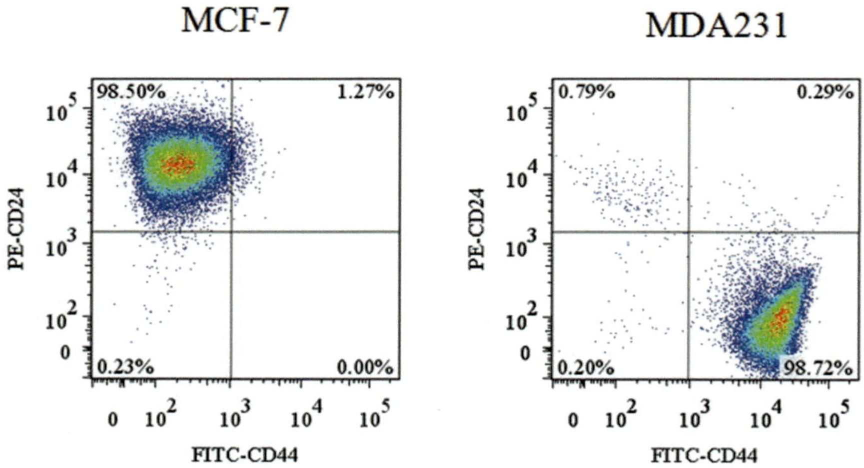

Fig. 2. Expression of CD44 and CD24 in established breast cancer cell lines.

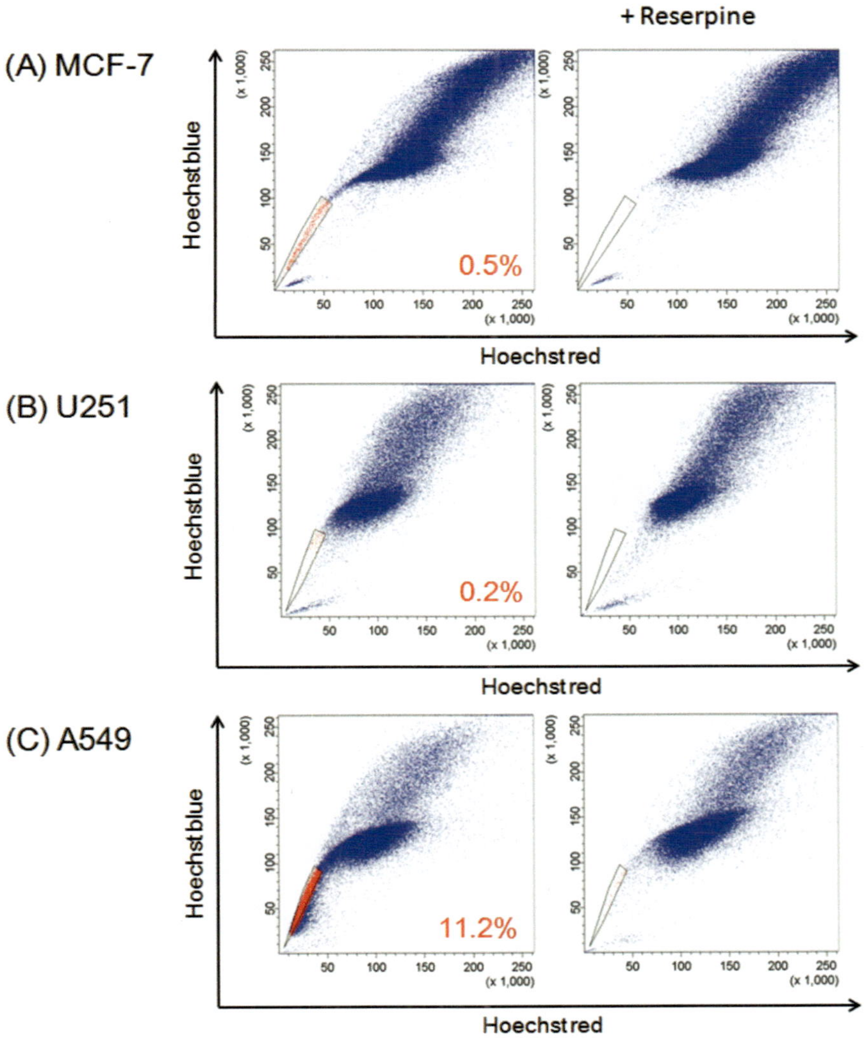

Fig. 3. Side population assay in human cancer cell lines.

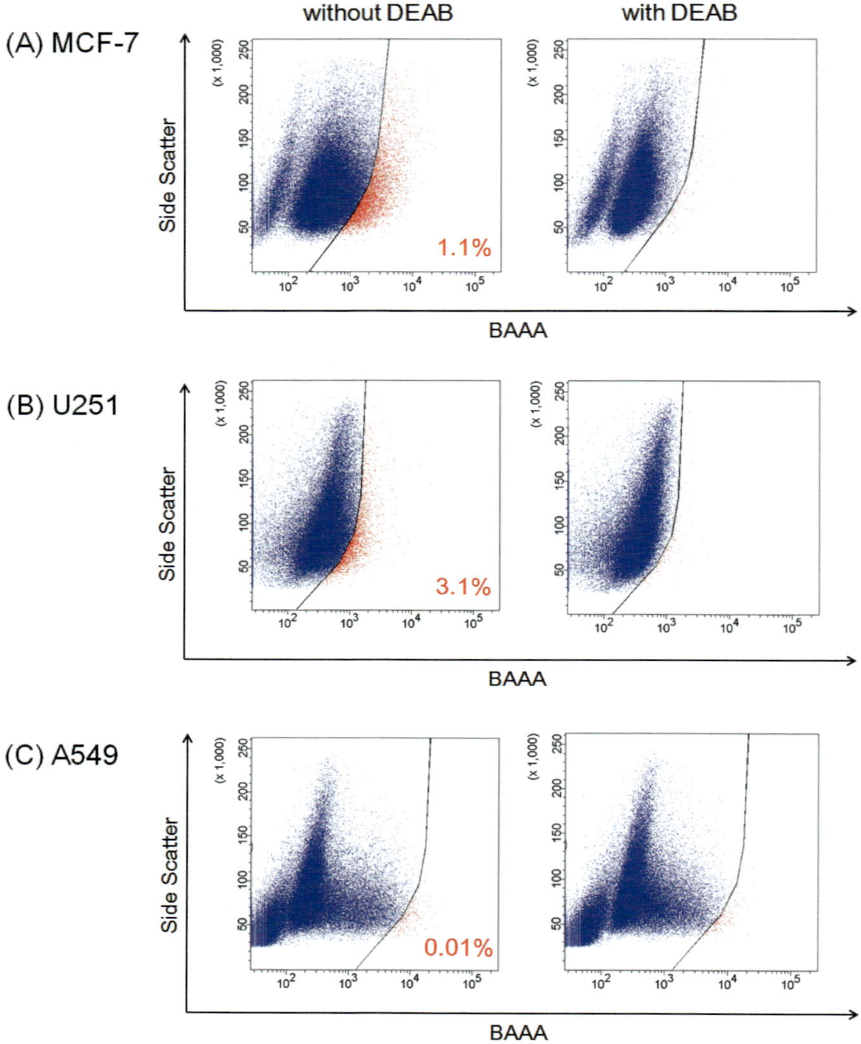

Fig. 4. ALDEFLUOR assay in human breast cancer cell lines. MCF-7 (A), U251 (B), and A549 (C) cancer cell lines were incubated with ALDEFLUOR substrate BAAA alone (left) or in the presence of the ALDH inhibitor DEAB (right), and then analyzed by flow cytometry. DEAB was used to establish the baseline fluorescence of these cells (blue) and to define the ALDEFLUOR-positive region (red).

Chapter 2

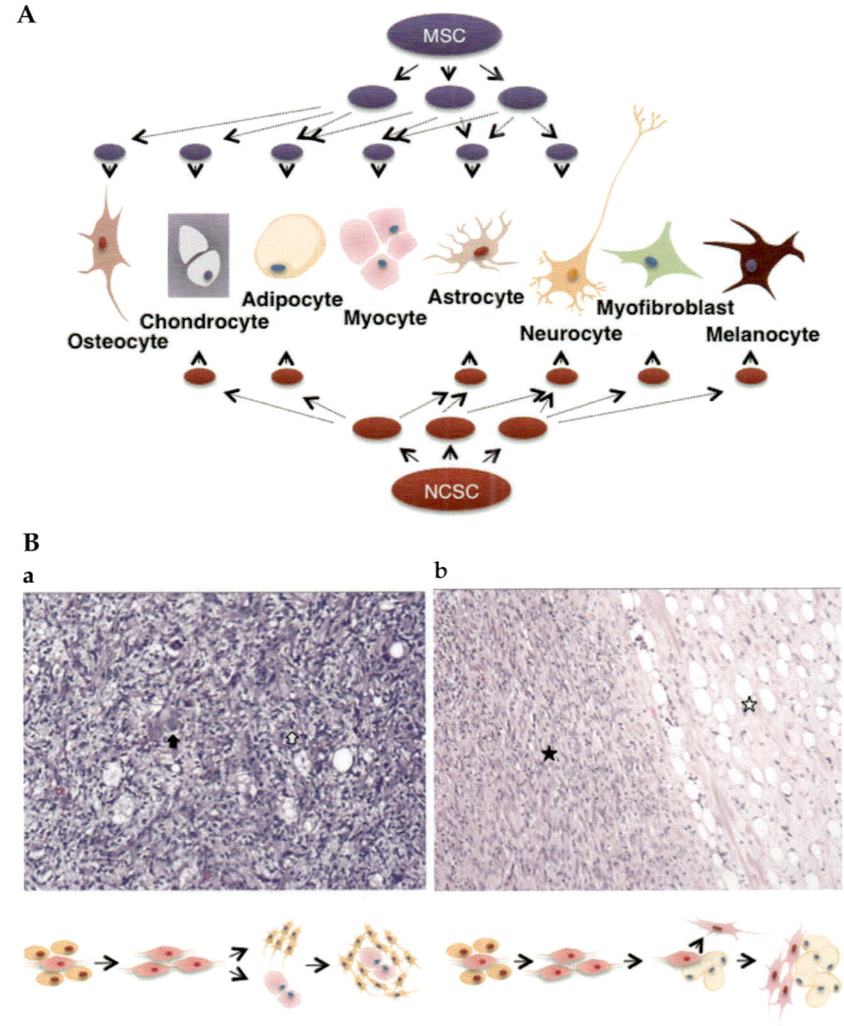

Fig. 1. Considerations for the origin of sarcoma. (A) Multi-differentiating potential of the mesenchymal stem cell (MSC) and neural crest stem cell (NCSC) (adapted from Toguchida J. Sarcoma-initiating cells and tissue stem cells. pp. 65–71. *In:* H. Esumi, N. Takakura, K. Miyazono, M. Mori [eds.] 2009. Experimental Medicine. 27. Yodosha, Tokyo, Japan) (B) Schemes of sarcomagenesis suggesting that tissue stem cells might be the origin of sarcomas. a. Malignant Triton tumor. Microscopically, the tumor cells exhibited pleomorphic (Schwannian) spindle tumor cells (white arrow) with focal rhabdomyosarcomatous differentiation (black arrow). b. Dedifferentiated liposarcoma. An abrupt transition from conventional well-differentiated liposarcoma (☆) to dedifferentiated liposarcoma composed of pleomorphic spindle cells (★).

Chapter 4

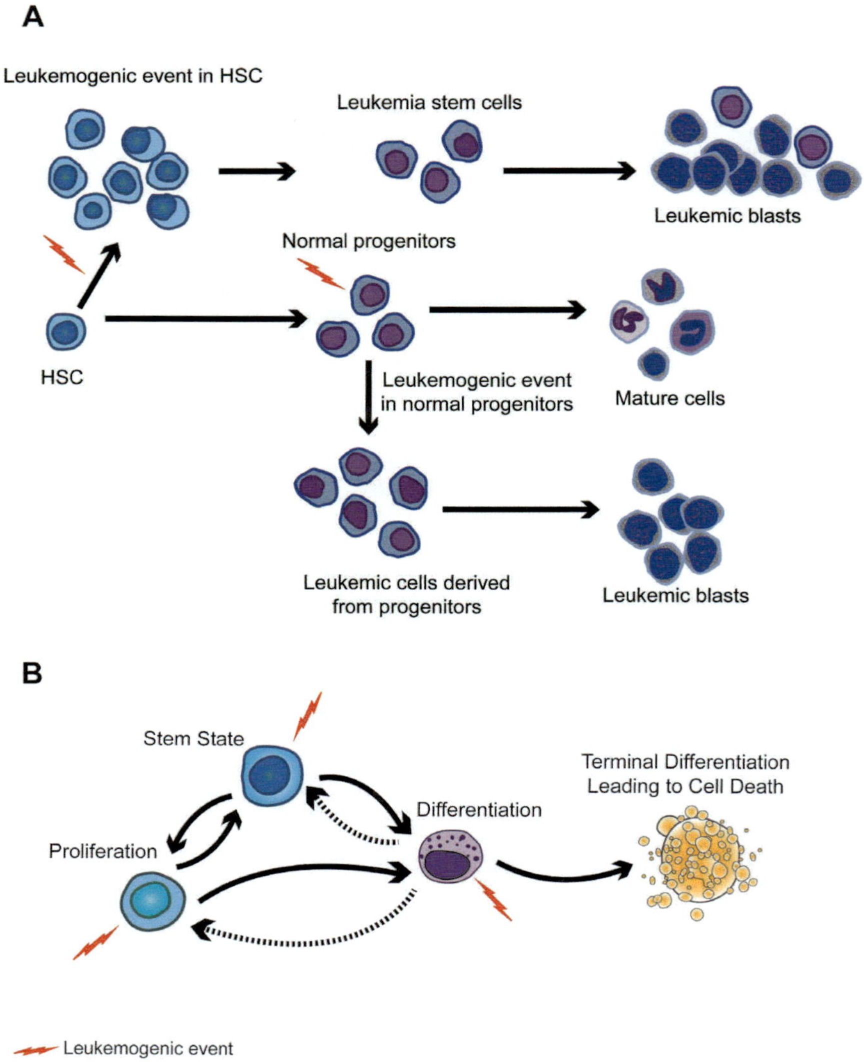

Fig. 1. Models of leukemogenesis. (A) The leukemogenic event ("hit") can occur in the HSC or in a hematopoietic progenitor. When a "hit" occurs in the HSC, part of the leukemic progeny carry stem cell characteristics. The transformed cells give rise to the leukemic blast population, some of which harbor LSC capacity (upper path). If the leukemogenic event occurs in a progenitor cell, none of the leukemic blasts carry LSC capacity (lower path) (Bonnet 2005, Dick 2008). (B) The link between stem cells and proliferating cells is not linear but rather bidirectional. The leukemogenic event can occur in a quiescent stem cell or in a proliferating cell. Any cell, unless destined to die, is capable of regaining stem cell capacity (Zipori 2004).

Chapter 5

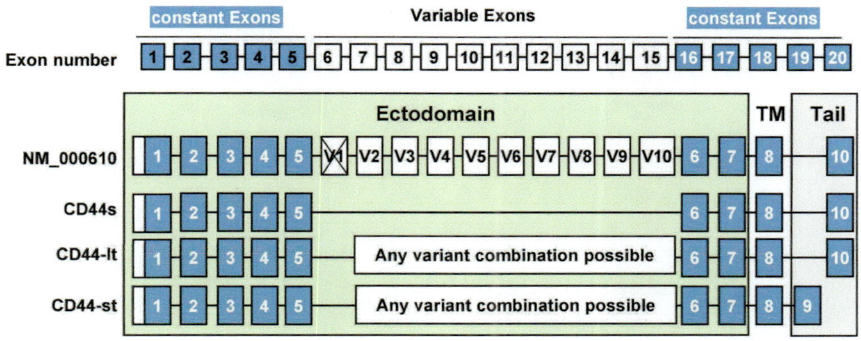

Fig. 1. Genomic structure of the human CD44 gene and its splice variants.

Top panel: The CD44 gene contains 20 exons and 19 introns. CD44s contains the constant exons 1–5 and 16–20 (blue boxes). This region is also included in all CD44 mRNAs and proteins. The variable exons 6 to 15 encode for v1 through v10 (white boxes). They are alternatively spliced to generate CD44 variant proteins containing one or multiple variable exons in tandem. An exception is human v1 which contains an in-frame termination codon and is therefore not included in any human CD44v molecules. In other species, however, CD44v could contain v1.

Bottom panel: Examples of alternatively spliced CD44 isoforms. Note that exon 19 is included in short-tail CD44v molecules whereas exon 20 is included in long-tail CD44v molecules.

Chapter 6

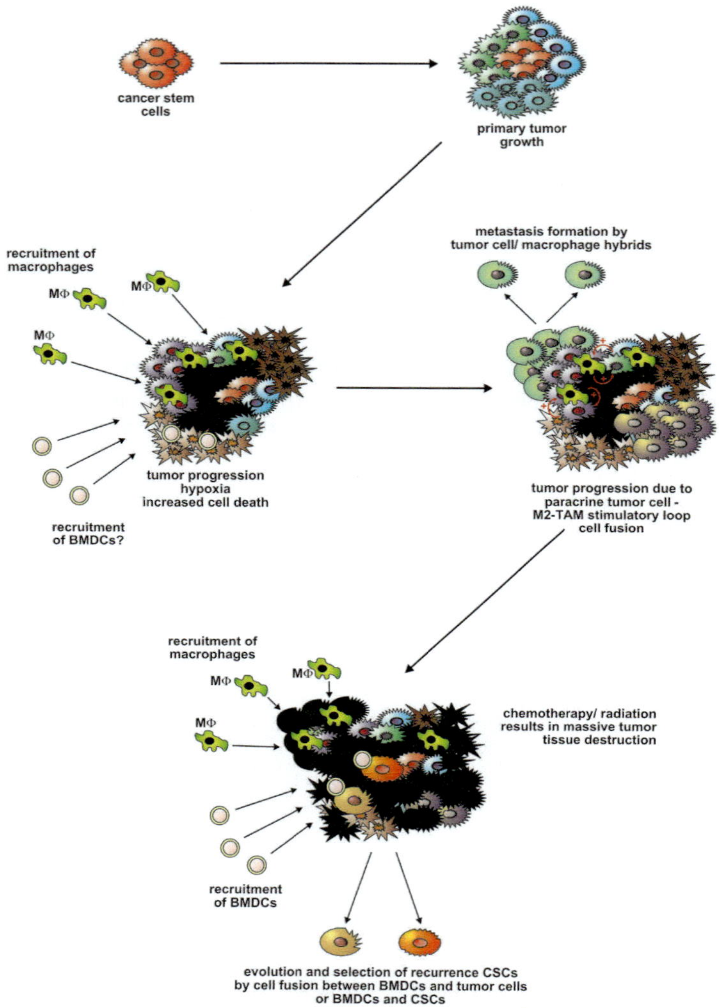

Fig. 1. Schematic overview about the cancer cycle. A primary tumor originates from CSCs. Due to increased tumor proliferation hypoxia concomitant with an enhanced loss of tumor cells occurs in the growing tumor leading to the recruitment of macrophages and BMDCs. Particularly M2-TAMs trigger tumor (cell) proliferation due to a paracrine stimulatory loop. M2-TAMs and tumor cells can fuse, thereby giving rise to hybrid cells exhibiting an enhanced metastatogenic behavior. Conventional cancer therapy, including chemotherapy and radiation, leads to a massive tumor tissue destruction (but spares CSCs) resulting in increased inflammatory conditions concomitant with an increased recruitment of BMDCs and macrophages. BMDCs might fuse with tumor cells and/or cancer stem cells, thereby giving rise to rCSCs—a process, which is possibly supported by the selection pressure conventional cancer therapy. Recurrence CSCs exhibit an "oncogenic resistance" phenotype.

Chapter 8

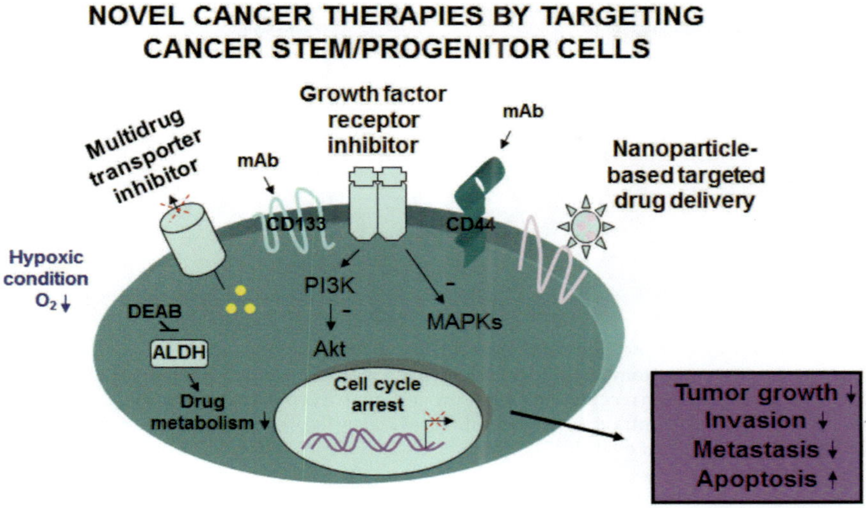

Fig. 2. New cancer therapies by molecular targeting distinct signaling pathways involved in aggressive behavior and multidrug resistance phenotype of cancer stem/progenitor cells. The inhibitory effects induced by distinct pharmacological agents such as the selective inhibitors of growth factor receptors or monoclonal antibody (mAb) directed against CD133 or CD44 stem cell-like surface marker, which may lead in an inhibition of PI3K/Akt and MAPK pathways in cancer stem/progenitor cells are indicated. The targeting of ABC multidrug transporters or aldehyde dehydrogenase (ALDH) by diethylaminobenzaldehyde (DEAB), which may result in the intracellular accumulation of chemotherapeutic drugs, is also shown. Moreover, the possibility to target the EMT process- and drug resistance-associated molecules as well as the signaling elements involved in the metabolic process under hypoxic condition are also indicated. The potential targeted delivery of drugs into cancer cells by using novel nanotechnology-based approaches is also illustrated. Hence, multitargeted strategies are likely to be effective to inhibit tumor growth, invasion, metastases and overcoming treatment resistance.

Chapter 9

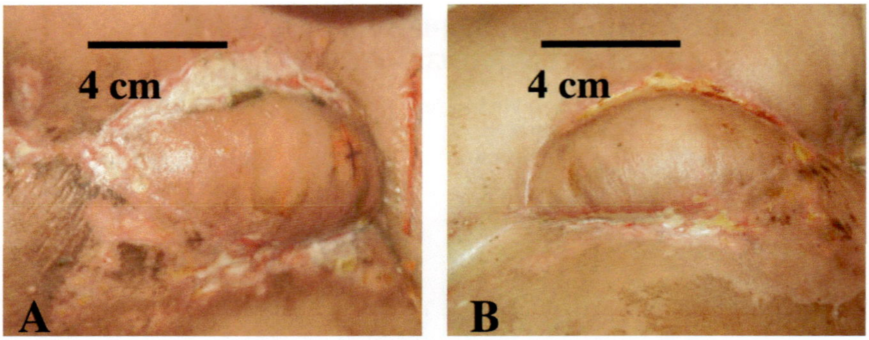

Fig. 2. Breast cancer metastasis regression by salinomycin. Orthotopic subcutaneous metastasis after mastectomy in a 40 year-old female with metastatic ductal breast carcinoma, triple negative (estrogen receptor: negative, progesterone receptor: negative, HER2: negative). (A) before, (B) after 12 intravenous applications of 200 μg^*kg^{-1} salinomycin every second day (adapted from Steinhart et al. 2012).

Chapter 11

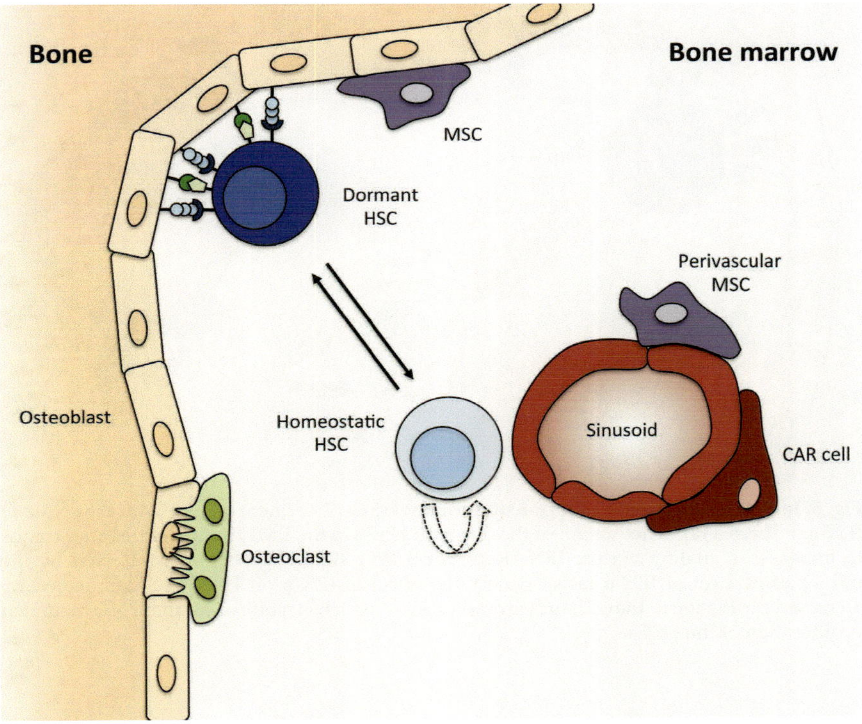

Fig. 1. A model for dormant and homeostatic hematopoietic stem cells in the bone marrow. Hematopoietic stem cells reside in the bone marrow in close proximity to the osteoblasts. Whereas dormant HSCs are in an osteoblastic niche composed of mainly osteoblasts and MSCs, homeostatic HSCs are within perivascular niches closer to sinusoids, CAR cells and perivascular MSCs.

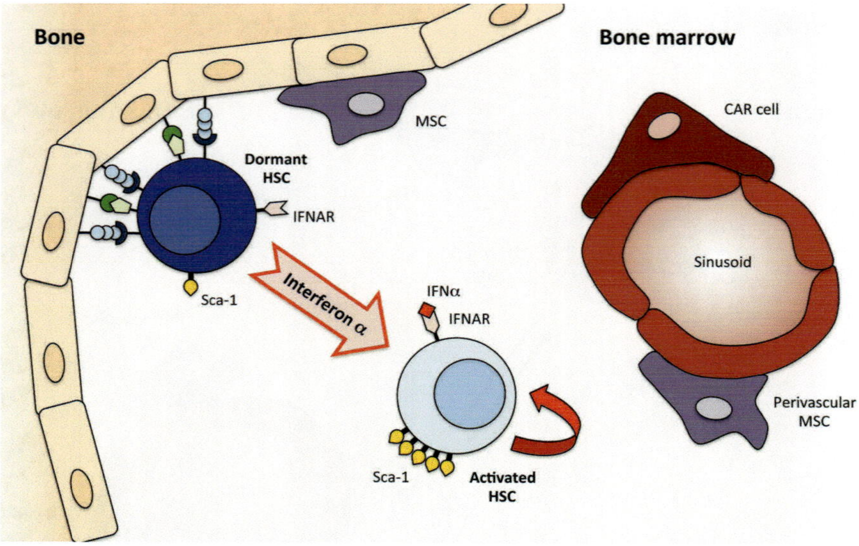

Fig. 5. Interferon-induced activation of dormant HSCs. Treatment of mice with the cytokine IFN-α induces a rapid activation of the dormant HSCs in the BM. The IFN-α treatment leads to increased signaling via the IFNAReceptor-STAT1 signaling pathway, followed by the induction of a set of IFN-α target genes. The effect of IFN-α on HSCs is accompanied by increased expression of the cell surface marker Sca-1, which is required for the IFN-α mediated proliferation of the HSCs.

For Product Safety Concerns and Information please contact our EU representative GPSR@taylorandfrancis.com Taylor & Francis Verlag GmbH, Kaufingerstraße 24, 80331 München, Germany